SHARING CITIES

共享城市

黄鹤　张烨　［荷］和马町
张悦　［新加坡］陈德钦　著

清華大學出版社
北　京

图书在版编目（CIP）数据

共享城市 / 黄鹤等著. — 北京：清华大学出版社，2019.12
ISBN 978-7-302-49695-3

Ⅰ. ①共… Ⅱ. ①黄… Ⅲ. ①城市规划 Ⅳ. ①TU984

中国版本图书馆CIP数据核字（2018）第035202号

责任编辑：张占奎
封面设计：陈国熙
责任校对：刘玉霞
责任印制：杨 艳

出版发行：清华大学出版社
网 址：http://www.tup.com.cn, http://www.wqbook.com
地 址：北京清华大学学研大厦A座 邮 编：100084
社 总 机：010-62770175 邮 购：010-62786544
投稿与读者服务：010-62776969, c-service@tup.tsinghua.edu cn
质量反馈：010-62772015, zhiliang@tup.tsinghua.edu.cn
印 装 者：小森印刷（北京）有限公司
经 销：全国新华书店
开 本：165mm × 230mm 印 张：20.75 字 数：619千字
版 次：2019年12月第1版 印 次：2019年12月第1次印刷
定 价：98.00元

产品编号：076883-01

CONTENTS

FOREWORD I

by Tsinghua

This book is one of the outcomes of English Program of Master in Architecture (EPMA) in Tsinghua School of Architecture. The EPMA of Tsinghua University's School of Architecture was aimed to create a new platform for teaching master courses of architecture for international students. We intended to enlarge the reach of our school's teaching philosophy by integrating a critical understanding of regional conditions with a global and multi-disciplinary network, and to have this teaching environment be engaged with the international dialogue.Since the establishment of the program in 2008, we have successfully attracted 91 international students from more than 30 countries in all five continents in the world.

Within our international program, we combine the critical academic atmosphere with real-world issues that arise from the current process of China's rapid urbanization. Students are encouraged to be critical, have be able to think reflexively in analyzing the real situation, and to create from a specific cultural context. The program includes 4 theory courses with 4 design studios. The first theme for theory course and design studio is local, with the theory course of Chinese local archicteture and the design studio of Architecture of space sprite. The second theme is contemporary, with the theory course of world contemporary architectural design trend and design studio of new design method including parametric design. The third theme is urban, with theory course of development of chinese cities and the urban design studio. The fourth theme is technology, with the theory course of green building and green building design studio.

Urban design studio is the course facing the urban issues among the above studios. It is aimed to set up a cross-culture platform of the urban topics discussion for the international students. In recent years, Knowledge City, Ecology City, Safe city and Edge City are selected as the themes of the studio. In 2017, Sharing City is chosen as the theme for the joint studio set up by School of Architecture in Tsinghua University and School of Design & Environment in National University of Singapore. The "Sharing Cities" studio aims to provide solutions to emerging concept of sharing, and responds to the idea of public space sharing and sustainable urban development from social, economic and humanitarian perspectives. The success of Tsinghua-NUS joint studio pushes forward the international cooperation process of urban design studio and other studios in EPMA.

We hope the improving EPMA and urban design studio would act as a window for international exchange students from all over the world to get access to the Chinese architectural and urban knowledge base and practice.

Director of English Program of Master in Architecture Prof. LI Xiaodong
Director of Urban Design Studio Prof. ZHANG Yue

本书是清华大学建筑学院英文硕士项目（EPMA）的教学成果系列丛书之一。建筑学院英文硕士项目旨在为国际学生搭建一个建筑学硕士课程的教学平台，致力于加深对于整合区域环境和全球性多学科网络的批判性理解，并注重实现国际对话中的教学环境。该项目自2008年创建以来，已成功吸引了来自5大洲30多个国家的91位国际学生。

英文硕士项目将重要学术领域和当前高速发展的城市化进程中的现实问题结合起来。项目鼓励学生开发批判性思维，能够独立思考实际情况，并能在特定文化背景下有所创造。主干课程设置包括4门理论课和4门设计课，相互对应。第一组课程主题为“乡土”，包括中国本土建筑介绍和具有场所精神的本土建筑设计课程；第二组课程主题为“当代”，包括当今建筑设计思潮介绍和包括参数化设计在内的新设计方法课程；第三组课程为“城市”，包括当代中国城市发展介绍和城市设计课程；第四组课程为“技术”，包括绿色建筑理论及实践介绍和绿色建设设计课程。

城市设计课程是上述课程中面向城市问题的设计课。课程旨在建立一个跨文化的平台，让不同国家的学生就城市发展的相关话题进行交流探讨。近年来课程选题涉及“知识城市”“生态城市”“安全城市”“边缘城市”等。2017年城市设计课程是与新加坡国立大学设计与环境学院共同进行的联合教学，选定“共享城市”为主题，旨在为正在广泛兴起的共享理念提供空间解决方案，并从社会、经济和人文角度回应公共空间的共享与城市的可持续发展。此次联合教学取得了良好的成效，为城市设计课程以及其他英文硕士项目课程的进一步国际化合作奠定基础。

我们希望，不断探索完善中的英文硕士项目及城市设计课程，能成为世界各地的国际交换生的窗口，使他们得以接触和了解中国城市和建筑的基础情况和实践。

英文硕士项目负责人　李晓东 教授

城市设计课程负责人　张　悦 教授

FOREWORD II

by NUS

The rapid pace of change in the cultural and economic landscape of China compels us to rethink the nature of the built environment afresh. While modes of production and consumption undergo breakneck speeds of transformation, the city and buildings remain unmoved, and innovations in building technology and perception crawl in snail pace by comparison. With the advent of smart phones ten years ago and the tens of millions of Apps that were developed since, urban living had changed dramatically. Is it possible for the physical environment to re-conceive in tandem with such changes in lifestyle, habit, social cohesion and other pertinent issues. How can we conceptualize the city of today and perhaps of tomorrow?

These are the questions grappled by architecture students from Tsinghua University and National University of Singapore in the Sharing Cities workshops and studios. The collaboration not only brings students from different international backgrounds together, but also young talents to imagine the future. They share the concerns of the current development in Beijing and Singapore as two of the most exciting and rapidly developing mega cities in the world and also project the future of urban and building designs in these cities that would embrace co-sharing as a practice for urban life. Studio is the key arena for postulations, debates, expressions and substantiations of ideas and collaborative studio, such as this one presented in the book, deepens the experience further.

The choice for the topic and format of the studio are relevant to the issues at hand, the real life situation that we are living in, and the method for eliciting and producing innovative ideas and proposals. The students have been investigating the nature of co-sharing as well as the myths surrounding this mode of consumption. They have explored different options and provided design solutions that would facilitate various forms of sharing: culture, heritage, transport, to name a few. Through research, case studies and design activities, the students collectively and individually have gained awareness and insights in this fast growing living pattern. They are able to hone their design skill too to facilitate and inspire the increasing popular lifestyle. This book bears witness to their achievement.

Head of Department of Architecture
Prof. Puay-Peng HO

新加坡国立大学

中国文化和经济格局的快速变化促使我们重新思考建成环境的本质。相较于生产和消费方式经历的急速变革，城市和建筑仍然止步不前，建筑技术和观念的创新也进展缓慢。随着十年前智能手机的出现，以及随之而生的数以千万计的应用程序，城市生活发生了巨大的变化。物质环境是否可以随着生活方式、习惯、社会凝聚力和其他相关问题的变化而重新建构？我们又应该如何概念化今天或许明天的城市？

这些问题正是清华大学-新加坡国立大学建筑系的共享城市合作设计课程和设计工作坊的关注所在。这一合作不仅集结了不同国际背景的学生，并且为这些年轻人才提供了一个畅想未来的机会。以世界上发展最活跃、最迅速的两座特大城市——北京和新加坡为对象，他们共同关注城市当前的发展状况，并探讨了未来的城市和建筑设计如何承载并推动共享作为城市生活的一种实践。设计课是思想形成、论辩、表达及深化的重要平台，而正如此书所呈现的，联合设计课恰恰将这一经验进一步深化。

设计课主题和形式的选择脱离不开我们当前需要面对的问题和所处的现实生活，同时也关系到激发和产生创新想法和方案的方法。学生们针对共享的本质以及围绕这一消费模式的诸多疑惑进行了深入的研究。他们探索了不同的可能性，并提出了能够促进不同类型如文化、遗产、交通等共享活动的一系列方案。通过研究、案例学习和设计探索，无论是在整体还是个人层面，学生们都强化了对这一迅速普及的生活模式的认识。他们也磨炼了自己的设计技能，并以此来推动这一日益流行的生活方式。这本书见证了他们取得的成绩。

建筑系系主任　何培斌 教授

1

INTRODUCTION 介绍

SHARING

Jonit studio introduction

By He Huang

“共享城市”联合教学介绍

黄鹤

Along side with the urban development nowadays, the continual emerging of new urban phenomena globally reminds the urban designers and architects to face the challenges from new perspectives and deal with these accordingly.

Recently, the extensive rise of sharing economy taking Uber, Airbnb, Wework as typical representative, introduces a new utilization pattern of existing urban resources around the world. Sharing economy, together, with other ways of sharing city, lead to sustainable city future. In the global city development process, this trend requires urban planners and architects, to carefully looking at the local social, economic and cultural background, to explore design strategies which meet the needs of sharing city in future.

Thus School of Architecture in Tsinghua University and School of Design & Environment in National University of Singapore set up a sharing city joint studio, in which 17 Tsinghua students from 12 countries and 17 NUS students from 3 countries participated. Students with different cultural background collected and analysed sharing city cases all over the world, integrating into the design task in Beijing and Singapore sites, explored how urban regeneration replies to the sustainable-oriented sharing urban space utilization from social, economic and cultural perspective.

CITIES

在世界范围内，随着城市的发展，新的城市现象不断出现，使得城市设计者和建筑师要不断以新的视角来审视这些现象，面对挑战并解决问题。

近年来共享经济广泛兴起，以Uber、Airbnb、Wework等为典型代表的共享经济实践在全球引发了城市资源共享共用的新模式。共享经济作为共享城市的一种体现方式，和城市资源共享的其他途径一起，指向了可持续发展的城市未来。在全球性的城市发展变化进程中，这样的指向，使得不同地区内的城市规划者和设计者需思考应当怎样立足于本地区的社会、经济、文化背景，以怎样的设计策略来回应面向未来的共享城市的发展需求。

基于此，清华大学建筑学院与新加坡国立大学设计与环境学院联合设立了“共享城市”联合教学，清华大学来自12个国家的17名学生与新加坡国立大学来自3个国家的17名同学参与了此次联合studio。来自不同文化背景的同学们，搜集分析了全球范围内共享城市的案例，通过北京和新加坡城市地段的设计任务，从社会、经济、文化等多视角探讨城市更新如何回应空间共享共用、促进城市可持续发展的需求。

共享城市

DESIGN

Urban regeneration and the practice of design research cooperation through education.

By Martijn de Geus

设计研究教学合作的城市更新实践

[荷] 和马町

This article describes a joint design research collaboration titled “Sharing Cities”, between Tsinghua University and the National University of Singapore. The article firstly explains the conceptual framework of a broader trend in educational design research, with secondly a case study review of how this framework was applied in the “Sharing Cities” joint studio. Together with the conclusion in the third part, the author hopes this review can benefit others in similar (urban) design studios and research collaborations.

The first part of the article covers aspects of educational design research, the underlying knowledge framework and the design studio’s methodology. It explains how the design research collaboration utilizes an underlying holistic design approach towards urban design in its approach towards design education.

The second part of the article takes Tsinghua University’s part of the joint-studio as a case study. This urban design studio is part of the school’s English Program for Master in Architecture, which takes the White Pagoda Temple district in Beijing Baitasi as its main intervention area. It explains how the design research framework is adapted for this particular studio to link emerging concepts of the sharing economy with the topic of urban regeneration. In addition, the studio utilized the cultural diversity of participating students, who came from over ten countries in the world as integral part of the educational design research. For instance, the studio started its design research with a case-study analysis into aspects of “sharing” expressed in an urban or architectural model, with cases from all home-countries of participating students.

In short, this paper describes the set up of a joint urban design studio, the design research framework, the student grouping, how the student’s output relates to the overall holistic design strategy, etc, and distills conclusion that could be helpful in establishing similar cases of (joint) educational design research.

Keywords: design education, design research, international cooperation, urban regeneration, urban design studio

RESEARCH

本文介绍了清华大学与新加坡国立大学设立的“共享城市”教学。文章首先解释了教学型设计研究更广泛趋势的概念框架，随之介绍了该概念框架如何应用在“共享城市”联合教学之中。最后在结论中，作者希望这样的工作对类似的城市设计联合课程和研究有所帮助。

本文的第一部分概括了教学型设计研究的基本知识框架和教学方法。它解释了设计研究协作如何运用一个基本的整体设计方法来达成城市设计的教学。

文章的第二部分以清华大学的部分联合教学为案例。该城市设计课程是建筑学英文硕士班的组成部分，以北京的白塔寺地区作为设计地段介绍了如何将设计研究框架运用于这个特定的设计课程，将新出现的共享经济概念与城市更新的主题联系起来。此外，来自世界十多个不同国家及文化背景的学生参与该设计课程。因为这些学生的不同背景，设计教学以“共享”元素表达至城市或建筑中的案例分析开始，而这些案例是来自参与学生自己的国家。

简而言之，本文介绍了联合城市设计教学、设计研究框架、学生分组，以及学生整体性的设计策略等过程，希望所得结论对正在建立的类似（联合）教育设计研究有所帮助。

关键词：设计教育、设计研究、国际合作、城市更新、城市设计课程

共享城市

1. Introducing Educational design research

The first part of this article covers aspects of educational design research, to explain how the Sharing Cities studio is part of an effort to embed design research into graduate level design education. In addition, this part explains the underlying knowledge framework and the design studio's methodology.

To start, in their 2012 book Conducting Educational Design Research McKenna and Reeves outline how the term "educational design research" is not a fixed definition, but instead can be used to describe a family of approaches that "strive toward the dual goals of developing theoretical understanding that can be of use to others while also designing and implementing interventions to address problems in practice." Other common ways of describing a similar concept include "design based research" (Kelly, 2003) or "development research" (Van den Akker, 1999). In order not to confuse the concept of this paper with the frequently used "design-based research," which more commonly concerns the aspect of research as part of a design process, or "research-informed design," the chosen term "educational design research" highlights the link suitable for this paper in which design research is part of the education of architects and urban designers. Although this of course does not exclude the possibility that the research can also be part of the design process.

Quite particular in design education is the relation reciprocity between research and practice. This considers the output of research, but also the position that tutors take in the educational process, since in design-based education the instructors and experts included might include researchers and practitioners. Wagner (1997), considers this relation a vital part of defining the educational design research, and terms it the "social design of research projects". He considers design research a platform for the social elements

1. 教学型设计研究介绍

本文第一部分概括了教学型设计研究的各个方面，说明了共享城市联合设计课是如何将设计研究嵌入到研究生的设计教育。此外，该部分解释了基本学术框架和设计课的方法。

在2012年的《实施教学型设计研究》一书中McKenna and Reeves勾勒了"教学型设计研究"这个没有固定定义的名词，但是可以用一组连贯的解决方案描述为"达到发展理论性解释的双重目的，既可以为他人所用，也可以解决实践中存在的问题。"其他常见的类似概念描述包括"基于研究的设计"（Kelly，2003）或"渐成研究"（Van den Akker，1999）。为了不混淆本文的概念与常用的将研究视为设计过程的"设计研究"或"以研究为依据的设计"，"教学型设计研究"强调了研究应当是建筑师和城市设计师教育的一部分。当然这并不排除研究成果可能成为设计过程一部分的可能性。

设计教育的独特性在于研究与实践的相互关系。这包括了研究成果，也包括导师在过程中的作用，因为在设计教育中的导师和专家也应当包括研究人员和实践从业者。瓦格纳（1997）认为，这种关系是界定教学型设计研究的一个重要成分，并称之为"研究项目的社会学设计"。他认为设计研究是社会因素在研究人员和从业者之间形成关系的平台。在他的研究者与实践者的三种形式合作中，我们在这个联合教学中使用的合作类型可描述为研究者、实践者和学生之间的"共同学习"。这种合作的特点是各方之间的合作与讨论。所有参与方都通过教育过程中的行动和讨论，例如通过头脑风暴、研讨会和工作营的形式参与其中。在应用合作成果的不同要素方面，教学型设计研究方式旨在通过合作研究获取知识，并将其传递至设计教学课程完成后，学生、研究者和从业者的个人研究和实践之中。

in forming relationships between researchers and practitioners. Following his "three forms of researcher-practitioner cooperation," the type of cooperation we utilize for this joint-studio can be best described as a "co-learning agreement" between researchers, practitioners and students. This cooperation is characterized by its ongoing collegial, reflexive inquiry between all parties. All parties participate through action and reflection in the process of education, for instance in the form of brainstorm sessions, seminars and workshops. To arrive at implementing elements of the outcome of the cooperation, the change model aims to draw on knowledge gained through co-operative research that is then dispersed into the individual research and practice of students, researchers and practitioners after the completion of the design studio.

In addition to these intangible outcomes, possible output of educational design research include theoretical and practical contributions. McKenna (2012) writes how as an example of theoretical contribution, educational design research can contribute to the development of theories that are used to describe, explain or predict certain phenomena. In addition the research can lead to prescriptive insights, often referred to as design principles: that recommend how to address a specific class of issues in a range of settings. For the practical contributions we do not necessarily mean only the output that has been applied in practice; when considering educational design research it mostly considers the development of solutions to practical problems. Whether or not without an actual implementation, the design interventions are "not merely hypothetical concepts" since they are considered and designed for an actual use. They are site-specific, designed with the goal of solving real problems .

除了这些无形的成果外，教学型设计研究的潜在成果还包括理论和实践贡献。麦克纳（2012）提到，作为理论贡献，教学型设计研究可以有助于用来描述、解释或预测某些现象的理论的发展。此外，这项研究可能会指向通常被引用为设计原则的一些规范，例如如何在一系列设置中强调特定类别的问题。从实际贡献而言，我们并不只指设计方案在实际中的应用，它主要考虑解决实际问题方法的发展。无论是否实施，设计方案都不只是假设概念，而是基于特定地点解决实际问题的。

1.1 Integrated Design Eduction

The Sharing Cities-studio is part of a graduate level program in architecture, however the design education content is focused on urban design. These two disciplines of architecture and urban design are conventionally disconnected, both in research and practice. In addition architectural design itself usually involves a series of hand-offs from owner to architect to builder to occupant. This path does not invite all affected parties into the planning process, and therefore dose not take into account their needs, areas of expertise or insights. In some cases, using the conventional method, incompatible elements of the design are not discovered until late in the process when it is expensive to make changes. This also considers cases where an urban master plan, designed by an urban designer, can be completely ignored by an architect later designing a building in this plan.

In contrast, for the Sharing Cities studio, the co-learning type agreement of cooperation mentioned before is followed through to an integrated studio design process that requires multidisciplinary collaboration, including interviews with local citizens, lectures by the local government, developer, design professionals, etc, from conception to completion. This integrated design method is a collaborative method for designing which emphasizes the development of a holistic design. Decision-making protocols and complementary design principles must be established early in the process in order to satisfy the goals of multiple stakeholders while achieving the overall project objectives.

Having utilized this approach for several years in our graduate program, we find that the necessary design research collaboration in the making of an urban design in this way makes the students familiar with a holistic approach towards urban design that benefits their architectural education. The studio considers cities to be living organisms, consisting of

1.1 整合的设计教学

共享城市联合设计教学是建筑学研究生课程的一部分，该课程内容侧重于城市设计。建筑学和城市设计这两门学科通常在研究和实践上都是相互割裂的。此外，建筑设计本身通常涉及从业主到建筑师、到施工方、到居住者的一系列交接工作。这过程并没有让所有受影响方参与其中，因此没有考虑到他们的需要、专业领域或见解。在惯用方法下，设计中的不相容元素有时直到过程后期才会被发现。类似状况也体现在城市规划中，建筑师在设计其中的建筑物的时候，可能会完全忽略之前城市设计师的工作。

相反，在共享城市联合设计教学中，之前提到的联合学习模式整合到设计教学过程中，包括多学科合作、与当地居民访谈、地方政府、开发商、设计人员举办的讲座等。这种整合的设计方法是一种强调整体设计开发的协同方法。决策制定工具和补充性的设计原则应当在过程早期确定下来，以确保实现整个项目目标的同时满足多个利益相关者的目标。

这种教学方式在我们的研究生课程中实行多年。我们发现，必要的城市设计研究协作方式使学生了解和熟悉整体的城市设计方法，有利于他们的建筑教育。城市设计课程认为城市是有生命的有机体，由各种不同的但同样重要的生命层组成。整个项目被看作是一个相互依存的系统，而不是独立组件（基地、结构、系统和使用）的积累。关照整个系统的目的是确保它们协调工作，而不是相互对抗。

various, equally important, layers of life. The whole project is viewed as an interdependent system, as opposed to an accumulation of its separate components (site, structure, systems and use). The goal of looking at all the systems together is to make sure they work in harmony rather than against each other.

1.2 Studio Method

The Sharing Cities course is primarily a studio course, consisting of 8 weeks (Tsinghua) or 12 weeks (NUS) of design studio sessions. The primary method of instruction involves students solving a predetermined, urban design problem. The problem space is defined in a particular manner to promote the learning objectives. Students were "coached" by both researchers, and practitioners on how to going about solving a complex, urban design problem, making use of individual and group crits, exercises, demonstrations, group discussion, and occasional lectures and/or site visits. In addition to these studio crits, the overall studio structure can be broken up in the following six components:

(1) Comparative case study research As theoretical ground work, the research used local cultural background knowledge and the cultural diversity of participating students, who came from over ten countries in the world as integral part of the educational design research. The comparitive case-study analysis took over 20 cases related to aspects of "sharing" expressed in an urban or architectural model, with cases from all home-countries of participating students, including fifty precent Chinese cases.

(2) Joint brainstorm charettes As part of the co-learning type cooperation, and to benefit the integrated design ideology, in the first phase of the joint studio, two short, intensive design charettes were held in Beijing and Singapore respectively. Students, researchers and practitioners from Tsinghua University and National University of Singapore collaborated

1.2 设计教学方式

共享城市课程主要概括了8周（清华大学）或者12周（新加坡国立大学）设计教学课，主要的教学方式针对学生解决预先确定的城市设计问题。问题空间以特定的方式定义以促成学习目标。学生由研究人员和从业人员联合教学，去解决一个复杂的城市设计问题，利用个人和团队的评判、练习、演示、小组讨论、偶尔的讲座和（或）现场考察。除了这些设计课程的部分，整个设计教学的结构可分为以下六部分：

（1）作为理论基础的案例比较研究。该部分包括来自世界各地十多个国家的学生所提供各自国家的案例。案例研究分析了超过20例的“共享”相关的全球建筑或城市案例，其中50%来自中国城市的实践。

（2）联合头脑风暴讨论，作为合作学习的一部分，有利于整体化的设计思想形成。在第一阶段的联合教学中，在北京和新加坡分别举行了为期一天的设计研讨会。来自清华大学和新加坡国立大学的学生、研究人员和从业人士相互交流，并分成不同小组，根据北京和新加坡的特定地段进行了为期1~2天的设计工作坊。工作坊结束后，各自的小组将在各自的学校完成自己的设计，包括比较性案例分析、场地分析及最终的城市设计方案。

and were divided into a several mixed groups, then carried out a 1-2 day workshop according to a specific site in Beijing and Singapore. After the collaboration, the respective groups completed their own design work at their home institutions, including comparative case studies, site analysis, and the eventual urban design proposal.

(3) Shared city forums and joint reviews In order to help participants in conceptualizing the "sharing cities" problem, two cross-institutional forums were held parallel to the charettes. In addition, the forum enabled the exchange of ideas and teaching pedagogies.

(4) Comparative site visit analysis Beijing/Singapore To highlight the importance of differentiating between global issues and local context, all participants visited sites in both Beijing and Singapore, connected to the possibility of urban regeneration and the emergency of the age of sharing .

(5) Student grouping in sub-themes Another part of the integrated design effort included the forming of teams for the Tsinghua side, in which 8 groups had particular focus in terms of topic and related urban area, but all discusses these in relation to the development of one overall masterplan. Students were grouped according to programmatic sub-themes that overlapped and influenced each-others urban designs.

(6) Sharing and publication of studio results Finally, a "Sharing City: Sharing Economy and Urban Regeneration" forum and final design presentation from both university was held in Tsinghua University's School of Architecture; and the conclusions and final student works were shared with the global architectural community at a forum and exhibition at the 26th UIA (International Union of Architects) World Congress in Seoul.

（3）共享城市的论坛和联合评图帮助学生在归纳“共享城市”问题方面起着积极作用，同时联合教学过程中举办了两个跨国的专家研讨会。论坛举办激发了相关讨论和教学方法的交流。

（4）北京/新加坡的比较性调研突出强调了全球性问题与本土性环境的区别，所有参与的学生到北京和新加坡的场地调研，使得他们直接面对各种城市更新，和共享时代兴起等现象。

（5）学生以不同的次主题形成分组。联合设计中清华大学部分分为8组不同专题并针对不同地区，但同时紧密围绕整体规划的讨论。学生分组依据不同专题，但相互重叠并相互影响。

（6）联合教学设计成果的分享与出版。最后，“共享城市：共享经济和城市再生”论坛和双方的最终评图在清华大学建筑学院举行；学生作品在第二十六届UIA（国际建筑师协会）论坛及展览中展出。

2. Studio Theme: Sharing Cities

The second part of the article takes Tsinghua University's part of the joint-studio as a case study, which takes the White Pagoda Temple district in Beijing (baitasi) as its main intervention area. It connects concepts of the emerging sharing economy, that consists of new concepts of for-profit sharing of basic everyday commodities, with urban regeneration. The implementation of these shared concept has changed the ways we commute (bike- and car sharing), shop (alibaba/ amazon) or work (co-work spaces) and redefines our living (shared-housing, air-bnb), and learning (e.g. open-online courses edX and MOOC) experiences. It redefines the need for urban citizens to own everything we want to use. Instead of everyone needing to have private access do daily necessities, with the emerging sharing concept, a larger group of users, for a smaller individual economic burden, can use more resources. This principle was used in the studio process as starting point of an urban regeneration model; According to John's (2016) the term "sharing" is a keyword for our times. But, as each of these examples show, the meanings of sharing vary wildly across different contexts. Arguably, booking an Uber ride is not "sharing"—it is a purchase of a service—and yet Uber, Lyft, and Airbnb have been hailed as part of a new "sharing economy." . Thus, when we talk about sharing we implicitly or explicitly engage with a set of values, not always with the same meaning. John claims that three basic spheres of sharing enacts different sets of values. However, for our studio, we stick with his core definition, in which, when we talk about sharing, we are talking about purportedly prosocial behaviors that promote, or are claimed to promote, greater openness, trust and understanding between people. This fits well with the conceptual set-up of the studio as a co-learning agreement between various researchers, students and practitioners. In addition, the studio themes

2. 联合教学主题：共享城市

文章的第二部分以清华大学部分的联合教学为案例，以北京白塔寺地区为设计基地。设计教学结合新兴的共享经济概念，包括日常基本商品的共享，将其和城市再生结合。这些共享概念的实施改变了我们通勤（共享自行车和共享汽车），购物（阿里巴巴、亚马逊）或工作（共享办公）方式重新定义了我们的生活（共享居住和AirBnB）、学习（例如网络开放课程EDX和MOOC）等方面的体验。随着共享概念的兴起，共享经济重新定义了城市居民拥有使用物品的必要性。随着共享概念的兴起，并不是每个人都需要拥有日常必需品，而是以较小的个人经济负担，在更大的用户群中使用更多的资源。这一原则在教学过程中作为城市更新模式的出发点。根据约翰（2016）的说法，“共享”是我们时代的关键词。但是，正如这些案例所示，“共享”的含义在不同的语境下有很大的差别，例如Uber、Lyft、AirBnB被称为共享经济的一部分，但是预订一个Uber车辆并不是“共享”，而是购买一个服务。因此，当我们谈论共享时，我们隐含地或明确地涉及不同价值体系，并不总是具有相同的意义。约翰声称，共享的三个基本领域体现了不同的价值观。在我们的设计教学中，我们坚持自己的核心概念，即当我们谈论共享，是促进社会行为，或是宣称要促进，以更大的开放促进人与人之间的信任和理解。这符合教学课程的理念设置，作为不同研究人员、学生和实践者之间的共同学习协议。此外，设计教学结合了共享主题与城市再生，并探讨了“共享”的生活方式的可能场景，及其对城市建成环境的影响。特别是，其关键问题在于新经济是否能给予历史文化街区新的愿景，包括自下而上的更新动力。如果能，那设计者和规划师在这一过程中该如何发挥作用。

combines this sharing topic with the notion of urban regeneration, and explores the possible scenarios for emerging "shared" lifestyles and its impact on the built environment of cities. Specifically, the key question is whether the new economy can give rise to new perspectives that include a bottom-up approach to regenerating historical urban areas and, if so, how architectural and urban design can be employed in this process to adapt the built environment.

2.1 Site: White Pagoda Temple historic district

For the site of this case, the White Pagoda Temple District was chosen, which is a historical conservation area in the Old City of Beijing. It largely consists of traditional hutong areas covering 37 hectares and is located just across Beijing's Financial Street District on the west second ring road. The district presents itself as a cultural oasis, a peaceful enclave among newly developed business areas. And with its profound historical heritage and rich cultural connotations, the district is home to significant cultural and or heritage sites representing various periods in the development of Beijing, and hence considered as one of the cultural hearts of the capital city. In addition, the area already has a long tradition of "sharing," in different times, with remnants of former shared Communist housing, or still utilized shared toilet facilities. Recently, reflecting on the unsuccessful large-scale demolition and redevelopment of the historical areas in Beijing over the past two decades, a new approach to urban regeneration was initiated, as part of which our studio-theme is very fitting. The general master plan strategy now aims for soft-strategies of development, taking into account today's network-based, digitally-shaped economies of sharing to enable and enhance public participation and communal engagement in urban re-generation. This approach has been successfully experimented in Baitasi District over the past two years, whereby positive impact of many creative initiatives on micro urban regeneration starts

2.1 场地：白塔寺历史文化地区

白塔寺历史文化地区被选作本次教学的场地。它主要由传统的胡同区域组成，占地37公顷，紧邻北京市西二环的金融街。这个场地是一个文化绿洲，是新开发商业区围绕的一片安逸领土，以深厚的历史底蕴和丰富的文化内涵，被视为首都文化的核心地区之一。此外，该地区早已有了“共享”的悠久传统，保留的社会主义大楼或仍然共用的公共厕所设施可作为证据。针对过去二十年来北京历史地段大规模拆除和重建的失败，在此次教学设计中我们提出了一种新的城市更新方法。现在的总体规划战略致力于软战略发展，考虑到现今以网络为基础的共享经济，促进公众参与和社区参与到城市更新之中。该方法已在过去的两年中成功于白塔寺地区实行，其中在微城市再生方面许多创造性的举措得以成功实施。这些成功经验为上述设计探索提供了充分的基础。

to emerge. Therefore, this site and in particular the initial outcomes provide a fertile ground for the above-mentioned design exploration.

2.2 Group findings

The "Sharing Cities" studio itself aimed to provide solutions to emerging concept of sharing, and sustainable urban development from social, economic and humanitarian perspectives. Therefore, eight sub-groups of two to three students each were established with a particular focus in terms of topic and related urban area, but all discusses these in relation to the development of one overall masterplan. The studio brief was designed to discuss the idea of sharing in eight different themes; shared housing, shared workspace, shared transportation, shared education, shared culture, shared heritage, shared commerce and shared infrastructure. The following is a short review of the concept of sharing as expressed in the urban design in each of the eight groups:

- Shared Infrastructure

This group focused on improving the outdated infrastructural conditions on which the community relies for conducting daily activities. Eg. the community lacks critical hygienic functions such as sinks for washing, toilets in public washrooms and evenly distributed showering facilities, or a comprehensive sewage systems. The proposal suggests a phased strategy, starting with replacing existing electrical infrastructure, to be replaced with a solar energy system which consists of roof (Hutong) mounted photovoltaic panels and solar powered street lamps. Then, after an analysis of existing functions and contextual conditions, public washrooms will be converted into service hubs. And thirdly, a mobile application is designed to create a map of all washrooms and their related features, facilitated by an existing digital database of public washrooms.

- Shared Transportation

The transportation group suggests to combine a park with parking and a multi-modal transfer system, to resolve the integration of various transportation modes and social issues in

2.2 分组工作

“共享城市”设计课程本身旨在从社会、经济和人文角度为新兴的共享概念和可持续的城市发展提供解决方案。因此，由每组两到三个学生形成的八个小组关注不同的分专题，涉及场地内的不同地区，但都与整体方案密切相连。八个分专题分别是：共享居住、共享办公、共享交通、共享教育、共享文化、共享遗产、共享商业和共享基础设施。以下是对八个分专题的简要说明：

共享基础设施

该小组致力改善社区里居民依赖但陈旧的日常基础设施。白塔寺地区缺乏重要的卫生设施如盥水槽、公共厕所、淋浴设施和综合污水处理系统。方案建议采取分阶段的战略，首先取代现有的电力基础设施，由屋顶（胡同）安装的光伏板和太阳能路灯组成的太阳能系统取而代之。然后，在分析现有的功能和条件后，公共厕所将转化为服务中心。最后，基于现有的公共厕所数据库基础，移动应用程序展现所有洗手间和其相关功能的地图。

共享交通

该小组建议将停车场与多功能换乘系统结合起来，解决历史街区各种交通方式和相应的社会问题。该方案的四个目标是：解决停车混乱问题、提供完善的多功能换乘系统、设计友善的行人街道、提供适当的交流空间。

the historic district. The four targets of the proposal are: to resolve the parking chaos, facilitate easy multi-modal transfers, design pedestrian friendly streets, and to provide suitable spaces for social activities.

- Sharing Heritage

Considering the situation of the urban design proposals in the historic district, this gropu found that heritage is not limited to a particular time in history. In fact, the White Pagoda Area is remarkable in its layered heritage. It consists of old temples and historic courtyard houses from the imperial era, but also has some key heritage objects from the early days of the Republic of China, and an icon of Soviet architecture, with the FuSuiJing building. In their proposal this group aims to integrate these three main components into on spatial heritage narrative.

- Sharing Culture

The culture group focused on the idea of sharing contemporary and historic culture, without creating disturbance to either. Part of the proposal included a Baitaisi skywalk, to recreate lost space for the local activities and to share it using a parallel path for outsiders to appreciate their cultural development without disturbing the local living.

- Shared Commerce

Although home to many cultural treasures, the study area did not have much commercial activity at present. After interviewing dozens of local business owners, this group realized that the Baitasi area offers a variety of goods and services from private-owned and public-owned businesses. In their proposal the group aimed to utilize local qualities and business owners as a starting point to develop a shared business platform spatially expressed through a curated, connected commercial corridor that links to surrounding commercial areas.

- Shared Housing

The area already being familiar with the concept of a shared courtyard with multiple families living in the adjacent houses, this group coined the idea to facilitate a shared public space that could expand the living

共享遗产

考虑到历史地区的城市设计方案特点，该小组发现遗产并不限于一个特定的历史时段，而是体现出在不同时期中积淀的显著特征。此地段拥有保留至今的古老寺庙和历史悠久的四合院，也有一些新中国早期建设，如具有社会主义象征性的福绥境大楼。在该小组的方案设计中，其目标综合考虑这三个主要遗产建筑。

共享文化

该小组致力于共享当代文化和历史文化，但彼此并不干扰。设计部分包括了白塔寺天空步道，重建当地遗失的活动。外来者使用该步道，可在不干扰当地生活的同时感知当地生活。

共享商业

虽然该地段拥有许多文化资源，但到目前为止没有多少商业活动。采访当地数十名商业运营者后，该小组意识到白塔寺地区由私营和公有商业为当地提供商品和服务。该小组的方案目的是利用当地特征和商业运营者为出发点，建立一个共享商业的平台，并通过走廊串联周边商业区域。

共享居住

白塔寺地区的居民已经熟悉了多个家庭居住在相邻的房子里并共享庭院的方式，该小组提出了以共享的公共空间扩大居住空间的想法。此外，为改善当地生活条件，公共街道生活区域被引入方案之中。居住环境的改善主要集中在房屋产权分析和访谈、拆除非法建筑、增加共用庭院空间，以及建立一些庭院与主要道路的连接。

area. In addition, areas for public street life were introduced, combined with regenerated housing that improves the local living conditions. The improvement of the living environment focused first on property analysis and interviews, removal of illegal buildings, addition of shared courtyard spaces, and better connecting some courtyards to the main pathways.

- Shared Education

This group realized that all educational program in the area is currently located behind strongly gated walls in which kids have to sit still and listen in uninspiring concrete boxes. With their proposal they tried to design an alternate educational typology, in which they redefined the purpose of a school, as a place not merely to absorb knowledge, but a platform that promotes social interaction, develops personal communication skills and character while the students interact among each other. By breaking the conservative and traditional learning environment toward a more vigorous learning environment, the new school becomes an educational systems that connecting the class room spaces to the urban life, and continues all the way to the hutong residential courtyards that creates a journey of learning that goes beyond the boundary of the schools.

- Sharing Workspaces

Lastly, in order to strengthen the community and identity of the area, this group, focused on co-working, wanted to provide spaces where users can share and exchange work and ideas. Following the diverse character of two distinct sites within the larger plan, their proposal includes two different typologies, which contrast and complement each other. First, the Fusuijing building, a former communist shared housing block, will work as a flexible hub. Secondly, a smaller scale intervention is focused on the historic courtyard typology with more permanent small offices and workshops, through a combination of shared and public courtyards.

共享教育

该小组意识到在该地区的所有教育局限于厚实的院墙之内，而孩子们必须坐在无趣的混凝土空间里听课。方案试图设计另一种教育类型，重新定义学校不仅仅是单方面的吸收知识，而是一个促进社会互动、培养个人沟通技巧和性格，以及促进学生彼此之间互动的平台。为打破保守和传统学习环境并营造更积极的学习环境，新的学校方案连接教室空间和城市生活，使其成为一个整体，并延续至胡同住宅庭院，从而形成一个超越学校边界的学习之旅。

共享办公

为加强社区归属感和本地形象，该小组聚焦共享办公，希望提供用户分享交流工作及想法的空间。方案涉及两个不同的地段和两种对比和互补的共享办公类型。福绥境大楼将作为一个灵活的办公中心，而较小规模的办公环境主要集中在典型的四合院里，结合院落空间，建立更为持久的小型办公室和工作室。

3. Strategic overlay

When distilling the individual group findings with their respective location focus, we can extrapolate the effect of their respective conclusions into the overall strategic master plan, as shown in the corresponding map. Now, this map shows in various colors the different groups' focus area. It finds that the different programmatic approaches have overlapping areas of interest and intervention, and that certain strategies complement desired development directions. It also shows that the initial idea of an organism that consists of several layers of life, is clearly expressed in these overlaps.

3.策略层面

提取各小组的调查与地段后，我们可以将相应结论扩展到总体战略计划中，如下图所示。这张地图中的不同颜色展示了不同小组关注的区域。不同的方案有重叠的设计领域，在这些重叠中，清晰地表达出若干城市生活方面的整体方案。

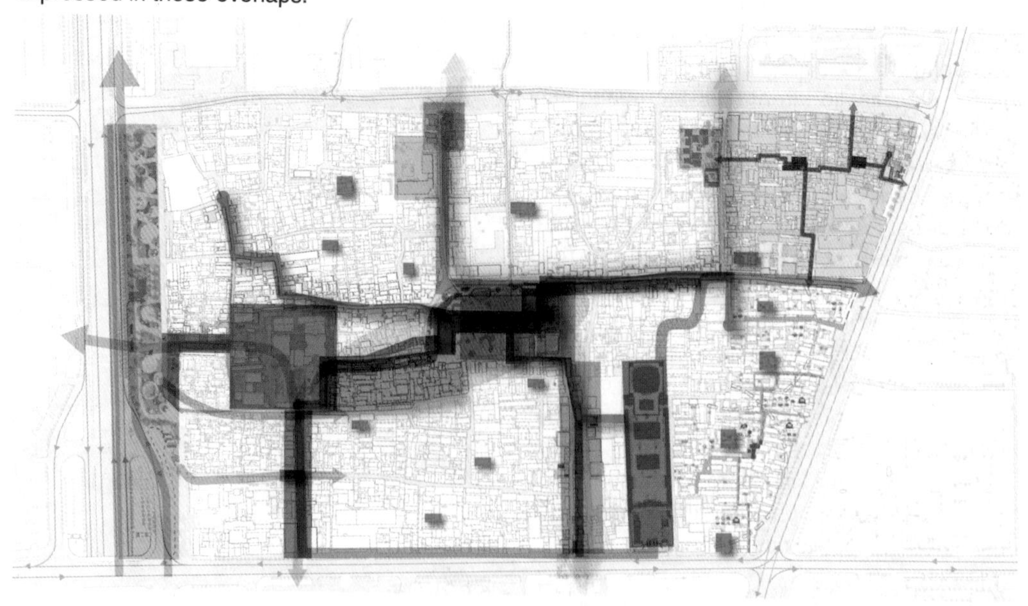

4. Conclusion

From the case study on Tsinghua's Baitasi site, we can conclude that the distribution into groups, while maintaining a collaborative, integrated design approach, can lead to a design strategy in which different individual plans complement each other. The process of integrated urban design can have graduate students design as teams, in which the cultural diversity of the participants' background enriches creative team work. In addition, through the starting point of teaching integrative urban design as part of the educational design research, a new generation of designers could be more prepared for designing sustainable urban environments, integrating cities and architecture.

4. 结论

从清华的白塔寺地区城市设计教学案例研究，我们可以得出这样的结论：任务分配到组，同时保持协同综合设计方法，可以使总体设计策略与个人计划相辅相成。城市设计一体化的过程可以使研究生设计成为一个团队，参与者自身的文化多样性丰富了团队创造性工作。此外，通过将城市设计一体化作为教育设计研究的一部分，新一代的设计师可以为城市可持续的环境设计、城市与建筑的整合做更充分的准备。

REFERENCES / 参考文献
1. George, Cherian. 2000. Singapore: The Air-conditioned Nation. Essays on the Politics of Comfort and Control, 1990-2000. Singapore: Landmark Books.
2. Harvey, David. 2012. Rebel Cities: From the Right to the City to the Urban Revolution. Verso.
3. Lee, Hsien Loong. 2014. "Transcript of Prime Minister Lee Hsien Loong's speech at Smart Nation launch on 24 November." National Infocomm Awards. Singapore: Prime Minister's Office Singapore.
4. McLaren, Duncan, and Agyeman, Julian. 2015. Sharing Cities: A Case for Truly Smart and Sustainable Cities. Cambridge: The MIT Press. Regional Research 116-137.

2

CASE STUDIES 案例

SHARING CITY CASES ALL OVER THE WORLD
世界范围内的共享城市案例

SHARING HOUSING
共享居住

1. Pole intergenerational of Etterbeek, Brussel, Belgium
2. Xiaomi You+ Youth Community, Beijing, China

SHARING WORKSPACE
共享办公

3. Hubud, Bali, Indonesia
4. Draft Atelier, Halles Pajol, Paris, France
5. We+, Beijing, China

SHARING TRANSPORTATION
共享交通

6. Metro Cable Caracas, Caracas, Venezuela
7. Car Sharing, Singapore
8. Bike Sharing, Beijing, China

SHARING EDUCATIO
共享教育

9. School Bridge, Xiashi, Village, Pinhe, Zhangzhou, Fujian, China
10. Booku Library, Kuala Lumpur, Malaysia

审图号：GS（2019）3121

cases introduced in the following pages

SHARING CULTURE
共享文化

Micro Yuan Library&Art Center, Beijing, China

41 Fleet Street, Kingston, Jamaica

Urban Remediation and Civic Infrastructure Hub, Sau Paulo, Brazil

SHARING HERITAGE
共享遗产

14. Revitalisation of Berzeit Historic Center, Birzeit, Palestine

SHARING COMMERCE
共享商业

15. The Corn Exchange, Brixton, London, United Kingdom
16. Etsy, Dumbo, Brooklyn,USA

SHARING INFRASTRUCTURE
共享基础设施

17. Water Reservoir Public Park, Medellín, Colombia
18. Repair Cafe(online), Cambridge, MA, United States

SHARING HOUSING / 共享住宅

The house is the most basic building unit in the city. A large amount of houses in the city have clear properties right, owned and used by families.

Sharing housing provides a new concept of living, it can be shared among different people in different time. It can also be sharing of some part of the original housing to a public sharing space, such as kitchen in the youth apartment, living room, or even bathroom that provide more opportunities for young people to interact.

住宅是城市之中最为基本的建筑单元。城市中相当多的住宅都是产权明晰，被固定的家庭拥有和使用的。

共享住宅提供了一种居住的新形态，它可以是不同人群在不同的时间中分享住屋，也可以是原有住宅中的一些部分更加公共地分享，例如青年公寓中的厨房、客厅，甚至卫生间，为青年人提供更多的交流机会。

SHARING WORKSPACE / 共享办公

Sharing workspace is currently one of the most common type of sharing in the sharing economy. The original architectural space is divided according to different working unit or office desk, in response to different building users in different periods of office needs. This kind of sharing workspace is commonly occupied by the individual entrepreneurs, new start-up of small institutions. Sharing workspace began to expand to tourist area, to provide temporary office space to travelling workers.

In these sharing workspace, the conference rooms, reception desk, cafes and some other public facilities belong to the operating company, and thus when the small institutions or individual office share these facilities, it reduces the operating costs.

共享办公是当前共享经济中最为普遍的一种模式，它将原有的建筑空间划分为若干办公单元或者办公桌，回应不同用户不同时段的办公需求，特别是一些个人创业者、初创时期的小型机构等，这样的共享办公方式也开始拓展到一些旅游地区，给在度假期间还需要进行办公的旅游者提供临时办公场所。

在这些共享办公空间中，会议室、前台、咖啡室等原来属于每个公司内部的功能空间被统一设置，那些小型机构或者个人办公者可以共享这些资源，降低其运营成本。

SHARING TRANSPORTATION / 共享交通

Car sharing, bike sharing and other sharing transportation are booming in recent years. Due to the need of providing transportation infrastructure for a city (including roads, parking lot areas etc.) and one needs a high costs to own a car, car sharing and bike sharing in a city not only bring convenience to the urban inhabitants, it also reduces the city's energy consumption, and promote a more sustainable development of the city.

In addition, some railway public transportations are no longer act as commute tools, they start to share the commercial, cultural and more functions, and play an active role in the overall development of the region.

共享汽车、共享自行车等共享交通方式近年来蓬勃发展，在机动车交通需要城市为其提供大量的交通基础设施（包括道路、停车场等）的地区，或者个人需要支付高成本才能拥有汽车的一些地区，共享汽车、共享自行车为城市居民出行带来方便的同时，也进一步降低了城市的能源消耗，促进了城市的可持续发展。

此外，一些轨道公共交通站点不再仅仅只是交通功能，它们共享商业、文化等多种功能，为地区的整体发展起着积极作用。

SHARING EDUCATION / 共享教育

The education system is becoming more and more diverse, and the traditional classroom space and teaching start to expand to more possibilities. Some multi-purpose public spaces became classrooms that impart and communicate knowledge, while some of the original classroom space transformed into a multi-functional public space for different periods of time and for different groups of people. Education is transforming into a sharing of experience and knowledge among different group of people, and the development of internet teaching and various media channels have greatly expanded the field of education for people to teach and learn.

教育方式正在变得日益多样化，传统的课堂式教学的空间模式和教育方式都有了很大的拓展。一些多用途的公共空间正成为传授知识、交流沟通的教室，而一些原有的课堂空间也正在转变为不同时间段、不同人群使用的多功能公共空间。教育正转变为不同人群间经验和知识的共享，并且网络教学的繁荣发展、多种媒介途径都大大扩展了教育的领域。

SHARING CULTURE / 共享文化

The cultural infrastructure brings convenience to people of different ethnics, ages and culture to interact. It responded to the needs of different groups, where it provides a space for them to share everyday life experience. Cultural infrastructure has to be able to adapt to the needs of a variety of cultural activities, and to be able to reflect the cultural features of the original area, and also to be able to accommodate a new culture.

文化基础设施的建设为不同宗族、不同年龄、不同文化背景的人们交流带来方便。因为要回应不同人群的需求，为他们分享日常生活、分享经验提供空间，文化基础设施要能适应多种文化活动开展的需要，既要反映出地区原有的文化特征，也要容纳新的文化。

SHARING HERITAGE / 共享遗产

Cultural heritage is a region's precious wealth, in terms of the tangible and intangible cultural heritage. These legacies are not ossified. They are the cohesion of past and to some extent, it is an evidence that influence today's life. Therefore, the ways of cultural heritage preservation should not be placing them in the museum, but intergrate them to everyday life, so that more people can share and pass it on to the next generation.

文化遗产是一个地区宝贵的财富，包括物质文化遗产和非物质文化遗产。这些遗产并不是僵化的，它们是以往生活的凝聚，并在一定程度上影响今天的生活。因此，对文化遗产的保护不应该是将其放到博物馆中，而是把它们和今天的生活结合起来，让更多的人在日常中感受分享，并把它们传承下去。

SHARING COMMERCE / 共享经济

Commerce activities start to transform in some urban areas. The rise of online shopping makes a great impact on the original business model, so some commercial activities began to transform to cater more leisure, cultural, and in a gathering mode for more activities running together. The original single commercial space has been transformed into a more diverse urban space for activities. On the other hand, the development of online shopping also enables more people to share the information of goods and provides new ways for shopping.

在一些城市地区，商业活动正在发生变化。网络购物的兴起使得原来的商业模式受到较大冲击，因此一些商业活动开始更多地和休闲、文化、聚会等体验活动结合在一起，原来单一的商业空间也日益转变为多样化城市活动的空间。另一方面，网络购物的繁荣发展也使得更多的人可以共享商品信息，为购物提供新的途径。

SHARING INFRASTRUCTURE / 共享设施

Conventionally, infrastructure such as water supply, waste disposal, and public toilets are considered negative facilities that affect the quality of the urban environment. In recent years, these facilities have been given new features, combined with education and exhibitions and other functional purpose, they become a new interest in the city, and even become a new icon of the city. In addition, similarly with the bike sharing, some small facilities, such as machinery, equipment and daily tools, have also begun a new mode of sharing.

传统意义上，例如供水基础设施、垃圾处理基础设施、公共卫生间等都被认为是影响城市环境品质的负面设施。近年来，这些设施被赋予了新的特点，结合科普教育、展览等功能，它们成为了城市中新的兴趣点，甚至成为城市新的标志。此外，如同共享单车一样，一些小型设施，如机械设备、日常工具等，也开始了共享共用的新模式。

SHARINGHOUSING / 共享居住

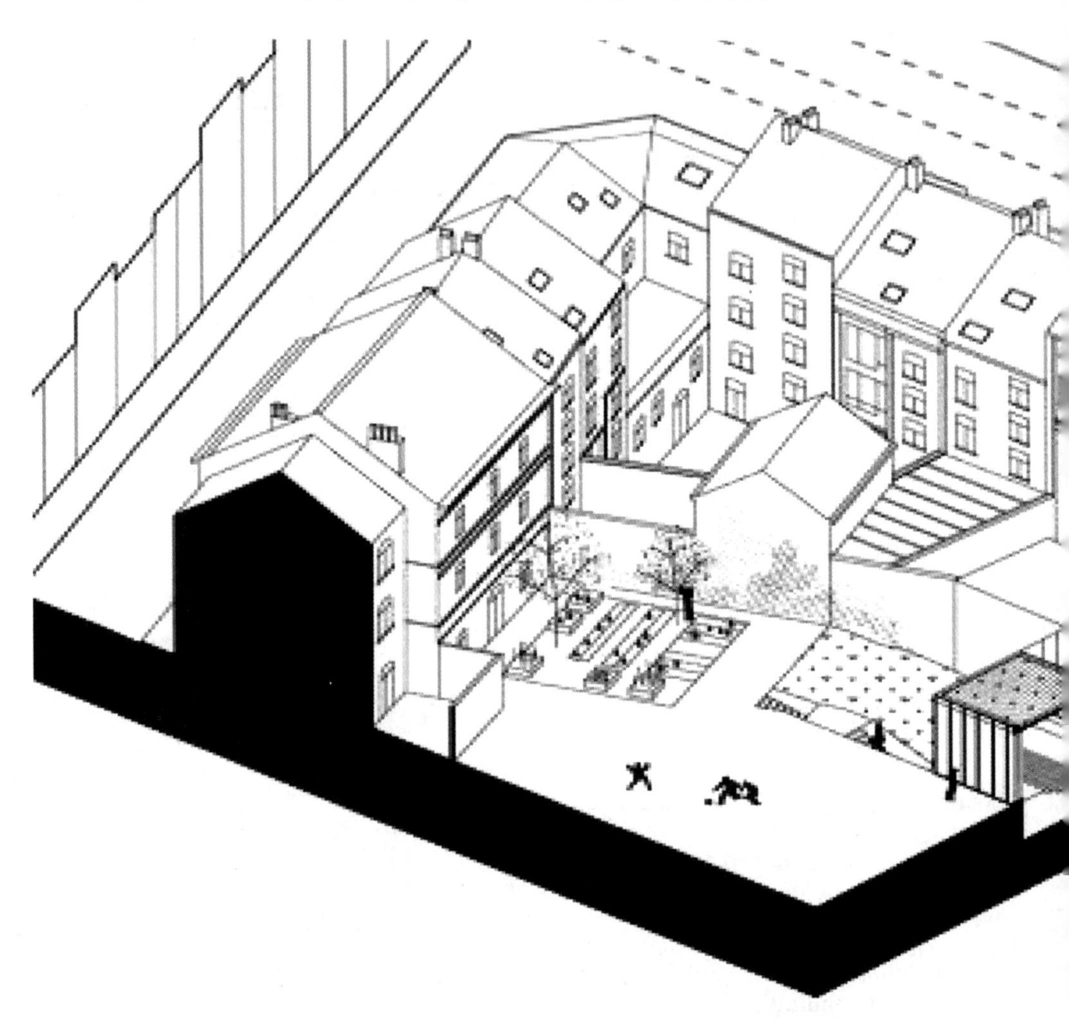

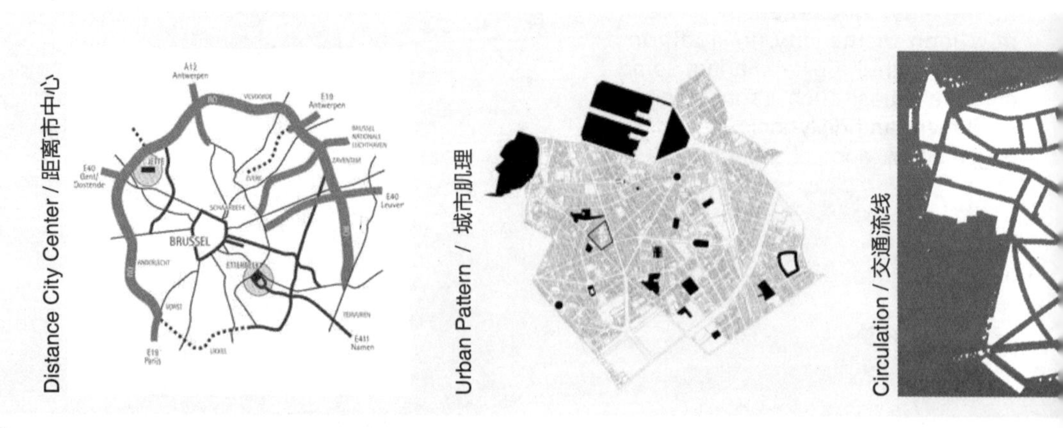

 Image source 图片来源：http://www.multiple.be/xfr/Etterbeek

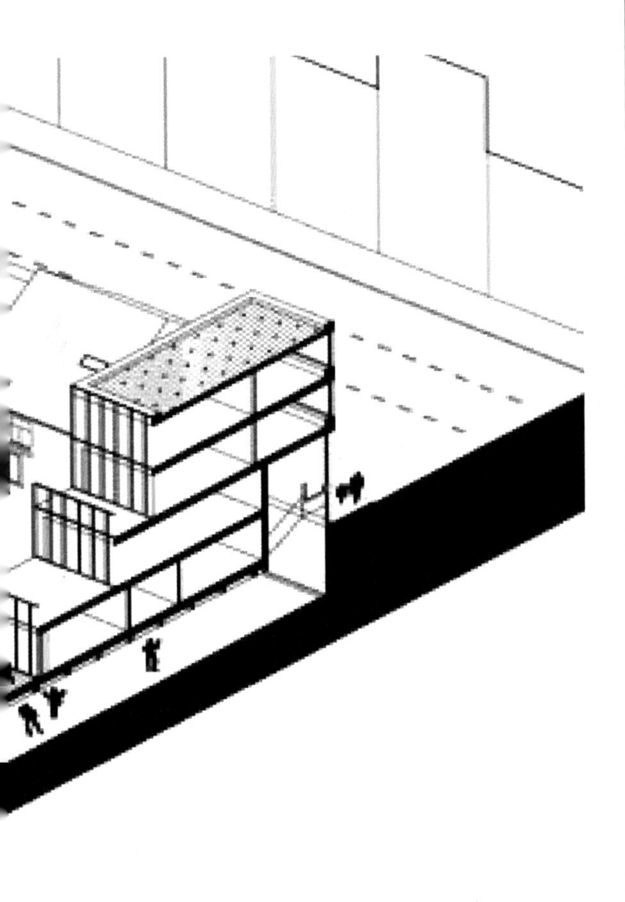

1. POLE INTERGENERATIONAL OF ETTERBEEK

ETTERBEEK, BELGIUM

Project owner : Commune d'Etterbeek
Area : 1'395 m²
Collaboration : Bureaux d'étude: JZH & Partners (Stabilité) MK Engineering (Techniques Spéciales & PEB) KAMAR (Coordination sécurité santé) STRAGES (consultance petite enfance)

The objective is to revitalize the district and to improve the lifestyle of the citizens. Housing is one of the essentiel elements.

Two major problems were identify:
- a multitude of houses are old and in bad conditions. It needs some renovation
- the proximity with the Europeen district made the land prize increase.

The program offer then, shared houses for different generation of people. But also activities as nursery, garden. In this way, the price of the land is decrease per person and the interection of the generation creat a community, revitilazing the district.

Zoning / 功能分区

1. 埃特尔贝克的国际极点住区

项目的目的是地区振兴和提升市民的生活水平，住房是其中的关键要素之一。
在项目中有两个显著问题被识别出来：
- 老旧房屋质量差，需要更新；
- 项目因靠近欧洲地区使得土地价格上涨。

因此项目为不同人群提供了共享住房、家庭共享幼儿园、花园等，通过这种方式，人均土地价格有所下降，人们之间的相互交往形成了良好的社区氛围。

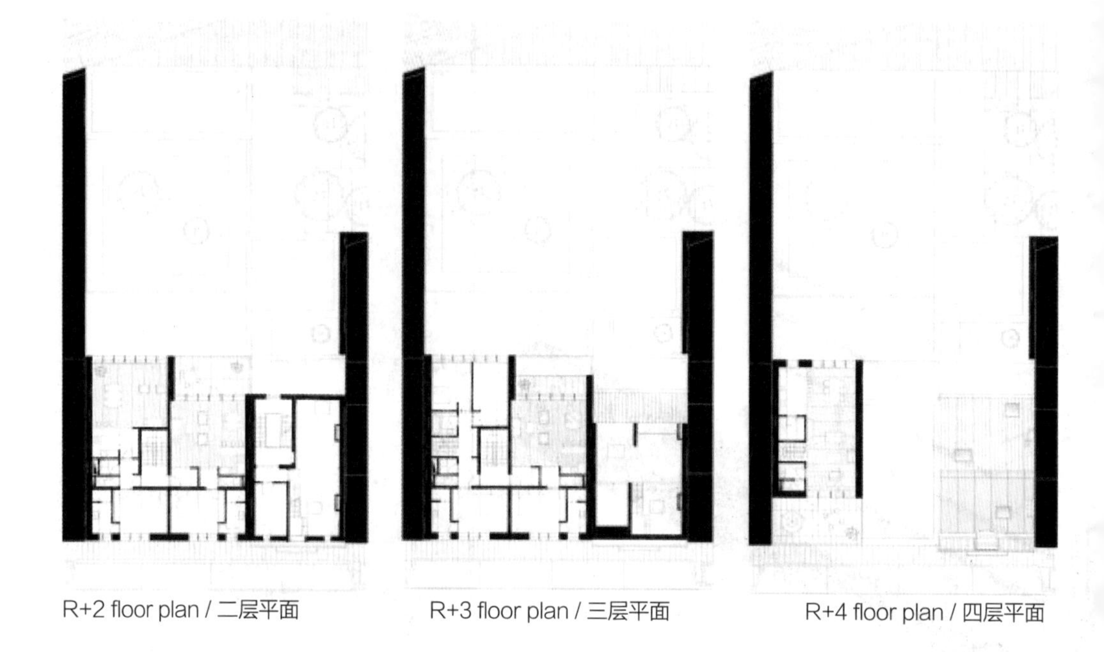

R+2 floor plan / 二层平面　　R+3 floor plan / 三层平面　　R+4 floor plan / 四层平面

Green space distribution in Brussels / 布鲁塞尔的绿地分布

Image source 图片来源：http://www.multiple.be/xfr/Etterbeek

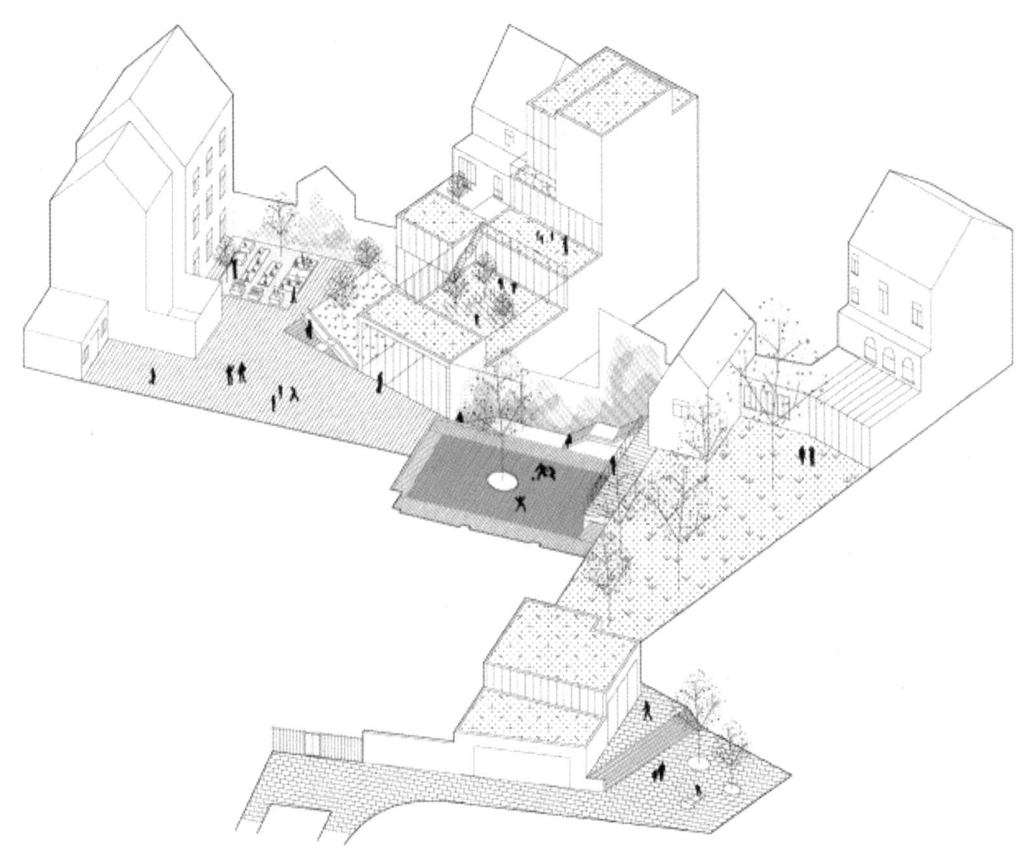

Isometric view / 轴测图

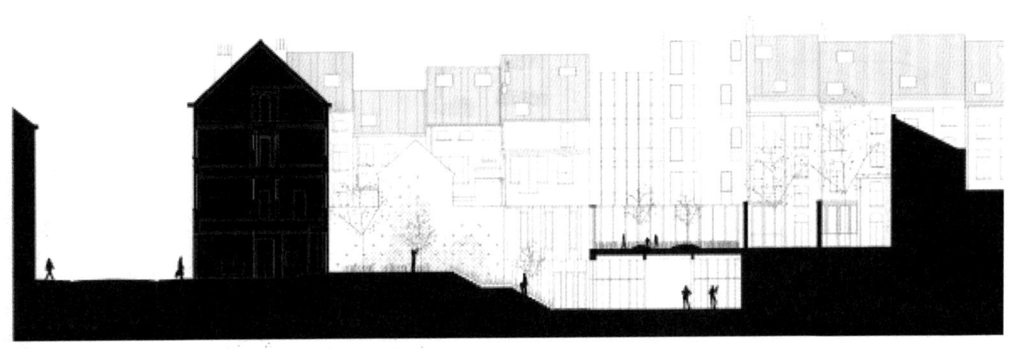

Section X—X / X—X 剖面

SHARINGHOUSING / 共享居住

Distance City Center / 距离市中心

Urban Pattern / 城市肌理

Circulation / 交通流线

 Image source 图片来源：http://map.baidu.com/

2. XIAOMI YOU+ YOUTH COMMUNITY

BEIJING, CHINA

Area: 1200 m²
Number of Rooms: 400

This apartment is all about the youth life and youth entrepreneurship. The company lease the apartment to the youth after renovation, especially targeting those who just started working. It has a nice public space and an interactive community atmosphere, which leads to better communication between young people. This apartment has three do not rent rules: first, do not rent to more than 45-year old's, do not rent to those who bring child, and lastly do not rent to those who are not interactive with others. The biggest characteristic of the design is to save a little space from each room to make a huge living room on the first floor.

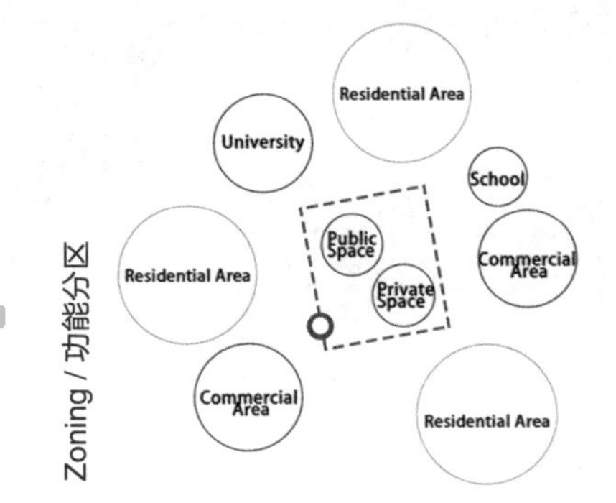

Zoning / 功能分区

2. 小米YOU+国际青年公寓

小米YOU+国际青年公寓的定位是给刚刚毕业、漂泊在外地寻求发展的年轻人提供性价比高的住处。住房经改造后出租给年轻人，小区公共空间和社区氛围好，便于年轻人的交流。公寓有三个不租的规定：不租给45岁以上的人，不租给带孩子的人，不租给不喜交往的人。设计的最大特点是从每个房间节省一点空间，在一楼设一个大客厅。

Exterior view of the hostel / 公寓外观

Public interactive space / 公共活动房间

chill out area / 放松休闲区域

Image source 图片来源：1. http://www.donews.com/net/201507/2897212.shtm 2. https://36kr.com/p/5059587.html 3. https://www.youplus.cc/commu

Interior view of the room with mezzanine floor / 带夹层房间内景

Interior view of the room with mezzanine floor / 带夹层房间内景

washroom / 盥洗间

SHARING WORKSPACE / 共享办公

 Image source 图片来源：1. Wikipedia (http://en.wikipedia.org/wiki/Valencia) 2. http://www.upv.es/ 3. Google map (https://maps.google.com)

3. HUBUD

BALI, INDONESIA

Area: 325 m²
Price: From 20$ to 275$ / month

Hubud, "Hub in Ubud" was created as a space which could embodied the paradigm of work and lifestyle.

As an international headquarter, Hubud provide different spaces where people from different business can work, meet, and share ideas, in a sense of a community and connectivity. They gathered re-purposed bamboo and worked alongside craftsmen to create the structure that balances play, pro-ductivity, and learning. They serve local coffee but also try to help the tourism-heavy economy, and to create opportunities where Balinese and the international community could engage with one another.

Hubud is about collaborating on the world's next big innovation and experimenting through some events and workshops.

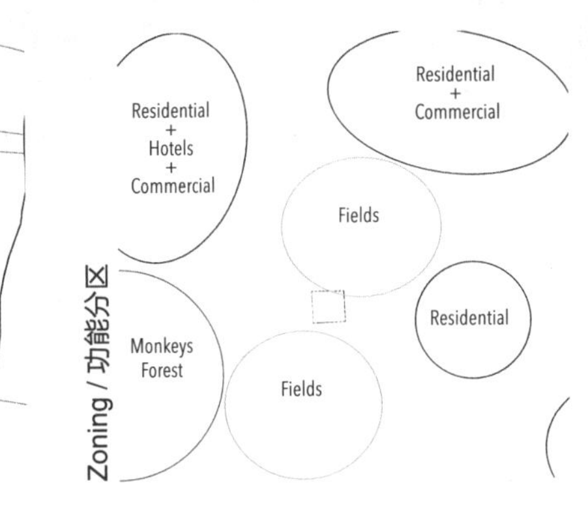

Zoning / 功能分区

3. HUBUD乌布节点

Hubud，巴厘岛的乌布节点，被创造为可以体现工作和生活方式的空间。作为国际性总部，Hubud提供给不同行业的人工作、会面和分享想法的场所，使人们感觉到社区感和彼此连接性。空间建造采用竹子，手工艺者协同一起创造了兼顾游戏、生产和学习等功能的空间。这里提供本地的咖啡，试图促进旅游业发展，并为巴厘岛和国际社会之间的交往创造机会。Hubud致力于通过一些活动和工作营，与世界范围下一项新的创新和实验相协作。

Hubud Coworking Spaces / 乌布节点共享办公空间

MAIN ROOM / 主厅

A dynamic, free-flowing space that encourages collaboration / 动态、自由流动的空间鼓励合作
- 129 m^2 / 129 平方米
- Flexible floor plan / 灵活布局平面
- Sound system / 声响系统
- Fan-based air circulation / 基于风扇的空气循环系统

For workshops, conferences, panels, team building, FGD sessions. / 可提供给工作营、会议、小组座谈、团队建设、小型讨论会等活动

THE LOFT / 阁楼

An intimate hideaway for in
之处
- 67m^2 / 67 平方米
- Flexible floor plan / 灵活
- Fan-based air circulation

For workshops, movie scre
FGD sessions, etc. / 可提供
建设、小型讨论会等活动

CONFERENCE ROOM / 会议室

- 28 m^2 / 28 平方米
- Air-Conditioner / 空调系统
- Live stream capabilities & equipment / 即时媒体系统
- Technical support / 技术支持

For meetings, workshops, brainstorming sessions, FGD sessions, etc. / 可提供给会议、工作营、头脑风暴、小型讨论会等

THE MEETING ROOM / 会

- 7.7m^2 / 7.7 平方米
- Air-Conditioner / 空调系统

For small workshops, one-o
提供小型工作营、一对一会

Image source 图片来源：https://www.hubud.org/event-venue-bali/

rings / 激发汇聚灵感的隐蔽

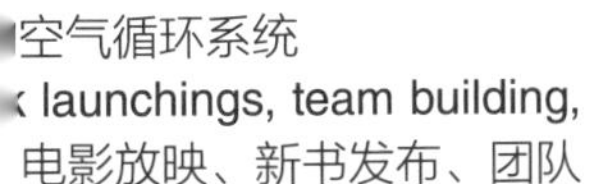

空气循环系统
launchings, team building,
电影放映、新书发布、团队

tings, running webinars / 可
讨会等

CAFE& GARDEN / 咖啡与花园

– 91 m^2 / 91 平方米
– Sound system / 声响系统
– Outdoor & Fan-based air circulation / 户外空间与基于风扇的空气循环系统
– Food + drink service available / 食物和饮料供应

For parties, evening events, workshops, movie screenings, book launching, team building, FGD sessions, etc. / 可提供聚会、夜间宴会、工作营、电影放映、新书发布、团队建设、小型讨论会等

SKYPE BOOTH / 视频电话间

– 1.2m^2 / 1.2 平方米
– Fan-based air circulation / 基于风扇的空气循环系统

For private video calls, crying alone, self contemplation or anytime you just need a minute alone / 可提供私人电视电话、独自哭泣、独自沉思或者任何需要独处的需求

SHARING WORKSPACE / 共享办公

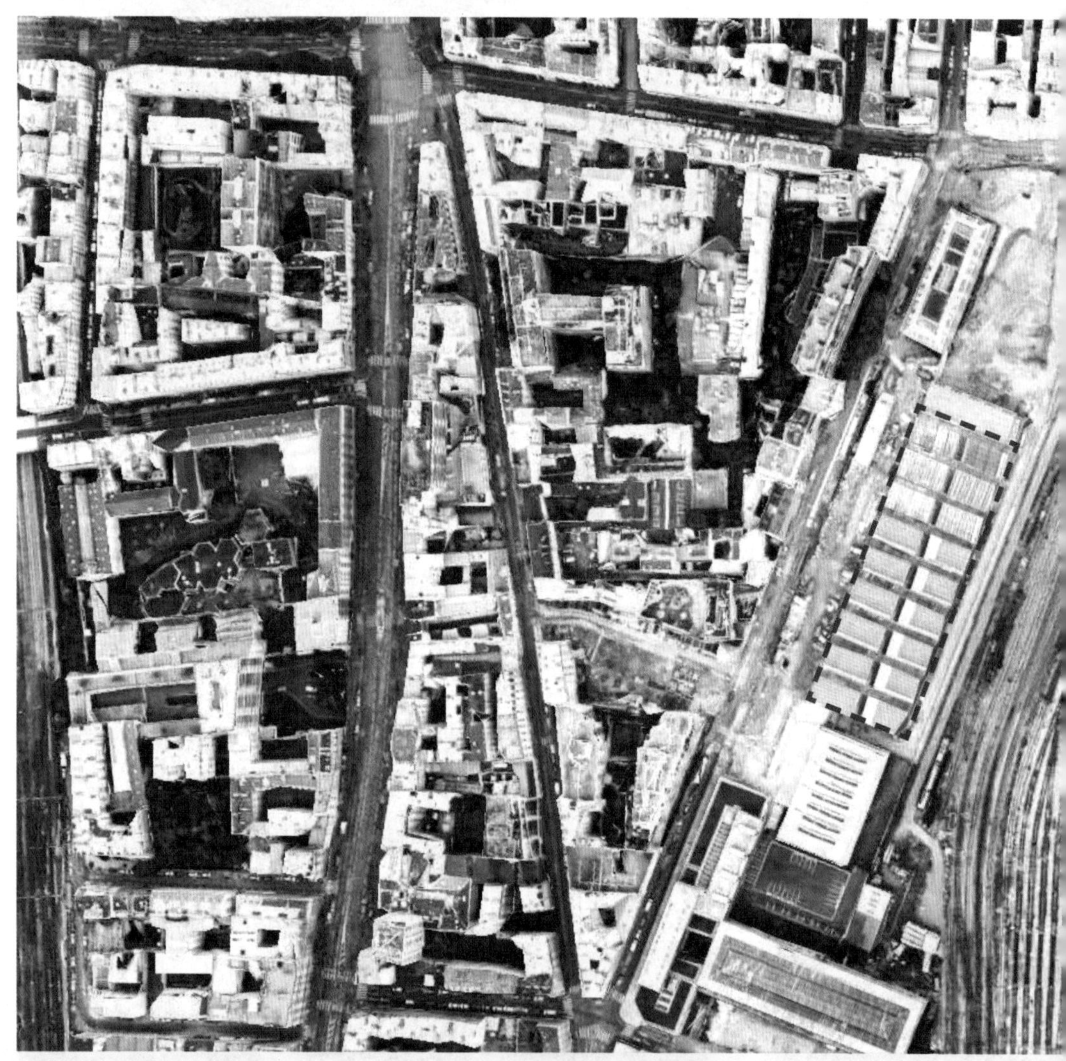

Distance City Center / 距离市中心

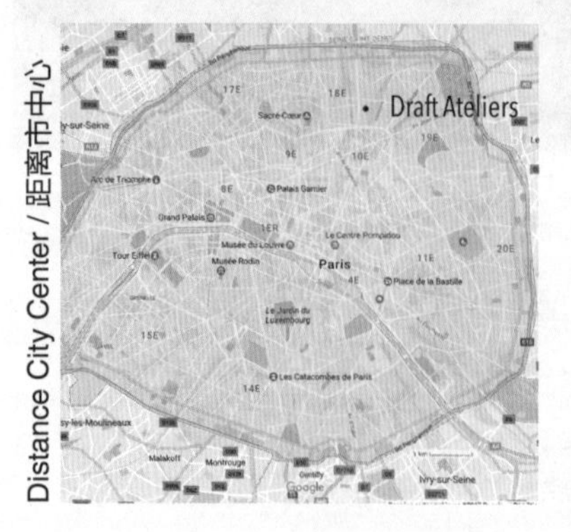

Urban Pattern / 城市肌理

Circulation / 交通流线

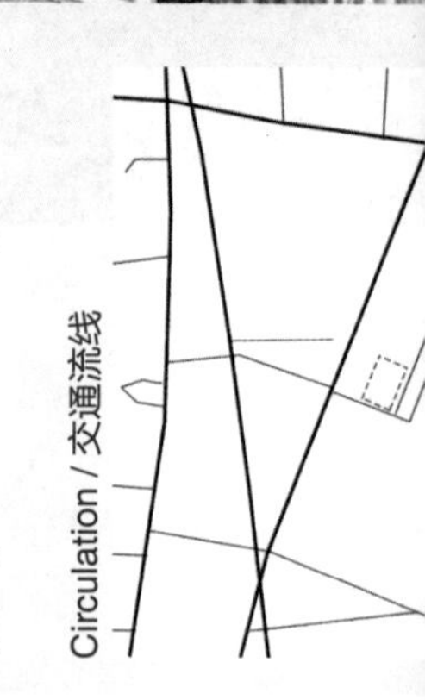

Image source 图片来源：1. Wikipedia (http://en.wikipedia.org/wiki/Valencia) 2. Google map (https://maps.google.com)

4. DRAFT ATLIERS
PARIS, FRANCE

Area: 130m²

Draft Atelier, "the connected workshops", located in the Halles Pajol, in Paris offers a colloborative space where students, contractors and companies can create and innovate together.

Numerical machines of prototypes or production are accessible for everyone in order to produce by ourselves our ideas. Draft Atelier give the chance to design, make and produce in a different way. Giving more priority to the use than the possesion, trying to define a new system of local production, testing new ways of personnalization and finally searching for a different way of consumption.

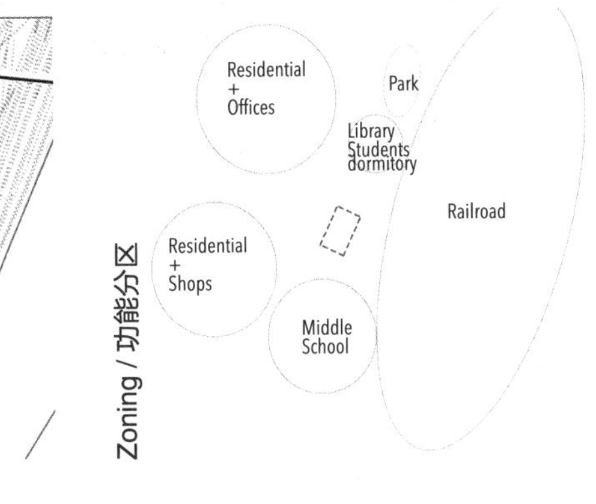

Zoning / 功能分区

4. 初步工作室

Draft Atelier初步工作室位于巴黎的哈尔斯帕乔尔，被称作"连接的工作室"。这里为学生、承包商和公司的共同创造和创新提供了合适的空间。

工作室提供大量的原型机器和机器产品供人们实现他们自己的想法。初步工作室让设计、制作和生产有了新的可能性。这里更重视空间和设备的使用而不是占有，并且尝试定义一种新的本土生产系统，旨在通过个性化的产出模式来发展新型消费方式。

Les Halles Pajol,Paris,

Coworking space in the Atelier, 130m²
共享办公空间，130 平方米

The Studio, 70m²
工作室，70 平方米

 Image source 图片来源：1. https://www.pinterest.com/pin/591801825790176 60/ 2. https://ateliers-draft.com/paris/concept#les-espaces

Wood Studio / 木工车间

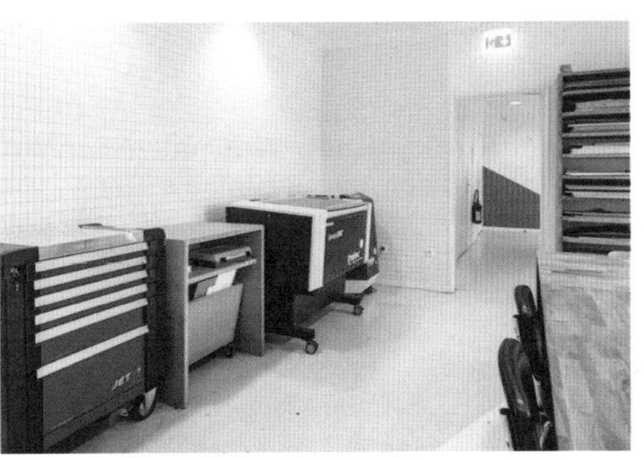

Laser cut / 激光雕刻室

SHARING WORKSPACE / 共享办公

WE+, China

WE+ co-working space is founded in May, 2015, now has 28 co-working office entities over 11 major cities within China, and USA, Finland. It is actively following the call from the central Chinese government for Innovation and Entrepreneurship. With a healthy political landscape and support, We+ focuses on providing affordable for SME (small and medium enterprises) and start-ups.

There are several major reasons for its fast growth.
(1) High demand for affordable office space targeting at new generation of millennials. New start-up companies limited resources in terms of money, talent, exposure to connections, all demand to way to group these SMEs together to create the agglomeration effect.

(2) Government support. The Central Government has made a clear stance for its full support of start-ups and SMEs.

(3) Location constraints and opportunities. We+, SOHO 3Q, Urwork, all of them are located in major city centres (Beijing Chaoyang), high-tech hubs (Bejing Zhongguancun), Universities and some even Airports. These location provide better opportunities for start-ups to grow, especially in attracting investments and talents.

(4) Technological innovation. China's networking app WeChat has a 93% penetration rate in China's Tier 1 cities. Along with We+'s own app, which is called WEPLUS (Fig.3), people can book a desk or a meeting room at We+ centre at ease.

WE+，中国

WE+合作空间成立于2015年5月，现有28个共同办事机构，分布于中国、美国、芬兰等国家的11个城市。WE+积极响应中国政府关于创新创业的倡导，凭借背景的支持，专注于为中小企业和初创企业提供可负担的服务。

WE+快速增长有几个主要原因。
（1）新的千禧一代对可支付办公空间的需求旺盛。新兴创业公司在资金、人才、接触风险等方面资源有限，这些中小企业集中在一起可形成集聚效应。

（2）政府支持。中央政府明确表示全力支持初创企业和中小企业。

（3）位置限制和机会。WE+、SOHO 3Q、Urwork都位于中心城区（北京朝阳）、高科技枢纽地区（北京中关村）、大学甚至机场周边。这些位置为初创企业提供了更好的机会，特别是在吸引投资和人才方面。

（4）技术创新。中国的网络应用程序WeChat在中国一线城市的使用率达93%。随着WE+自己的应用程序，即WEPLUS的应用，人们可以轻松地在WE+中心预订桌面或会议室。办公室通常位于市中心，方便大家见面。

 Image source 图片来源：1. https://www.weplus.com/ 2. https://www.douban.com/note/531752125/

5. WE+ CO-WORKING SPACE
CHINA

Number of Co-working space: 28
Total Area: 98800m²
Price: 1,200 ¥/ month

We+ is a co-working space brand, established in 2015 in China. In response to the centralising "public entrepreneurship and innovation". We+ provides a suitable office space for small and micro enterprises. At the same time, the open plan office environment enhance the interaction and promotes the multimedia sharing spirit.

5. WE+共享办公空间

WE+是一个联合办公空间品牌，成立于2015年5月，响应“大众创业，万众创新”的指导，积极打造适合中小微及创业型企业的办公场所，营造开敞的办公环境，倡导互联网共享精神。

Unlike We+, Hubud in Bali is tagging on something called "Coworkation". People from developed countries such as USA, UK, going on a co-working vacation to Bali Island. Ranging from a few weeks, to 3-6 months.

如果将WE+与巴厘岛的Hubud相比较，可以看到，与WE+不同，Hubud旨在体现“共享工作营”的方式。来自美国、英国等发达国家的人们前往巴厘岛度过从几周到3~6个月的共享办公假期。

Most of the people working here are doing freelance programing, independent designers, bloggers. "Digital nomads," they call themselves. These people cannot afford work-free holidays, so the combination of these two becomes a new lifestyle.

在Hubud工作的大多数人是自由编程者、独立设计师、博客博主。他们称自己为“数字游牧民族”。这些人不能负担完全不工作的假期，所以度假与工作的结合变成了一种新的生活方式。

Waking up doing yoga in the rice paddy, either go surfing or working in front of a field of green in the afternoon. Climb a volcano and talk about this cool business idea you have with your friend on the way down and work on it that night. The exotic work-living lifestyle and the cheaper living cost compared to NYC, is what drives people there. Joo chiat presents similar potential to expatriates and backpackers despite higher living cost, which can be reduced once coupled with self-production programs like urban farming and sharing by giving. The atmospheric setting in Hubud co-working space is even less formal than We+, if We+ is

醒来在稻田里做瑜伽，下午去冲浪或在绿林中工作，爬上一座火山，在与朋友下山的路上谈论你很酷的商业理念，并在夜间围绕该理念进行工作。与纽约市相比，这样的异国情调的工作生活方式和便宜的生活成本是驱使人们前往那里的原因。新加坡如切地区对外籍人士和背包客也具有类似的吸引力，尽管生活成本较高。但如果加上自主生产项目，如城市农业和资源分享，生活成本可以减少。

Hubud共享办公空间的环境比WE+更不正式，如果WE+像一个咖啡馆，那么Hubud更像一个家。人们穿着凉鞋往来其间，办公空间在白天完全是自然的通风和照明。Hubud的网站展示了组织的各种活动，以促进共同学习，从比特币投资到印度尼西亚新的IT技巧。这些共享学习是完全免费的。

WE+ CO-WORKING SPACE / WE+ 共享办公空间

Image source 图片来源：1. https://www.weplus.com/ 2. https://www.douban.com/note/531752125/

like a café then Hubud is like a home. People are wearing sandals, space is completely natural ventilated and lit during the day. Their website shows wide range of activities they organised to promote co-learning, from Bitcoin investment, new IT tricks to Bahasa Indonesia. To learn from each other without any cost.

Comparisons and Reflections

The mixture of these two groups of people may create even greater synergy and innovation from a business standpoint. The digital nomads, who are mostly independent, creative and constantly on the move around the globe, can inject fresh blood and vibrancy into SMEs who are more structured. The local start-ups can expose to, not only expatriates' expertise, but also digital nomads' ideas who are constantly on the move.

In terms of design, WE+ seems to be copying the West's model of urban co-working. There is little reflection of the place or the city the office is in. The space is modern, yet somehow lack a sense of belonging and familiarity. However, the solution appears to be local in Hubud's case. Honest use of material, exposure to natural elements, responding to local context and creating a life or narrative not just within the building but within the city.

Ultimately what differentiates a co-working space from short-term leasing offices, is its ability to build trust and connection among its workers, despite company entities or social class. It is empowerment the people to release their full potential. It is pertinent for businesses to see this potential. Moreover, it is important to increase the scale of this potential to the district, into the neighbourhoods to form an ecosystem of its own.

比较与反思

从商业角度来看，数字游牧民族和当地初创企业从业人员这两类人的混合可能会产生更大的协同作用和创新。数字游牧民族主要是独立的、创造性的，不断在全球范围内移动，可以为更具结构性的中小企业注入新鲜血液和活力。当地的初创企业不仅可以体现出外派人士的专业知识，还可以让数字游牧民族的观念不断提升。

在设计方面，WE+似乎正在复制西方城市的合作模式。办公室的地方性或所在城市的特点几乎没有反映。空间是现代的，但没有归属感和熟悉感。相对而言，Hubud的解决方案看起来更具地方性。真实的使用本地材料，暴露于自然元素之间，回应当地环境，创造在建筑内、城市内的生活或叙事。

共享办公空间区别于短租办公室的是它更有利于建立员工之间的信任和联系，无论是公司实体之间还是社会阶层之间。它赋予人们发挥自己的全部潜力，而且看到这一潜力可资发展相关的业务。此外，重要的是将这一潜力的规模扩大到地区内、进入邻里以形成自我的生态系统。

SHARING TRANSPORTATION / 共享交通

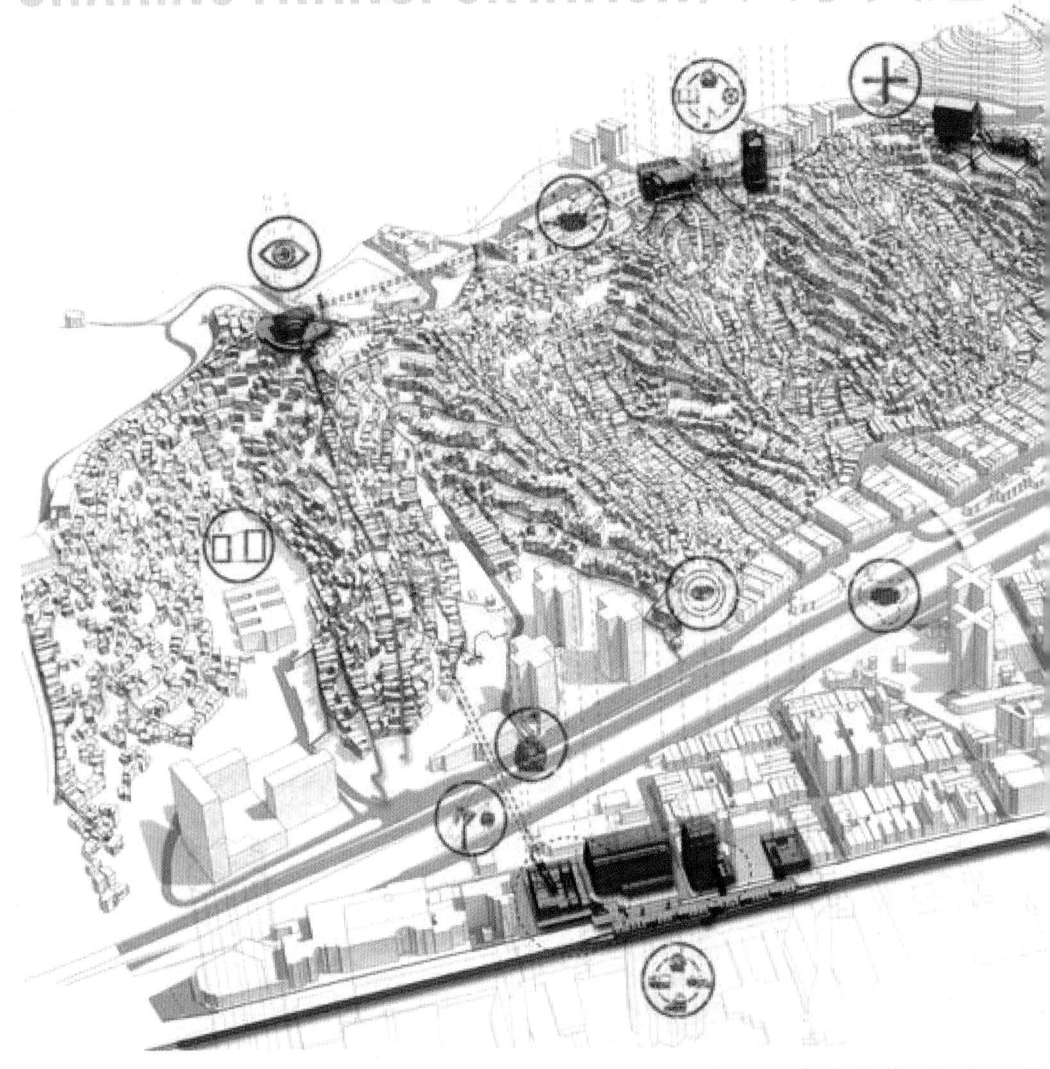

La Ceiba Station / LaCeia车站

Vertical gym integration / 整合的立体健身房

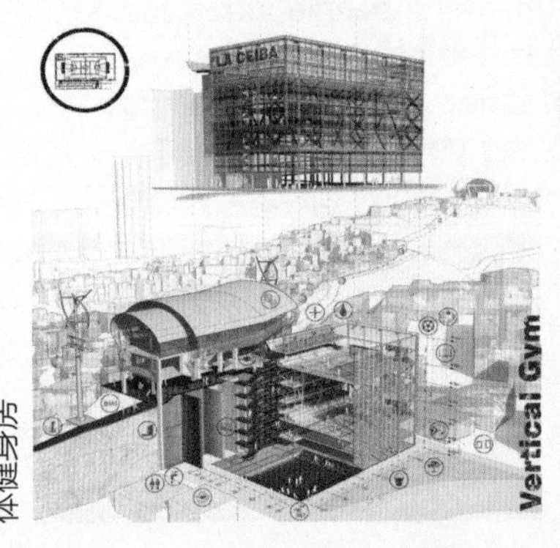

Growing House integration / 整合不断增长的住宅

Image source 图片来源：1. https://www.detail-online.com/article/simply-constructed-a-chance-for-our-urban-planet-14177/ 2. http://www.archdaily.com/429744/metro-cable-caracas-urban-think-tank

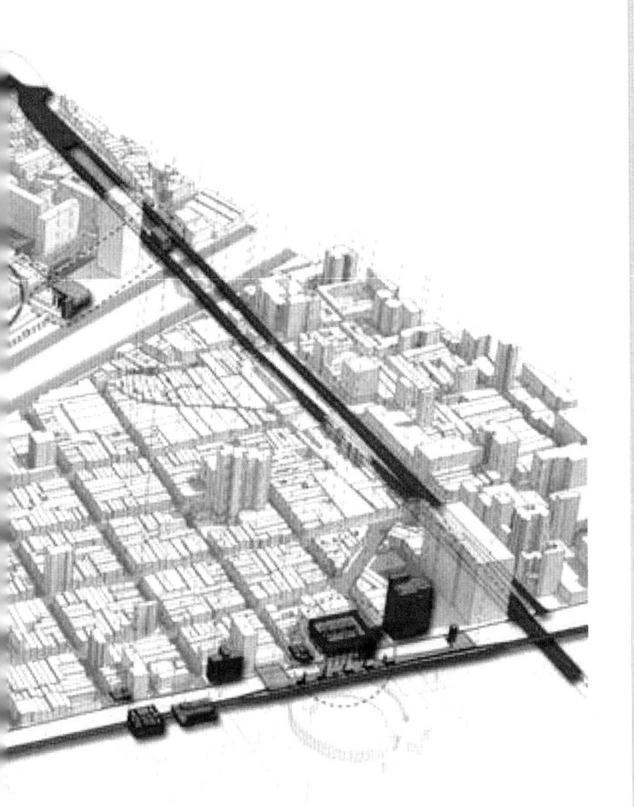

6. METRO CABLE CARACAS

CARACAS, VENEZUELA

Architect: Urban-Think Tank
Client: C.A. Metro de Caracas

Metro Cable Caracas was the first cable car system being designed just to serve the community of Caracas. Its aim is to integrate the slums with the rest of the city, by setting up five cable car stations, integrated with the existing Caracas Metro station. Where most of the slums are located at the hillsides, the strategic of choosing the location for the stations on the slums will be based on:

(1) Accessibility, adequate pedestrian circulation patterns.
(2) Constructive sustainability with minimum expropriation and demolition of existing houses.

The other 2 stations located on the valley serve as connection with the public transport system of capital. Facilities such as vertical gym, sharing housing and music factory are integrated as part of the Metro cable system at slum area.

6. 加拉加斯地铁缆车索道

加拉加斯地铁缆车索道是第一个为加拉加斯社区服务的缆车系统。项目设立五个缆车站并将其与现有的加拉加斯地铁站结合起来，实现了贫民窟与城市的整合。 大多数贫民窟位于山坡上，其中3个站点的选择基于以下两点：

（1）步行可达性；
（2）建设的可持续性：最小限度的土地开发和最低程度的房屋拆迁。

位于山谷的其他两个站点与首都的公共交通系统连接。立体健身房、共享住房和音乐工厂等设施被整合为贫民区地铁电缆系统的一部分。

Growing House

Music Factory

Music factory integration / 整合的音乐工厂

SHARING TRANSPORTATION / 共享交通

Current Service Providers
1. Smove
2. Car Club
3. Whizz Car

These companies all provide similar services and attempt to streamline the car sharing process and differentiate themselves from regular car rental firms through the following:
1. Keyless Access.
2. Phone App managed.
3. One-way service .
4. Short-term rental.
5. Pay based on usage (time/distance).

The process of car sharing in Singapore can be generalised to these few steps:
1. Online registration and verification of driver's license.
2. Reservation of car through web portal or app, up to 10-20 minutes prior to booking.
3. Check for defects and pick up of car at designated parking lots all around Singapore.
4. Return of car at any of the designated parking lots, payment to be automatically deducted.

The phone app for each company manages supply and demand in each district so that customers can be updated on the availability of cars. As quoted from Car Club's website, their fleet size is based on usage statistics, and each car serves up to 27–30 users. The 100+ hubs where the fleets are parked at are located at existing carparks of establishments/HDB (Housing Development Board) flats which the companies have liaised with. HDB has identified another 100 carparks with the potential to house these car sharing systems, as well as an additional 300 in the near future.

当前服务提供商有：
1. Smove
2. 汽车俱乐部
3. Whizz汽车

这些公司都提供类似的服务，并试图简化汽车共享过程，通过以下方式与普通汽车租赁公司区分开来：
1.无钥匙接入。
2.电话应用程序管理。
3.单程服务。
4.短期租赁。
5.根据使用情况（时间/距离）付款。

新加坡汽车共享的过程可以概括为以下几个步骤：
1.在线注册和核实驾照。
2.提前10~20分钟通过门户网站或应用程序预订汽车。
3.在新加坡各地的指定停车场检查车辆的缺陷和车辆。
4.在任何指定停车场返回汽车，付款自动扣除。

每个公司的电话应用程序管理每个区域的供应和需求，以便客户可以更新汽车的使用状态。引用汽车俱乐部的网站信息，其车队规模是根据使用情况统计的，每辆车最多可以为27~30个用户提供服务。车队停泊的100多个节点位于现有公司/屋宇署（房屋发展局）单位停车场，共享汽车公司与这些既有停车场单位建立了联系。房屋发展局已经确定了另外100个停车场，有可能安置这些汽车共享系统，未来将还有300个停车场应对共享汽车的停放问题。

Image source 图片来源：1. https://sg.carousell.com/p/uberflex-club-part-time-job-20-ph-lobang-104868430/ 2. https://vulcanpost.com/19851/smove-green-rental-cars-singapore/

7. CAR SHARING
SINGAPORE

Currently in Singapore, the combined fleet of shared cars consists of approximately 650 cars. Car sharing has recently gained ground in Singapore due to rising COE (Certificate of Entitlement, a license required to own a car in Singapore) prices, costs of car ownership as well as an improved public transport infrastructure. These factors combined have deterred residents from car ownership and turned them instead towards car sharing for the short periods of autonomous car usage that might pop up now and then.

7. 共享汽车

目前在新加坡，共享汽车共有约650辆。本地汽车拥车证价格的上调，使得拥有私家汽车的成本增加，以及改进公交系统的成本促使了共享汽车需求的增长。昂贵的拥车成本和交通设施的进步改变了人们对拥有私家车的渴望。

SHARING TRANSPORTATION / 共享交通

OFO

OFO was initiated in the school compound, until October 2016, it has reached 22 cities, more than 200 universities, and provide more than 400,000,000 bike sharing service. Now, it has solved the most of the campus traffic solution, bring ease to the college teachers and students, reduced carbon consumed and increase the efficiency of the bike sharing.
In the city, inefficient travel conditions unable to meet the fast-paced city life. However, OFO with the simple vision of "ride anytime, anywhere", provides people's short-distance travel needs and thus, bike sharing has became the new way to connect the people with the city.

OFO 小黄车

OFO 源起于校园，直至2016年10月，已覆盖了全国22座城市、200多所高校，累计提供超过4000万次共享单车出行服务。已成为中国规模最大的校园交通代步解决方案，为广大高校师生提供便捷经济、绿色低碳、更高效率的校园共享单车服务。
低效率的出行状况已经无法满足快节奏的城市生活，两点一线的生活半径和不断加速的城市改造，让外来人融不进来，也让原住民忘了城市本来的样子。正因如此，OFO共享单车怀揣"随时随地有车骑"的朴素愿景来到城市，试图满足人们短途代步的需求，成为联系城市中人们的新方式。

Mobike

Mobike created the world's first intelligent shared bicycle model. Its smart lock integrated GPS and communication module in conection with smart phone APP allow users to locate and use the bike, and ride to their destination, park at the roadside or any appropriate area, then lock them to achieve electronic payment settlement.

Mobike was established in January 2015, and officially launched in Shanghai. Mobike has launched in Shanghai, Beijing, Guangzhou, Shenzhen, Chengdu, Ningbo, Foshan, Xiamen, Wuhan, Dongguan, Kunming, Nanjing, Zhuhai, Haikou, Nanning, more than 30 cities in China and Singapore. Mobike is the world's largest intelligent bike sharing company. It has successfully make a comeback for bikes to the city and promotes green city with a smart technology solution.

摩拜

摩拜单车创建了全球首个智能共享单车模式，其专利智能锁集成了GPS和通信模块，采用了新一代物联网技术，通过智能手机APP让用户随时随地可以定位并使用最近的摩拜单车，骑行到达目的地后，就近停放在路边合适的区域，关锁即实现电子付费结算。

摩拜单车于2015年1月成立，2016年4月22日地球日当天在上海正式推出智能共享单车服务，并已先后进入上海、北京、广州、深圳、成都、宁波、佛山、厦门、武汉、东莞、昆明、南京、珠海、海口、南宁等30多个国内城市和包括新加坡在内的海外城市，稳居全球最大的智能共享单车运营平台。摩拜在所到的城市中掀起骑行的热潮，推动"让自行车回归城市"，为更多人的出行带来方便，也给城市倡导绿色出行提供了可持续发展的智能解决方案。

Image source 图片来源：1. http://www.ofo.so/about/ 2. https://mobike.com/cn/ 3. http://www.boxihao.com/blog/ofo-ofo 4. http://www.sohu.com/a/191363636_204550

8. BIKE SHARING
CHINA

At the end of 2016, the bike sharing companies started to provide a bicycle sharing service on a campus, subway station, bus station, residential area, commercial area and public service area by a low charging rate (from ¥0.50–1.50 per hour) . This bike sharing concept is a new type of sharing economy too. At the same time, due to its low-carbon travel concept, the bike sharing has attracted attention from the society too.

8. 共享单车

2016年底，共享单车企业以一种较低成本（0.5~1.5元/小时）的分时租赁模式在校园、地铁站点、公交站点、居民区、商业区、公共服务区等提供自行车单车共享服务。共享单车是一种新型共享经济。由于其符合低碳出行理念，共享单车已经引起人们的注意。

SHARING EDUCATION / 共享教育

 Image source 图片来源：1. Google map (https://maps.google.com) 2. http://www.archdaily.com/45409/school-bridge-xiaodong-li

9. SCHOOL BRIDGE

FUJIAN, CHINA

Area: 195 m²
Student: 40
Staff: 2

The bridge school was established in 2008 under the mission of the Xiashi Village in Pinghe. The main concept of the design is to enliven an old community (the village) and to sustain a traditional culture through a contemporary language which does not compete with the traditional, but presents and communicates with the traditional with respect. It is done by combining few different functions into one space — a bridge which connects two old castles cross the creek, a school which also symbolically connects past, current with future, a playground (for the kids) and the stage (for the villagers).

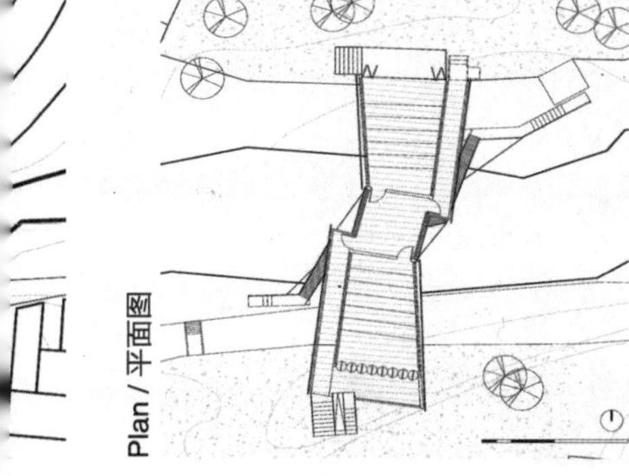

Plan / 平面图

9. 桥上书屋

桥上书屋于2008年完成，是一项由平和下石村筹募的项目。主要设计理念是激活一个老社区（村庄），在不与传统文化形成冲突的前提下，通过当代的语言实现与传统的交流。它由几个不同的功能整合为一个共同的空间——过溪桥梁连接两个旧城堡，象征学校连接了过去、现在和未来；连接了操场（为孩子）和舞台（为村民）。

The Bridge School is a two-classroom school in the small village of Xiashi, up in the mountains of the Fujian Province in China. It is so called because it bridges the two parts of the village that lie on either side of a small creek that runs about ten metres below the village. Suspended from the structure of the school and below it is a pedestrian bridge for people to use. There are two abandoned tulous on either side of the creek.The Bridge School is like a symbol of a truce.

The building is small and modern in design. It has no reference to traditional building styles but is set in place as if it always belonged there. The idea of a building as a bridge, although not unknown in other parts of the world, is a new concept here, and appreciated by the local community. It has a quiet and dignified presence and is striking for its simplicity. It has been able to transform the way the people of the village think about buildings and to introduce new aesthetic values to them.

桥上书屋是下石村一所有两个教室的学校，位于中国福建省的众山之中。它之所以被称为桥上书屋是因为它联系了位于一条 10m 宽的小溪两侧的两部分村庄。在书屋立体结构之下悬挂着供人们使用的步行桥，在小溪的两侧各有一座被遗弃了的城堡式的土楼，而桥上书屋看起来就像是某种停战的符号。

建筑尺度小但设计很现代，尽管没有采用传统的建筑形式但身处大环境下使得其体现出很显著的归属感。建筑的概念是作为一座桥，对本地社区而言是一个很新但被接受的概念。建筑安静、端庄，并以其简洁性打动观者。它改变了当地居民对建筑的看法并将新的美学观念引入其间。

Image source 图片来源：http://www.archdaily.com/45409/school-bridge-xiaodong-li

SHARING EDUCATION / 共享教育

Distance City Center/ 距离市中心

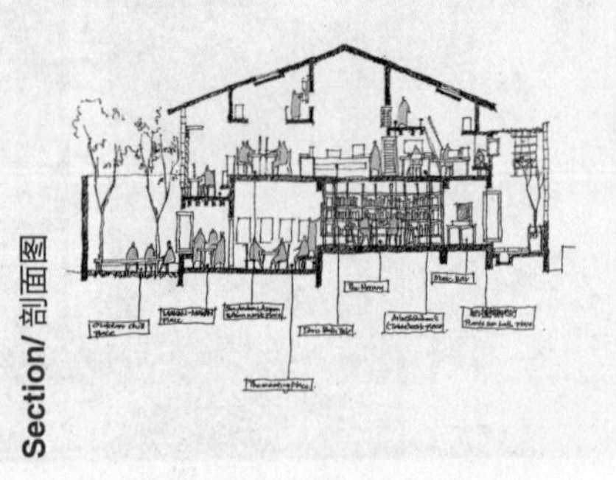
Section/ 剖面图

Front Elevation/ 正面图

Image source 图片来源：1. http://readbooku.wixsite.com/booku 2. Google map (https://maps.google.com)

10. BOOKU LIBRARY
KUALA LUMPUR, MALAYSIA

Architect: Tetawowe Atelier
Founders: 2015(-ongoing)

"Booku" is an architects initiated, non-profit library committed to the exchange of cultural knowledge within a multidisciplinary community. "Booku" is dedicated to promote literature whilst facilitating and showcasing architecture through exhibition, discourse and education.
Thus, Booku is a multi-functional library that fills the gap in access and opportunity by engaging discourse through literature, music, visual arts, performance and events.

10. Booku图书馆
马来西亚

Booku是由建筑师发起的非营利图书馆，目的在于形成多学科社区，促进文化知识的交流。Booku在促进文学交流的同时，致力于通过展览、交流和学习来推广建筑。因此，Booku是一个多功能的图书馆，促进了文学、音乐、视觉艺术、表演和各类活动的交流机会。

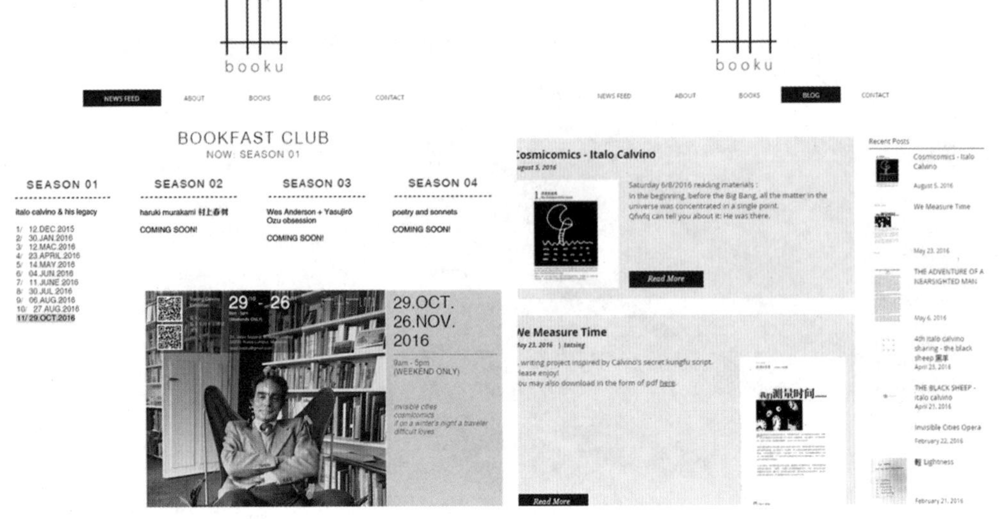

Website / 网页

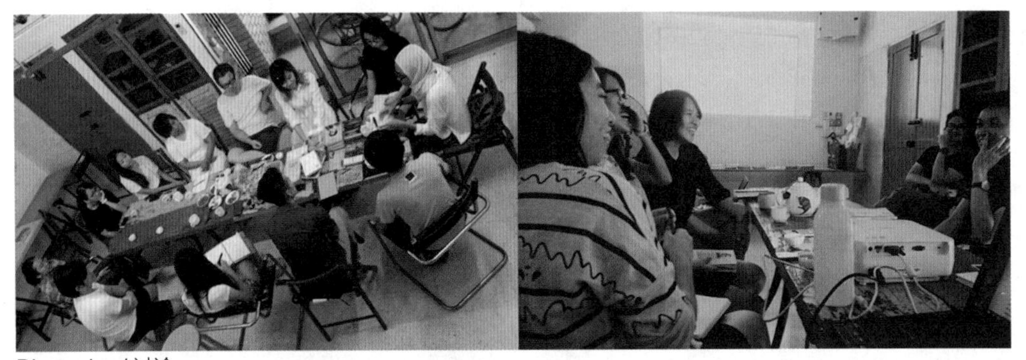

Discussion / 讨论

Reading Space / 阅读空间

Image source 图片来源：1.http://littleplanetlab.wixsite.com/lppl/copy-of-a3-jalan-sepakat-4 2.http://littleplanetlab.wixsite.com/lppl/tracingcalvino 3. http://readbooku.wixsite.com/

Outdoor Reading Space / 户外阅读空间

Exhibition Space / 展览空间

Exhibition Space / 展览空间

SHARING CULTURE / 共享文化

Distance City Center / 距离市中心

Urban Pattern / 城市肌理

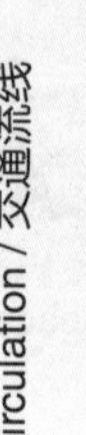

Circulation / 交通流线

Image source 图片来源：1. Google map (https://maps.google.com)

11. MICRO YUAN LIBRARY & ART CENTER
BEIJING, CHINA

Architects: ZAO/standardarchitecture
Area: 17.0 sqm
Project Year: 2013

The Micro Yuan'er is located in the Cha'er Hutong in Beijing China. It is located within the Dashilar district among other hutongs to the south of Tiananmen Square. Hutongs as we know are historical structures within dense areas with narrow streets and alleys. They are usually used as objects of tourism to illustrate their survival for over the many years past. Hutongs are generally in danger and most of them will vanish one day.

The Micro Yuan'er was created to introduce activities the local community had never experienced. It is mostly focused on children's activities which functions as a library and as a art/dance studio.

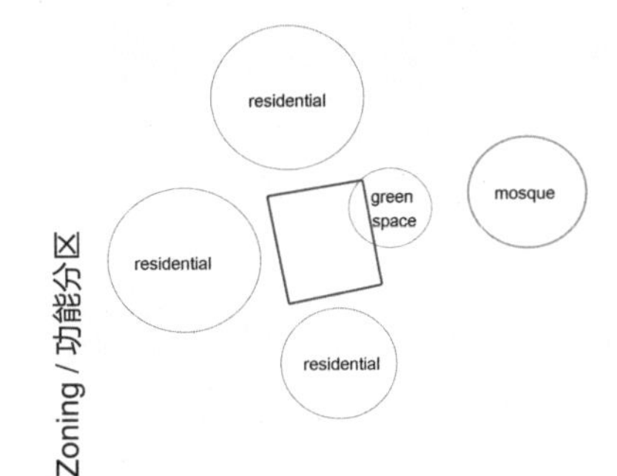

Zoning / 功能分区

11. 微杂院
北京，中国

微杂院位于北京旧城大栅栏街区茶儿胡同。众所周知，胡同是传统密集住区里的狭窄街道和小巷。在一定程度上，胡同展示了老北京人传统的邻里生活方式。今天，传统的胡同逐渐被城市道路所取代，面临着消失的困境。
“微杂院”的设计为传统社区引入了新的活动。设计聚焦在为儿童提供作为图书馆和艺术/舞蹈培训等的活动空间。

The space was renovated around a very old scholar tree and carefully designed within th existing structures. The main materials of the project are traditional grey bricks both new and recycled to make the project blend gracefully within its surroundings. The design maintains and improved the exisitng space composition using the exisitng infastructure systems as well.

The people of the hutong commuity has benefited from the library and its other facilities as it introduces them to activities from outside world and shares the knowledge of the arts (painting, dancing and reading). The center has also become a central area for the community to exchange knowledge among visitors and locals.

围绕一棵老树的空间被更新，并谨慎地同既有构筑物建立联系。项目的主要材料是传统的灰砖，一部分是新的，一部分是回收利用的，使得新的部分与其周边环境优美地混合在一起。设计维持并改善了既有空间构成，同时也维持并改善了既有的基础设施。

胡同社区的居民从图书馆和其他设施上获益，这些设施将外部世界的活动和艺术知识（如绘画、舞蹈和阅读）介绍给当地居民，该项目也已成为社区居民与外来参观者交流知识的中心地区。

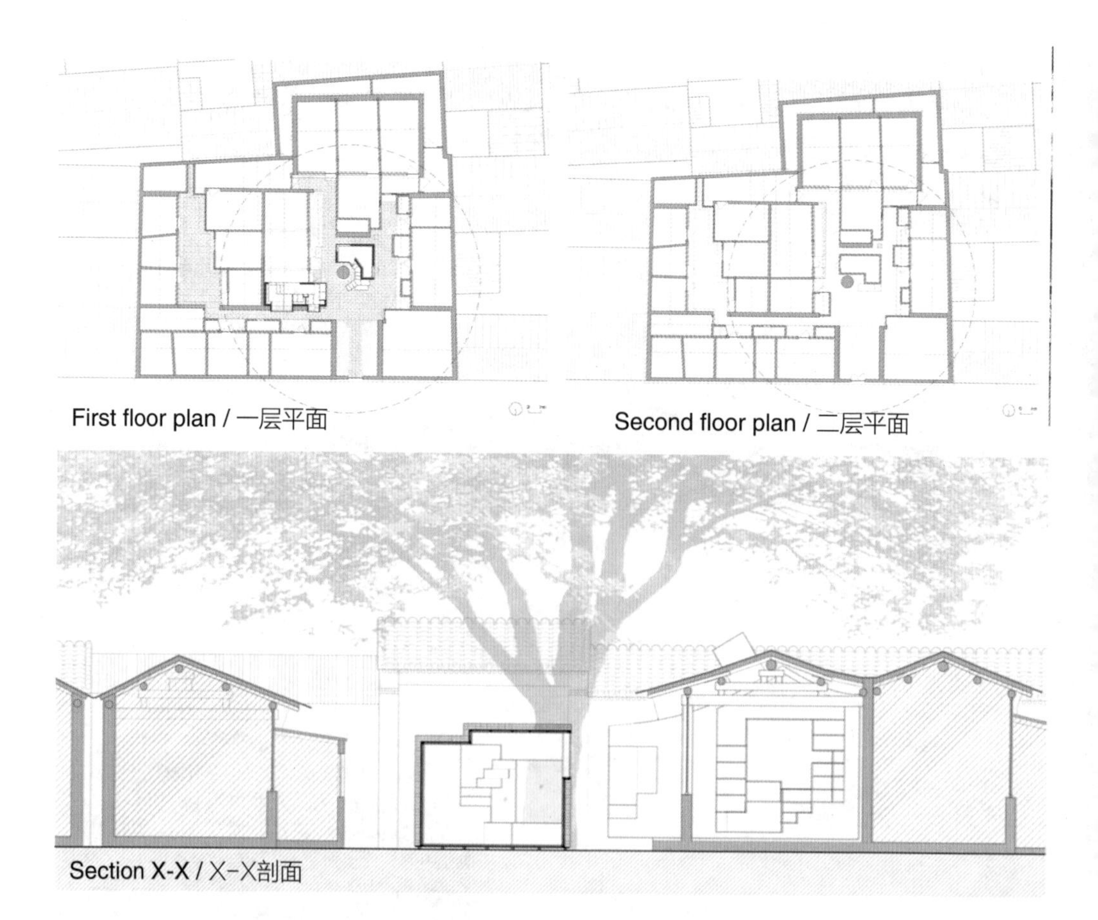

First floor plan / 一层平面

Second floor plan / 二层平面

Section X-X / X-X剖面

Image source 图片来源：1. http://www.core77.com/gallery/28202/beijing-design-week-2014/?channel_id=96#image=21
2. http://www.archdaily.com/775172/micro-yuaner-zao-standardarchitecturebooku

SHARING CULTURE / 共享文化

Distance City Center / 距离市中心

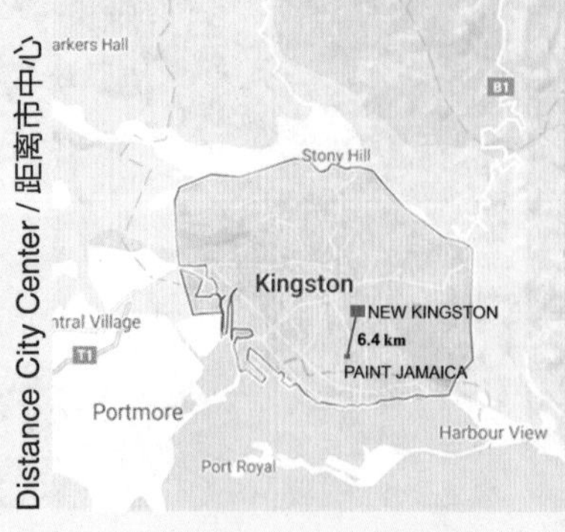

Urban Pattern / 城市肌理

Circulation / 交通流线

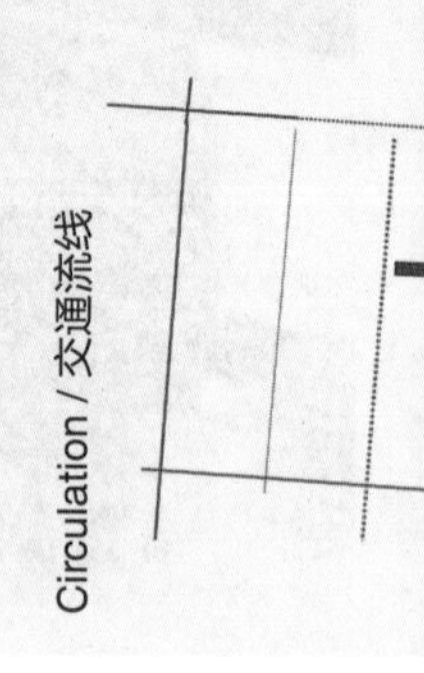

Image source 图片来源：Google map (https://maps.google.com)

12. 41 FLEET STREET
KINGSTON, JAMAICA

41 Fleet Street is located in Kingston Jamaica.The location was discovered by Paint Jamaica which was also established around the same time in 2014.

Paint Jamaica is a social and art intervention NGO established to transform surroundings through democratic art. It was created by Marianna Farag with a group of Jamaican artists, architectural consultants and volunteers.The idea of this project was to inspire and remind others in the community and beyond about the mazing Jamaican culture. The project started from warehouse alongside a school on fleet street. They renovated this old warehouse to provide a safe space for people to gather and improve the bonds among the area.

12. 41舰队街项目
金斯敦，牙买加

项目位于牙买加，最先在2014年由“描绘牙买加”发现并成立。“描绘牙买加”是一个社会和艺术的公益组织，通过民主艺术来改造环境。设计由组织者玛丽安娜·法拉格与一群牙买加艺术家、建筑顾问和志愿者共同创作。这个项目的想法是激发并提醒社区和其他人关于牙买加文化的思考。该项目从仓库沿着车队街道的一所学校开始，设计翻新了旧仓库，为改善区域间联系提供了安全空间。

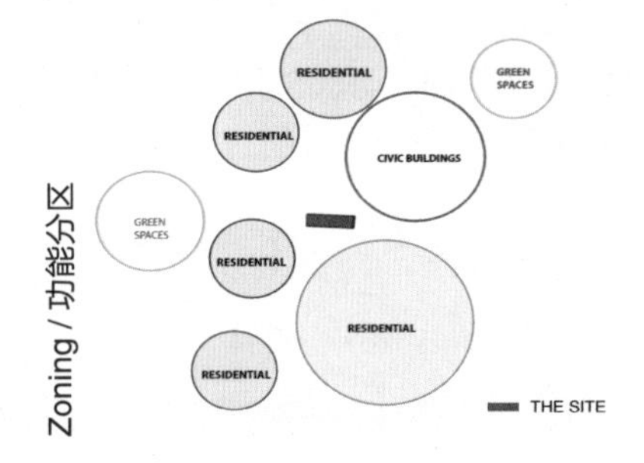

Zoning / 功能分区

This project is an interactive program which included the local community members and the team. The community learn more about art and even about each other. The redeveloped space attracted more new activities to the community and at the same time provided a space for daily activities. It is now considered one of the hot spots for concerts,games,excercise,food vendros and seasonal events. The space is even big enough to have more than one activities at the same time. The project was such a success, it has inspired a quest to redevelop other locations like this around the island.It will apply a similar interactive concept within the project redevelopment process with the support from the public sector companies.

这是一个涉及当地社区居民和团体的项目。当地社区居民从彼此间学习艺术、从事相关活动。重新改造的空间吸引了更多主办商给当地居民提供新的活动，同时也为当地居民提供了日常交流空间。现在，该空间被认为是音乐会、戏剧、运动、食物和季节性活动的热点地区之一。空间足够让好几个活动同时进行。该项目的成功，启发了岛内更多的重建项目。在政府部门的支持下，岛上也将以类似的互动概念引领重建项目。

Image source 图片来源：1.http://www.huffingtonpost.com/kirkanthony-hamilton/hope-lives-here_b_6269476.html 2. https://experiencejamaique.com/blog/street-art-kingston 3. http://www.dothereggae.com/portal/entrevista-a-jah9-por-supah-frans/ 4. http://jamaicansmusic.com/news/Culture/Paint_And_Plant_Jamaica_Returns_To_Parade_Gardens_With_One_Of_A_Kind_Project_Merging_Art_And_Nature

SHARING CULTURE / 共享文化

Context

Home to about 100,000 inhabitants, the Paraisópolis favela (a slum) in St. Paulo is one of the world's largest informal communities.

While mapping the area and interviewing residents, Urban Think Tank learned of a local music organization hoping to build a new practice and performance center.

Inspired by this story, U-TT developed a proposal that would transform an inaccessible void within the dense fabric, which had been cleared of homes after a severe mudslide, into a productive and dynamic community hub.

背景

圣保罗的Paraisópolis约有10万居民,是世界上最大的贫民窟之一。

U-TT在测绘当地和采访居民时，了解到当地的音乐组织希望建立一个新的表演中心。

受到这个故事的启发，U-TT制定了一项方案，将被严重泥石流后掏空的地域转化为一个富有活力的社区中心。

Image source 图片来源：1. https://www.lafargeholcim-foundation.org/projects/urban-remediation-and-civic-infrastructure-hub-so-paulo-brazi
2. http://www.archdaily.com/222057/global-holcim-award-2012-winners-announced

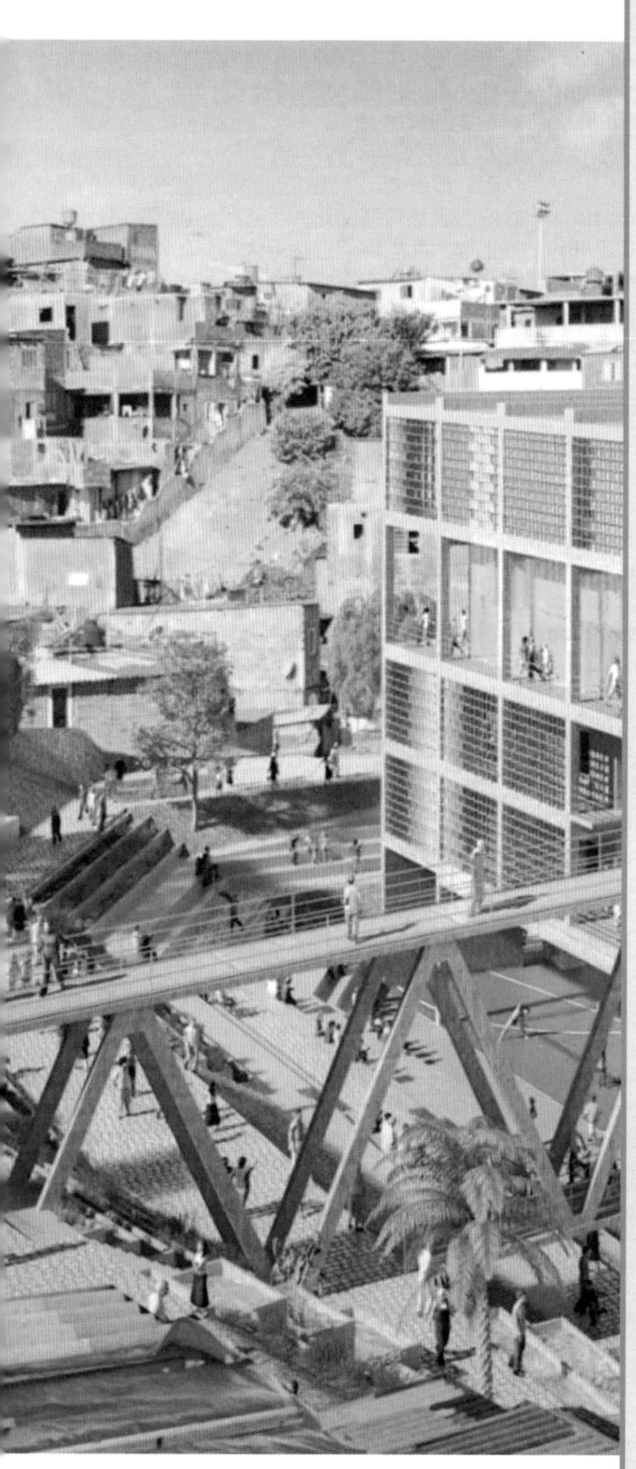

13. URBAN REMEDIATION AND CIVIC INFRASTRUCTURE HUB

St. PAULO, BRAZIL

Architect: Urban-Think Tank
Competition: Global Holcim Award 2012

This project transforms an eroded landscape into a productive zone and dynamic public space, preventing further erosion and dangerous mudslides on the steep slopes and provides social and cultural infrastructure. It includes a terraced public space with areas for urban agriculture, a water management system, a public amphitheater, a music school, a small concert hall, sports facilities, public spaces and transport infrastructure.

13. 整治后的城市公共设施中心

圣保罗，巴西

该项目将消逝的景观转化为生产区和公共设施，一方面防止陡坡进一步的泥石流，另一方面提供社会和文化基础设施。设施包括了可进行都市农业种植的公共空间、供水系统、露天剧场、音乐学校、小型音乐厅、体育设施，公共空间和交通基础设施。

Urban Farming

In this project, the agriculture is used as more than just a food source – it also plays a part in preventing further soil erosion, thus reduing additional structural interventions.

However, what the designers may not have taken into consideration, is the availability of residents to tend to the urban farm. Given that farming can be rather time consuming, and considering that the residents are in fact, struggling to make a living by working longer hours, perhaps it isn't suitable to incorporate urban farming.

Application

Given that the area is impoverished and public funding is limited, other sustainable measures such as hybrid airconditioning, solar panels, passive shading and facilitating cross ventilation help realise the economic resilience of such a project.

Though not immediately relevant to Joo Chiat at the urban scale, it is important to explore how the other sustainable measures that may benefit HDB flat residents can be integrated into the urban farming initiative.

都市农业

在这个项目中，农业不仅是食物来源，同时扮演着防止进一步的水土流失，从而减少额外建筑结构需要的角色。

然而，设计者可能没想到的是，基于居民的经济状况，农场可能面对人手问题。农业是相当耗时的工作，而居民们已耗尽大部分时间工作谋生，空余的时间实在不多。因此，城市农业在这个社区的可行性需进一步讨论。

应用

该地区十分贫困，公共资金有限，因此其他可持续发展措施方案有助于实现此项目，如混合空调、太阳能电池板、被动遮阳和交叉通风等。

尽管这区域与课程涉及的新加坡如切地段的规模并不太相似，我们仍能探讨如何将城市农业纳入新加坡的住宅区（HDB），让居民受益于其可持续计划。

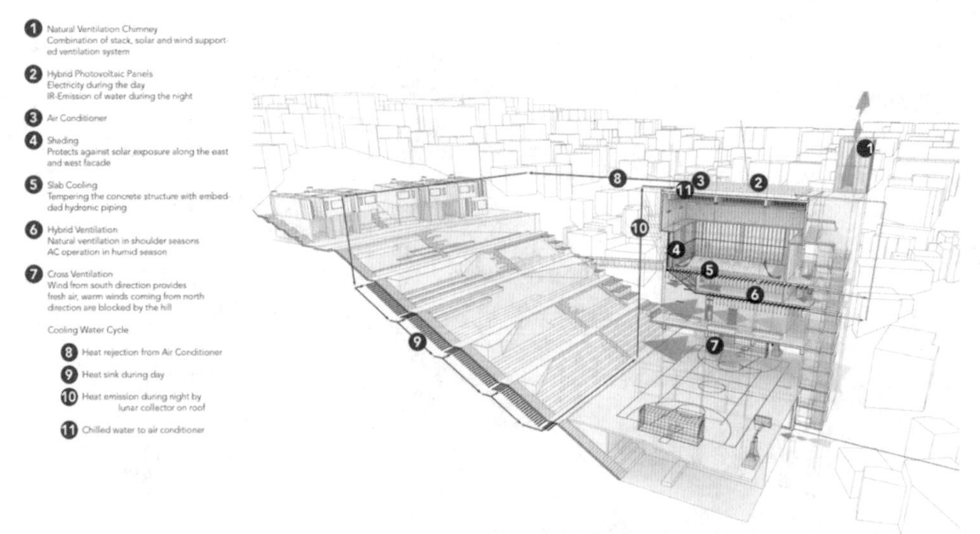

Image source 图片来源：1. https://www.lafargeholcim-foundation.org/projects/urban-remediation-and-civic-infrastructure-hub-so-paulo-brazi
2. http://www.archdaily.com/222057/global-holcim-award-2012-winners-announced

Programmes

The design of the open theatre acts as an inviting mitigating space connecting the flow of the interior of the building into the surrounding landscape. The spaces are multipurpose, with the open theater doubling as a soccer court, and the green areas as both urban farms and a park, and seats are built into the landscape to create social spaces.

However, one might consider the modern typology of the building and selction of programmes to be incongruent with the surrounding area, causing the residents to avoid the place.

设计方案

开放式剧院的设计将建筑物融合于周围景观。这些空间都属多用途的：开放式剧院可作为足球场，农场可以作为公园，座位也被融入周围环境中，以营造人们交流的空间。

然而，现代建筑似乎与周边地区不协调，可能导致居民不太喜欢这些地方。

SHARINGHERITAGE / 共享遗产

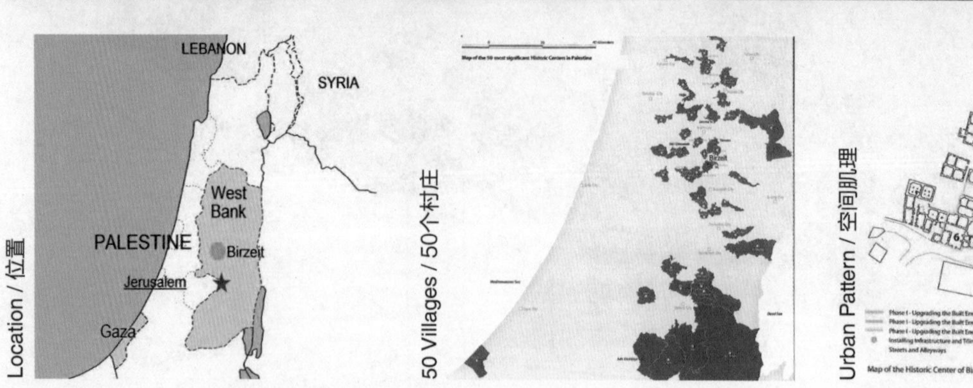

Image source 图片来源：1. Google map (https://maps.google.com) 2. http://arenaofspeculation.org/research/interviews/farhat-muhawi/ 3. http://www.akdn.org/architecture/project/revitalisation-of-birzeit-historic-centre

14. REVITALISATION OF BIRZEIT HISTORIC CENTER

BIRZEIT, PALESTINE

Architect: Riwaq - Centre for Architectural Conservation, Ramallah, Palestine
Design: 2007-2011
Size: 40640 sqm
Completed: 2009 - ongoing

The programme is driven by the desire to save the cultural heritage but takes on added meaning and significance in the context of the Israeli occupation.

Riwaq is one of the key actors lobbying for the protection of the cultural heritage of Palestine. Since its establishment in 1991, this NGO has pursued a strategy that includes documentation, conservation, revitalisation, community participation and activism, legislative reform, publicity, job training and public awareness programmes.

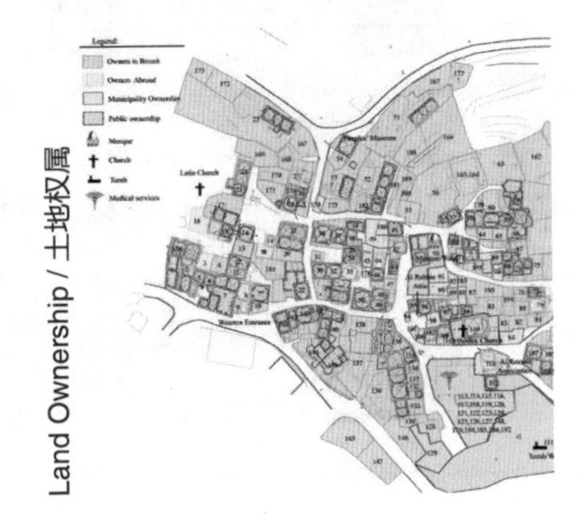

Land Ownership / 土地权属

14. 比尔泽历史中心复兴

比尔泽，巴勒斯坦

该项目缘于拯救文化遗产，在被以色列占领的背景下具有更深远的意义。

利瓦克是游说保护巴勒斯坦文化遗产的主要参与组织之一。自1991年成立以来，这个非政府组织一直奉行订立文件、保护、振兴、社区参与、行动、立法改革、宣传、职业培训和培养公众意识等策略。

Sharing Heritage:
Involving the local community in the process of rehabilitation

-transform the decaying town of Birzeit
-create employment through conservation
-revive vanishing traditional crafts in the process

Encourage the revival of vanishing traditional crafts as well as the use of local materials or locally produced supplies, contributing to the national GDP. Local blacksmiths create gates and grilles for several restored buildings. Riwaq also organised a joint Belgian-Palestinian initiative to exchange experience in smithing.

共享遗产
使当地社区参与遗产修复过程中
-改变衰退中的比尔泽老城
-通过历史保护创造就业
-在过程中复兴消失的传统手工艺产业

鼓励当地消失的传统手工艺复兴的同时，也鼓励使用当地材料或当地供应产品，可发展当地经济。当地铁匠为几座修复的建筑制作大门和烧烤炉具，利瓦克同时组织了比利时—巴勒斯坦联合提议以交流锻造技艺。

Image source 图片来源：1. Google map (https://maps.google.com) 2.https://archnet.org/system/publications/contents/2570/original/FLS3269.pdf?1384764 3. http://www.akdn.org/architecture/project/revitalisation-of-birzeit-historic-centre 4. http://www.riwaq.org/historic-buildings/projects/hosh-al-%E2%80%98etem-resedincy

SHARING COMMERCE / 共享商业

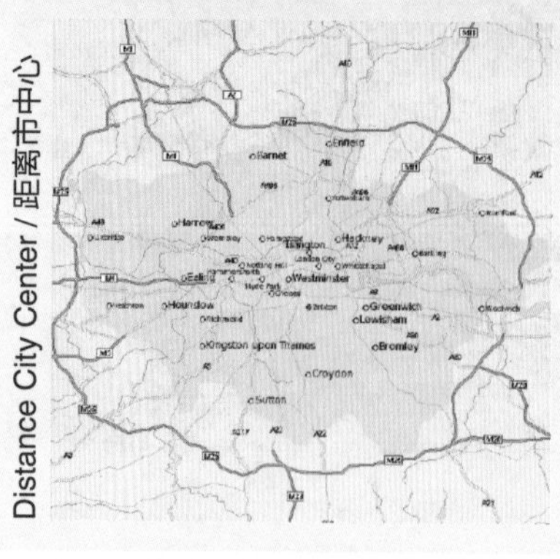

Distance City Center / 距离市中心

Urban pattern / 城市肌理

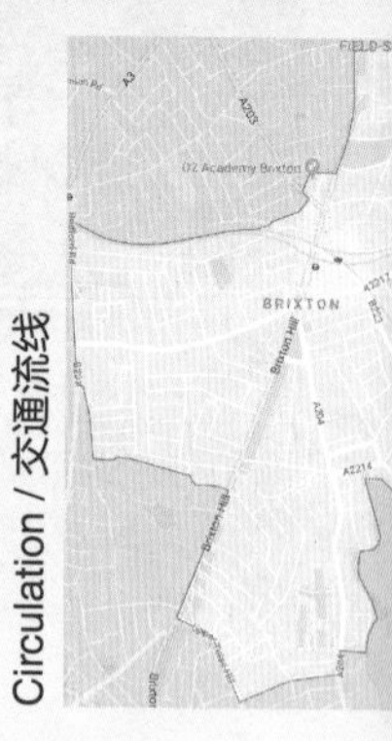

Circulation / 交通流线

Image source 图片来源：1. Google map (https://maps.google.com) 2.Google map (https://maps.google.com) 3.Google map (https://maps.google.co

Landuse / 土地利用

15. THE CORN EXCHANGE

BRIXTON, UK

The Brixton Pound (B£) is money that sticks to Brixton. It's designed to support Brixton businesses and encourage local trade and production. It's a complementary currency, working alongside (not replacing) pounds sterling, for use by independent local shops and traders. The B£ gives local traders and customers the chance to get together to support each other and maintain the diversity of the high street and strengthen pride in Brixton. The B£ makes money work for Brixton by supporting smaller shops and traders who are under threat from the recession and larger chains. It stays in Brixton and circulates, increasing local trade and community connections. Money spent with independent businesses circulates within the local economy up to three times longer than when it's spent with national chains.

15. 谷物市场

布克斯顿，英国

布克斯顿镑（B£）是布克斯顿的货币，用来促进当地的贸易和生产。这是一种作为补充的货币，和英镑一起使用。B£为当地贸易商和客户提供互相支持的机会，在保持大街多样性的同时，加强了布克斯顿的自豪感。B£通过支持受经济衰退及相关商业链条威胁的小型商店和商人，为Brixton带来经济利益。此种货币仅在布里克斯顿流通，增加了当地贸易和社区间的联系。B£在本地的经济流通中形成了高效的闭环系统。

5 five

5 BRIXTON £

Luol Deng (1985–) basketball player

10 ten

10 BRIXTON £

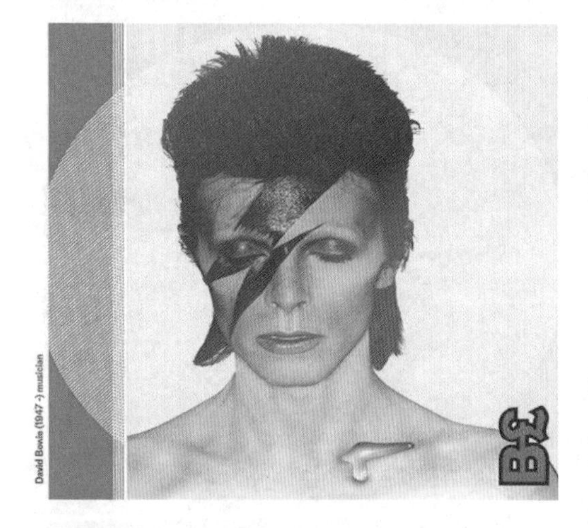

20 twenty

20 BRIXTON £

Image source 图片来源：1. Google map (https://maps.google.com)

What is Brixton Pound (B£) ?

It's a complementary currency, working alongside (not replacing) pounds sterling, for use by independent local shops and traders.

什么是布克斯顿镑

作为英镑的补充货币，为本地的小型商店和商人所使用

20 Storey

A Smile Dry Cleaners
Adornment
Alisha Humphrey medical& Sports Injury Thearapist
All Traders Builders Ltd.
Amaru
Amy Williams
Anna Searight's singing lessons
Ans Home Furnishings
Any Book
Aroma Nouceau
artYesart. com
Artz Designer Wear
Aurora Belly Dance
Avisa's
Axe DVD's
Azmarino Cafe
Back Home Foods
Baker Baby
Balance and Puerpose
Bambinos
Bamboula
Bella West
Beyond Future Focus UK(BFFUK)CIC
Binks' Gin Bar
BLEU
Blue Jay
Blue Turtle Oasis
Bombay Inn

■ Fully signed up traders who accept online and mobile transactions as well as paper Brixton Pounds / 同时接受线上支付和纸制布克斯顿镑的商铺

■ Traders with applications pending, you can spend paper / 能使用纸制布克斯顿镑的商铺

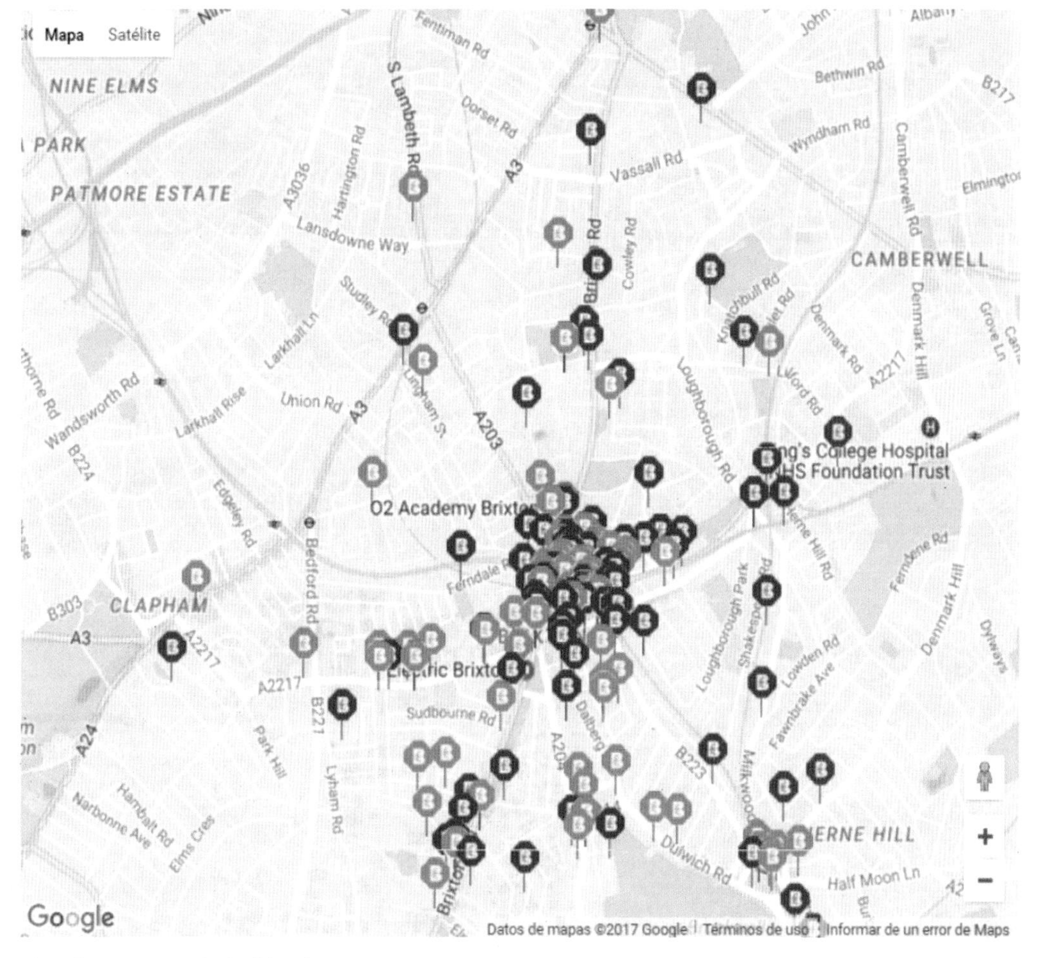

2. https://spacesyntax.com/project/brixton/

SHARING COMMERCE / 共享商业

Distance City Center / 距离市中心

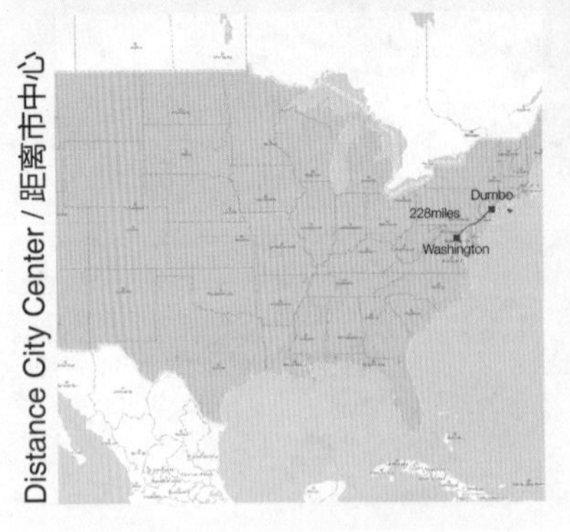

Urban Pattern / 城市肌理

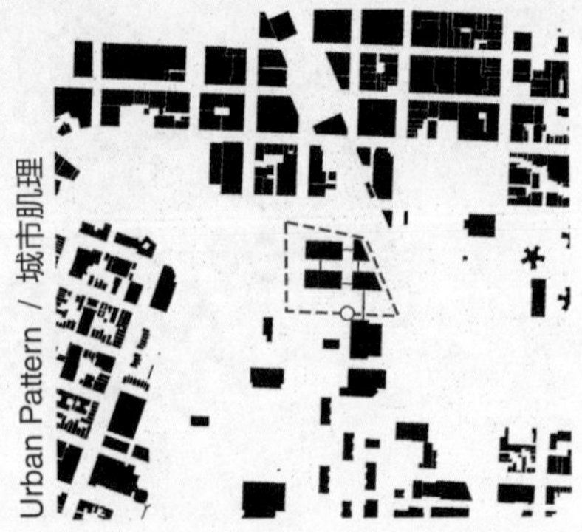

Circulation / 交通流线

 Image source 图片来源：Google map (https://maps.google.com)

16. Etsy
DUMBO, UNITED STATES

Area: 9,000 ft²
Staff: 921 (total)

E-commerce - Etsy is popular as a side-business as well as a place to buy goods made from recycled and upcycled materials, along with less expensive or more unusual versions of mass-produced items. The unique nature of many of the items for sale is part of their appeal to some shoppers. Product photos on Etsy tend to be editorial or artistic instead of commercial catalog style. Sellers can add tags to their products to help buyers find them, and buyers can choose to search for items available locally. Etsy staffers publish lists of featured items.

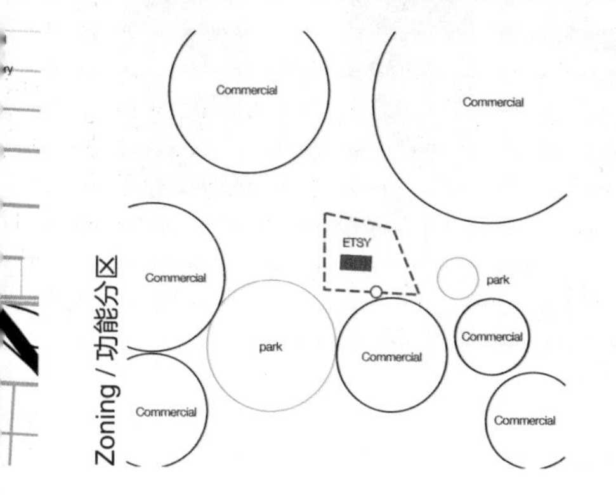

Zoning / 功能分区

16. Etsy网络商店平台
丹布，美国

电子商务—Etsy作为购买由回收和循环材料制成商品的平台，其低廉的价格和独特的批量产品生产方法受到广泛喜爱。出售物品的独特性是吸引消费者的重要因素。Etsy上的产品照片往往是精心编辑过的，而非简单的商业目录。卖家可以在自己的产品上添加标签方便买家寻找，买家可以方便地搜索本地商品。 Etsy工作人员还发布了特色项目清单。

Etsy

Image source 图片来源：1. https://officesnapshots.com/2017/01/23/etsy-offices-new-york-city/ 2. https://www.etsy.com/uk/?ref=lgo
3. http://www.businessinsider.com/inside-etsys-new-perk-filled-office-2016-6/#etsy-signed-a-10-year-lease-on-the-nine-story-building-2

Shop by category / 分类商店

SHARINGINFRASTRUCTURE / 共享设施

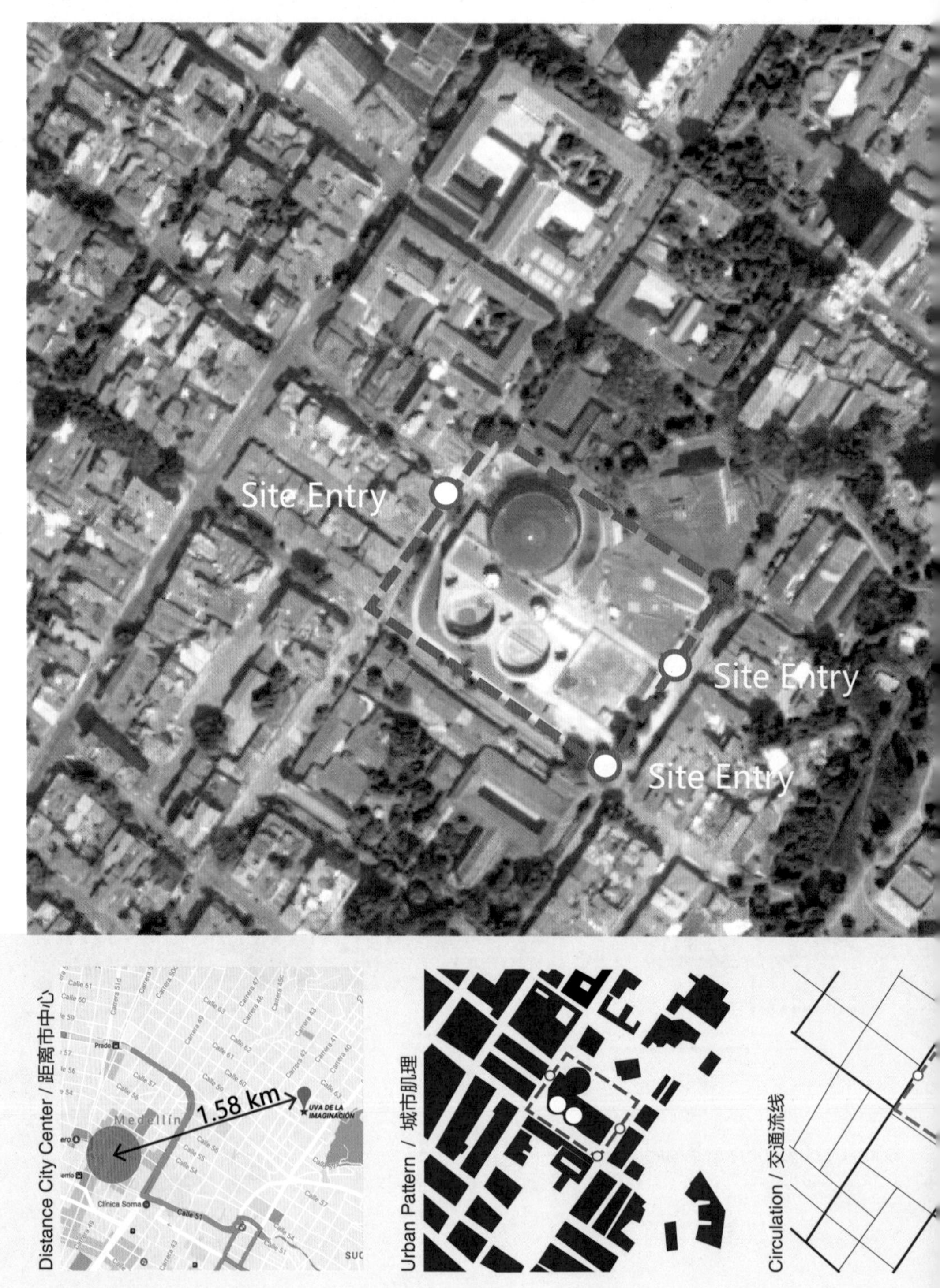

Image source 图片来源: 1.Wikipedia (https://en.wikipedia.org/wiki/Medellin) 2. Summary (https://www.lafargeholcim-foundation.org/projects/articulated-site
3. Google map (https://maps.google.com)

17. WATER RESERVOIR PUBLIC PARK
MEDELLIN, COLOMBIA

Architect: Colectivo720
Area: 1.54 ha

From a multidisciplinary vantage point, this project for a public park in Medellín, Colombia, centers on the creation of spaces around and above a series of water reservoirs. Tracing the site's history, the architectural form takes its inspiration from the surrounding topography as well as from the structure of the existing tanks and pools, resulting in an intervention with minimal environmental impact. Considering the infrastructural use of the site, special attention is given to water management, which utilizes recycling technologies that involve rainwater and grey water harvesting through simple systems for the irrigation of the park. In an interaction between nature and the urban landscape, the park seeks to improve the quality of life in the city.

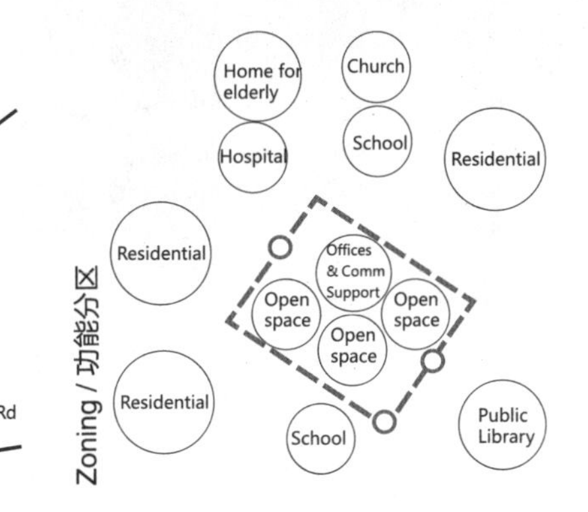

Zoning / 功能分区

17. 水库公共公园
麦德林，哥伦比亚

从多学科角度来看，哥伦比亚麦德林公园项目是围绕一系列水库建立的。追溯历史，其建筑形式的灵感来自于周围地形和水池的形状，设计采用对环境影响最小的干预措施。过程中考虑到场址基础设施的使用情况，注重水管理并采取循环利用技术，通过简单的灌溉系统收集雨水和中水，通过促进自然与城市景观的互动，致力于提高城市生活质量。

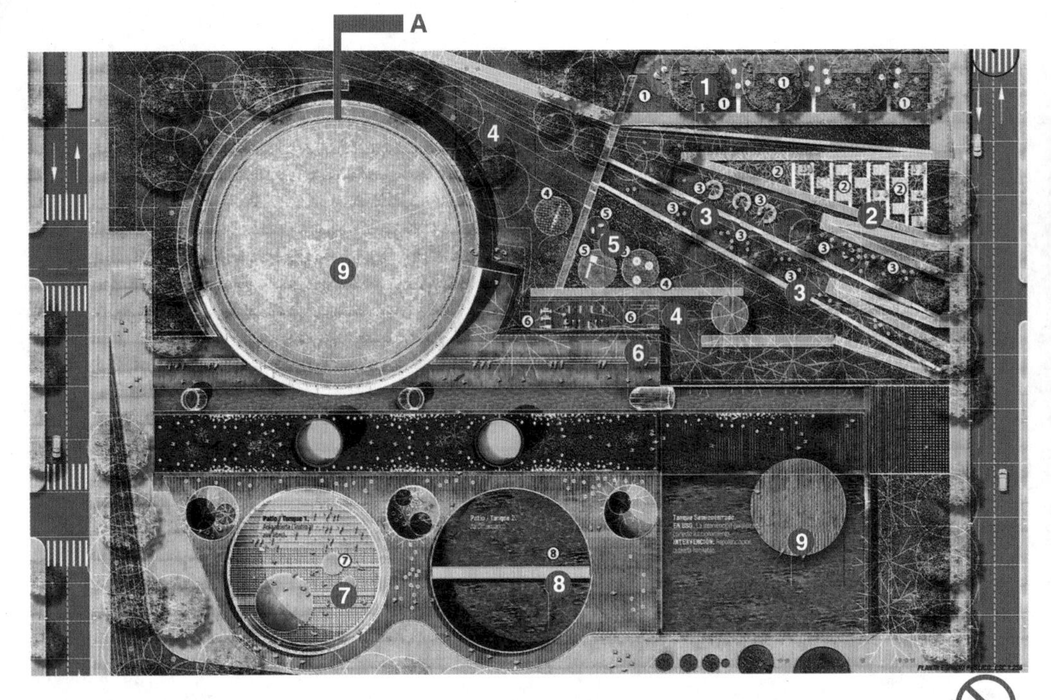

SITE PLAN / 总平面

1. Relaxation area; 2. Picnic area; 3. Meeting area; 4. Sand area; 5. Children's area; 6. Urban gym; 7. Open-air theatre; 8. Water garden; 9. Water tank

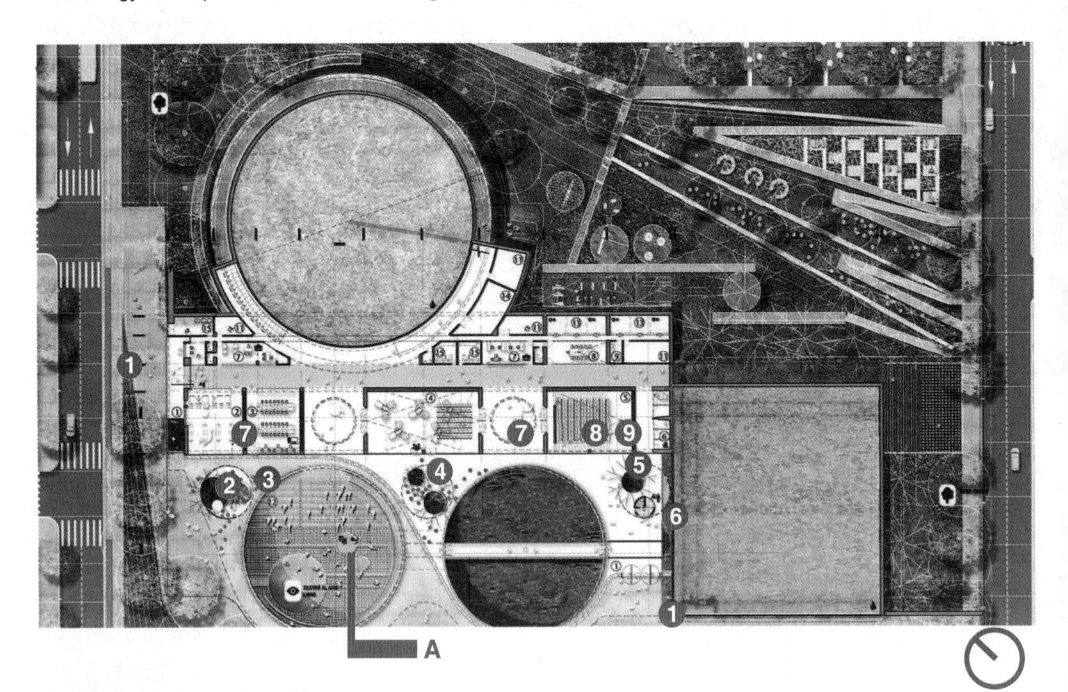

GROUND FLOOR / 地面层

1. Entrance hall; 2. Local businesses; 3. Internet cafe; 4. Classrooms; 5. Auditorium; 6. Private study rooms; 7. Offices; 8. Boardroom; 9. Kitchen

Image source 图片来源：1.Wikipedia (https://en.wikipedia.org/wiki/Medellin) 2. Summary (https://www.lafargeholcim-foundation.org/projects/articulated-sit
3. Google map (https://maps.google.com)

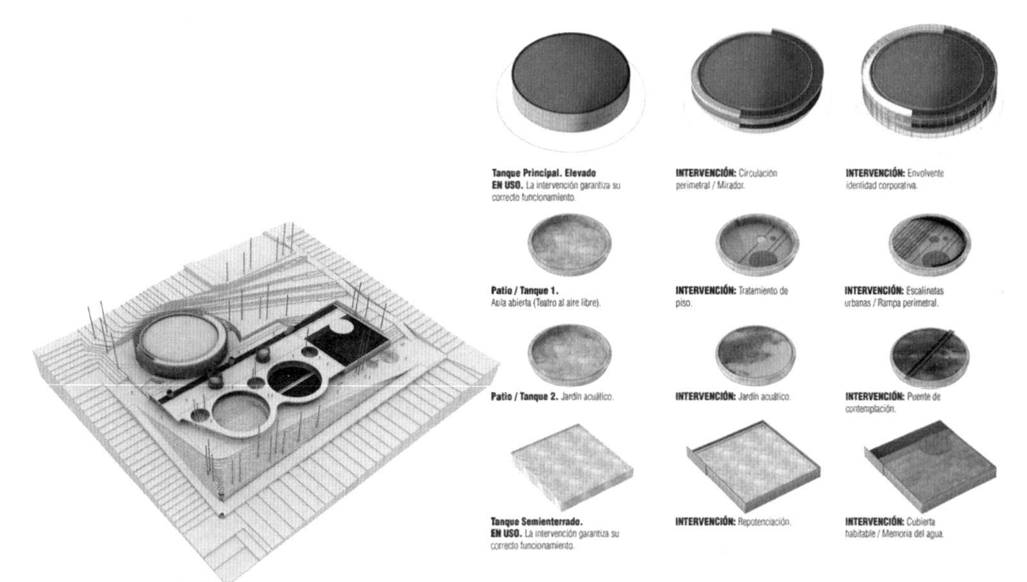

Site model / 场地模型

Diagram illustrating process of reusing existing site features to create new inhabitable spaces for users / 说明重新利用现有场地特点去创造服务使用者的空间分析图

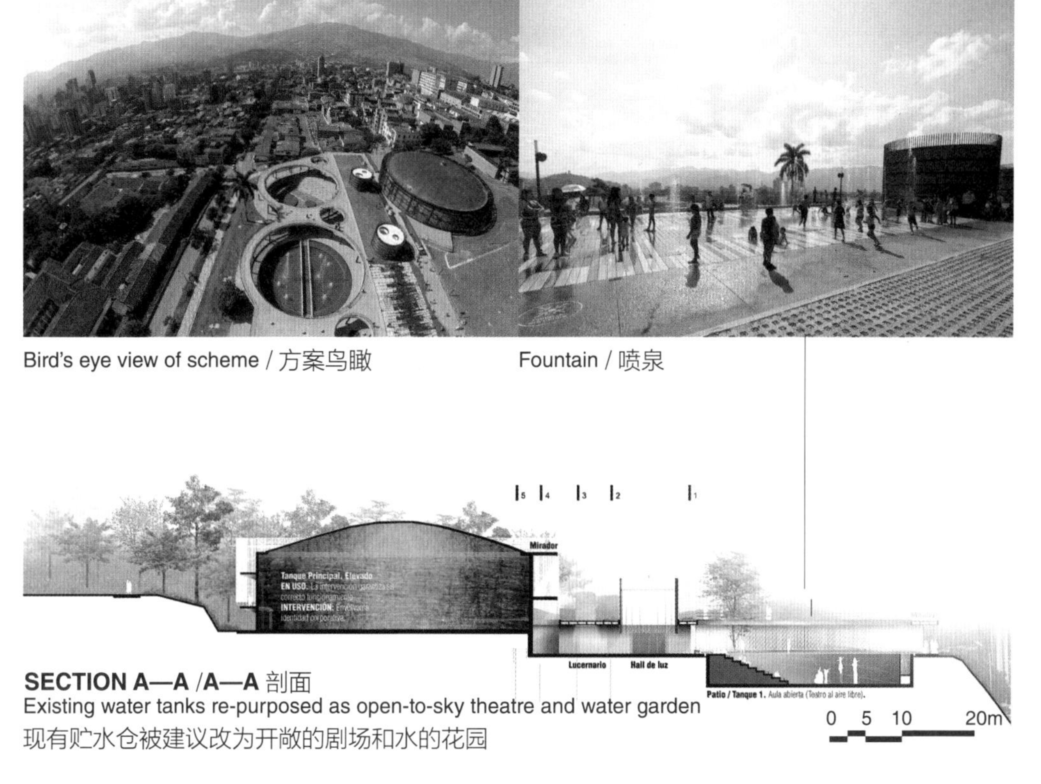

Bird's eye view of scheme / 方案鸟瞰

Fountain / 喷泉

SECTION A—A /A—A 剖面

Existing water tanks re-purposed as open-to-sky theatre and water garden

现有贮水仓被建议改为开敞的剧场和水的花园

SHARING INFRASTRUCTURE / 共享设施

Distance City Center / 距离市中心

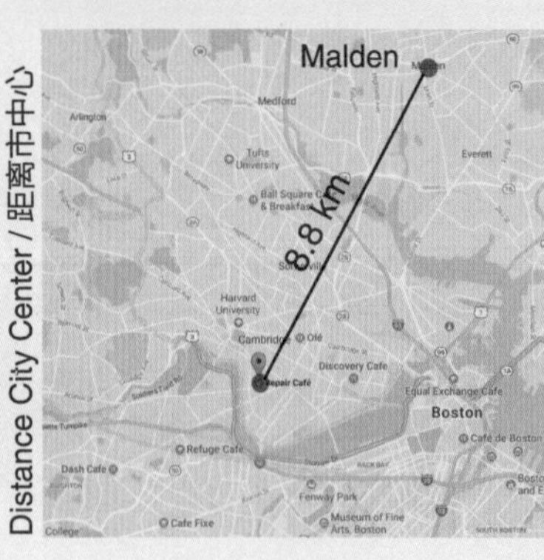

World Distribution / 各地分布

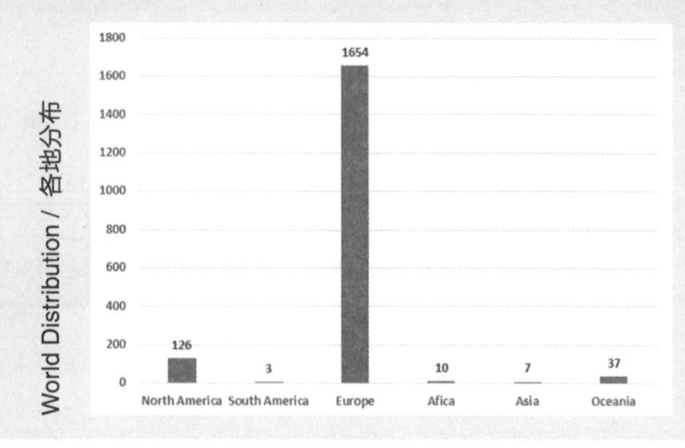

Image source 图片来源：1.Google map (https://maps.google.com) 2.https://repaircafe.org/en/community/ 3. https://repaircafe.org/en/shop/

18. REPAIR CAFE
CAMBRIDGE, MA, UNITED STATES

Founder: Martine Postma
Established: 2009
Number of cafes worldwide: 1223

Repair cafe is the creation of Martine Postma who started the first Repair Cafe in October 18, 2009. Repair cafe is a platform for the gathering of persons who wish to have their broken items (within reason) repaired as well as volunteers who have the expertise to fix these items.

The gatherings are held monthly, sometimes weekly, based on the demand of participants and the cafes are usually held at community centers or businesses that volunteer their spaces for the event. One of the main goals of these non profit events is to not only give a second chance to items that were meant for landfills but to also teach participants a way of thinking by recycling or reusing items with the hope that they can also transfer this way of living to others.

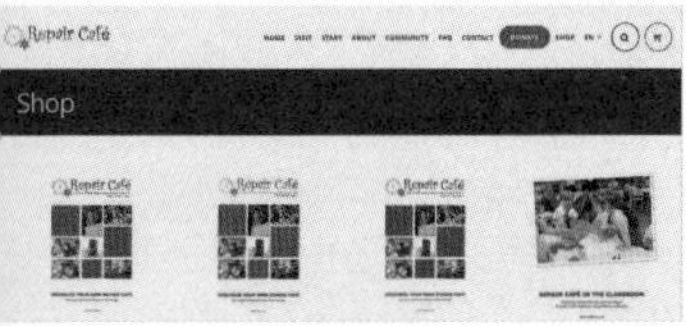

18. 维修咖啡馆
剑桥，马萨诸塞州，美国

维修咖啡馆由马丁·波斯特马创建于2009年10月18日。维修咖啡馆是一个为希望修复破碎物品的客户以及有专门知识的志愿者提供的平台。
活动根据参与者的需求每月或者每周举行，通常在社区中心或志愿提供活动空间的企业中举行。这些非营利活动的主要目标之一是，创造一个废物再利用的机会，同时唤醒参与者回收和再利用物品的意识，并将这种生活方式传播给更多的人。

PLATFORM POTENTIAL

The simple idea of people coming together fixing everyday broken items can be expanded by using the internet as an agent to increase the number of "cafes" The accompanying message that not everything has to be thrown away, but some things can be used longer can travel further. More technical volunteers could be invited for more assistance in the area of our everyday electronic items such as cellphones and laptops as voiced by Ms. Postma on repaircafe.org.

WHAT DOES THIS PLATFORM REPRESENT FOR THE CONCEPT OF SHARING ?

The platform at its shallowest level helps people fix their items but it's also an avenue for people to meet new faces, make connections and share their life's experiences. Repair Cafe can also be seen as a latent light therapy event for persons who are reintroducing themselves into society.

HOW CAN ARCHITECTURE PLAY A ROLE IN REPAIR CAFE'S DELIVERY OF SERVICE ?

For most activities that occur in a space, it can be improved by tailoring the space to fit the needs and habits of the users. Repair cafe spaces are usually held in a large room, but through design a space that is not only flexible but carefully divided with respect to layers of privacy, can add value to the work at hand and the experience of the space.

WHY IS REPAIR CAFE POPULAR IF THE INTERNET HAS MANY AVAILABLE RESOURCES ?

People prefer human interaction when doing activities. The instructions received by users may not always be simple to execute. The feedback component that some people seek is not always consistent or guaranteed. Repair Cafe however connects the people seeking repair information and the person giving the instruction directly. Face to face interaction eleminates the feedback fault of the internet which makes people more inclined to attend events such as this.

THE IDEA OF REPAIRING ITEMS FOR FREE IS NOT NEW

The internet is currently the largest platform to search and share information as well as connect people. The search engine Google gives users access to all relevant websites containing information related to fixing items. A dedicated website such as instructables.com, and media website Youtube.com among others have outlets to search , share and receive feedback and various repair topics.

· 潜在的平台

通过网络人们聚集起来修复日常破损的物品，这样的简单想法能够扩展成为咖啡增加顾客。随之而产生的信息是，并非破损的每样东西都不得不丢掉，一些东西可以修复后再续利用。通过维修咖啡网络中波斯特马女士的声讯，更多的技术服务志愿者能够被邀请为日常电器，如电话、笔记本电脑等的维修提供服务。

· 平台如何体现共享的理念

该平台最基本的层面是能够帮助人们维修他们的物品，更进一步而言，它为人们遇见新面孔营造场所，建立联系使得人们可以分享他们的生活经验。维修咖啡也被视作为人们将自己重新引入社会的潜在的轻微治疗活动。

· 建筑师如何在维修咖啡的派送或服务中起作用?

在空间发生的大多数行为，都可以被空间的裁缝通过空间的改进而满足使用者的需求。维修咖啡的活动通常可以在一个大空间中进行，不过通过非固定的灵活空间的设计，空间可分为不同私密性的层次来满足不同需求，能提升手工工作和交换经验的空间价值。

· 为何即便互联网有多样资源维修咖啡仍受追捧?

当人们进行活动时更倾向于交流互动。使用者接收到的指令并不总是简单的操作指令，一些使用者寻求的反馈信息也不总是能被承诺，尽管如此维修咖啡仍是直接地将寻求维修信息和提供指导的人直接联系在一起，面对面的交流能消除网络反馈中的失误，这使得人们更加趋向于参加活动。

· 免费获取维修信息的想法并不新

互联网现在是搜索和共享信息以及联系人们的最大平台，搜索引擎 Google 使得人们可以登录包含维修物品信息的网站，如 mstructables.com,Youtube.com 等都可以获取多样的维修信息与反馈。

Image source 图片来源：1.Google map (https://maps.google.com) 2. https://repaircafe.org/ 3. http://circularcambridge.org/events/cambridge-repair-caf

Repair Café

VICTORIA

CLOTHES* FURNITURE
*** ELECTRICAL APPLIANCES ***
TOYS * BICYCLES * ET CETERA

EXPERIENCED "FIXERS" WILL BE ON HAND TO HELP YOU REPAIR YOUR POSSESSIONS.

DON'T TOSS IT! REPAIR IT!!

SATURDAY, FEBRUARY 18, 2017

9:30 am - 12:30 pm

CENTRAL BRANCH OF THE GREATER VICTORIA PUBLIC LIBRARY
735 Broughton St, Victoria

visit us on facebook: www.facebook.com/repaircafevictoria

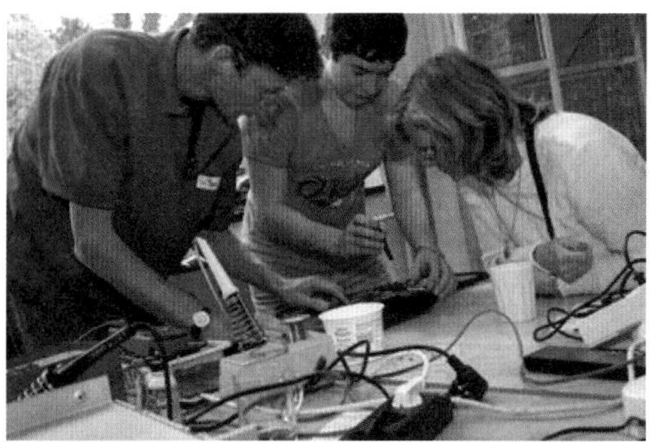

3

CONCEPT UALIZATION 概念

SHARING

Sharing and design for shareability

By Jeffrey CHAN Kok Hui and Ye ZHANG

为共享而设计

曾国辉，张烨

Designing sharing systems

Sharing today is both a social practice and an economic institution. As the former, sharing practices have been found throughout history and across the world in many different forms. For instance, the sharing of responsibilities and resources was seen to be fundamental for survival in the hunter-gatherer societies of prehistory (Harari, 2015). But as for the latter, sharing is a somewhat novel phenomenon in the so-called Sharing Economy. And sharing practices have been exemplified by the recent emergence of rapidly growing enterprises such as Airbnb, Uber and Mobike.

Nevertheless, the marriage between the time-honoured social practice of sharing and the enterprises found in this relatively new "Sharing Economy" has been an uneasy one. This is because many of the practices found in the Sharing Economy exploited the approbative image of these time-honoured practices of sharing—for instance, altruism (Tomasello, 2009)—even when they have little in common with actual (i.e., unremunerated) sharing practices (John, 2017). After all, are enterprises offering shared meeting spaces (e.g., https://www.sharedesk.net) or the shared car services (e.g., https://www.zipcar.com), about sharing, or more pointedly, more about for-profit rental services merely masked by a high-sounding rhetoric? But barring unproductive cynicism and presuming a more charitable reception, these new "sharing"

CITIES

设计共享系统

当今社会，共享是一种社会实践，同时也是一种经济模式。前者对于我们并不陌生，在历史的长河中，共享始终以不同的形式推动着人类的进程。譬如，在史前的狩猎采集社会，共享资源和分担责任被认为是人类生存的根本所在(Harari，2015)。作为后者，特别是在共享经济的语境下，共享则是一个新的现象，尤其突出地体现在一系列快速增长的、以共享经济为商业模式的创新产业中，如爱彼迎 (Airbnb)、优步 (Uber) 以及摩拜单车 (Mobike)。

然而，新的共享经济并不能被简单地理解为有着优良传统的共享社会实践和市场经济的成功对接。事实上，许多所谓的以共享经济模式运营的商业不过是以共享为名，绑架了作为社会实践的共享所蕴含的被长久认可的价值，比如利他主义(Tomasello，2009)，而其具体商业内容与真正意义上的无偿共享或许并无关联。更准确地说，究竟共享商业（无论是共享会议空间，如https://www.sharedesk.net，还是共享汽车，如https://www.zipcar.com）所推动的是共享实践本身，还是以盈利为目的的租赁服务，恰恰被一层浮夸的面纱所蒙蔽。如果抛开犬儒主义的观点从更具建设性的角度来看，那么这些作为共享经济主体的共享服务业很可能是未来城市发展的主要动力之一。也正是这个原因，建筑师、城市规划师和设计师需要对此给予特别的关注。由此也引出如下问题：我们应该如何设计更好的共享系统，从而可以有效地避免共享的积极意义被市场经济所绑架而更好地实现其真正的价值，同时使得共享系统能够不断自我完善并达到自给自足。简言之，我们应该如何设计一个原真的、合伦理的共享系统。

共享城市

services integral to the Sharing Economy are likely to become a major driver of future urbanism, and for this reason, demand serious attention from architects, planners and urban designers. In turn, this ought to beg the following important questions: how do we design better sharing systems that can effectively address the spurious notions of sharing exploited by the market economy on the one hand, and on the other, design better sharing systems that are self-reinforcing and therefore, also self-sustaining? In short, how do we design for authentic and ethical sharing systems?

The concept of sharing

In our research, we also discover that while sharing appears as the primary term of reference in literature on the Sharing Economy, the concept of "sharing", when found elsewhere, is almost always mentioned as a secondary attribute. For example, the urban commons has often been characterized as shared spaces (Stavrides, 2016). For this reason, any exposition on "sharing" has to first answer this question: why should we prioritize the concept of "sharing"?

To answer this question, it is important to begin with a definition of the word, "share". According to Oxford English Dictionary (OED), "share" as a verb basically means to have a portion of something with another or others, and it may take any of the following meanings: give a portion of something to another or others; use, occupy, or enjoy (something) jointly with another or others; possess (a view or quality) in common with others; (of a number of people or organisations) have a part in (something, especially an activity); tell someone about (something, especially personal); post or repost (something) on a social media website or application. Because of these different meanings, the actual meaning of "share" is primarily contextual and its usage will continue to expand with changes in society. In summarising his inquiry into the

共享的概念

我们在对共享的初步研究中发现，虽然在关于共享经济的文献中，共享总被作为首要切入点，但在相关其他领域，这一概念却常常被当作次要参考。比如，城市共同体（Urban Commons）的一种解释便是共享空间（Stavrides，2016）。因此，在剖析“共享”之前，我们必须先回答一个问题，即为什么要特别强调这个概念？

为此，我们有必要先探讨一下“共享”的含义。在英语环境下，基于牛津英语字典，共享作为动词的基本定义是：和他人共有某事或某物的一部分。而其具体含义可以是：给予他人某事或某物的一部分；和他人一起使用，拥有或者享受某事或某物；和他人持有相同的观点或者共有相同的特质；不同的人或者组织共同参与分担某事或某物，尤其是某项活动；告知他人某事，尤其是关于个人的某事；以及在社交媒体网络或应用上发布或转发某事。这些不同的定义也意味着“共享”的含义主要由具体的语境所决定，并且随着社会的发展而不断延伸。《共享时代》一书的作者尼古拉斯·约翰在总结他对于共享的含义的研究时也提出，这一词语的恰当定义和其用法随着时间的推移而产生了显著的变化（John，2017）。然而，无论是哪一重语境，共享这一词语都并完全没有贬损之意。换言之，共享是一个表示褒扬的积极的概念。

或许正是由于这个原因，尤其在共享经济的语境下和文献中，共享被反复强调。但是我们认为这并不是这一概念本身应该被重视的原因。共享之所以应该被优先探讨，是因为相比于其他相近的概念，它能够更加准确地揭示当今建筑、规划以及城市设计领域潜在的意义和可能性。换言之，如果不去强调这一概念，我们或许将面对一个双输的尴尬局面：一方面我们可能无法清晰地表述唯有共享一词才可能准确传达的含义，另一方面假设我们轻易使用其他相近的概念以替代共享，则可能带来一系列的误解和谬论。我们将在下文对此进行详解。

其一，与其他相近的概念不同，共享能够表达一些特别的含义。譬如，在某些语境下，共享常常与合作、共有以及参与交换使用。然而，与合作和参与所描述的集体过程或者社会过程不同，共享同时还更表达了对于有形资源的获

meaning of "sharing", Nicholas A. John, the author of The Age of Sharing (2017) suggests that the "proper" meanings and the way in which this word is used have changed quite dramatically over time (John, 2017: pp. 4). Importantly here, we suggest that "share" and "sharing"—in spite of these different contextual meanings—are rarely invoked with a pejorative undertone. In other words, "sharing" is mostly a positive, or an approbative concept.

For this reason, we argue that the concept of sharing has been prioritized—especially in the context of the Sharing Economy—because of this approbative undertone in spite of its many meanings. But this is not the reason why we think the concept of sharing ought to be prioritized. Instead, we argue that sharing ought to be prioritized today because this concept is more accurate, relative to other concepts proximate to it, to reveal meanings and possibilities in architecture, planning and urban design today. In other words if the concept of sharing is not prioritized, a double loss becomes inevitable: not only is it challenging, if not impossible, to articulate meanings that only sharing can convey, but also that the different substitutive concepts that would be applied in its place will result in both misleading interpretations and erroneous conclusions. Here, we further elaborate on why sharing ought to be prioritized.

Firstly, "sharing" is unique. Unlike the constellation of other concepts that approximate its meanings, sharing denotes its own special meanings. For instance, depending on the context, sharing is often used interchangeably with cooperating, commoning and participating. However, unlike cooperating and participating, which tend to suggest communal or social processes, sharing also denotes access to some (tangible) portions of resources. And unlike commoning, sharing offers a more precise—exact—logic of division and distribution. The remaining haziness between all these

得。共享和共有的不同则在于前者更准确地表达了切分或者分配逻辑的存在。客观而言，这一系列概念的细微含义差别在英文语境下要比在中文语境下更加模糊难辨。在中文语境下，颇为重要的一点是“共享”一词明确表达了“享受”的含义，而这是合作、共有以及参与等其他几个概念所无法表达的。在这个意义上，应该说共享所表达的享受、快乐的含义与其在英文语境中的褒义是一致的。换言之，我们共享的是美好而非邪恶，并且我们因此而快乐。

其二，共享与其他相近概念的本质不同还在于其带有一种情感价值。共享经济推崇者和实践者们恰恰充分利用了共享的这一情感内涵。比如说，如果可以分享并通过这一过程而拥有新的朋友，为什么谁还会选择占有事物并承受由此而来的负担呢？再有，如果可以通过爱彼迎（Airbnb）居住在当地人家并深入了解当地的风土人情，为什么还要选择居住在酒店呢？当共享的情感价向被放大并强化，这一概念便被赋予了一种十分正面的社会性，而这是其他相近概念所无法比拟的。进一步而言，共享能够唤起一种温暖且舒适的情愫，这对于设计一个活跃欢快且利于彼此交往的社区和城市是无法也不应该被轻视的。也就是说，与重在描述过程且相对冷漠的概念（如合作、参与或是共有）所不同，共享能够有效地拉近人与人之间的距离。仅此一点，共享就应该成为一个被建筑师、城市规划师和设计师重视的概念并在创造宜人的城市空间的实践中得以应用。

最后，共享这一概念更预设了一系列道德伦理价值的存在，例如信任、共同性、平等以及公正，这些价值也通常被认为是一个繁荣且可持续城市的先决要素。然而，在当今的城市中，面对着过度的商业化和消费主义、快速的科技更新以及愈发弱化的识别性和场所性，这些伦理价值已经萎缩甚至被废黜。简言之，全球化资本主义的统治性优势（Sklair, 2017）不断塑造并强化一种新自由主义的都市性，并且鼓励空间的封闭与隔离以及公共空间的私有化。然而讽刺的是，正是这些被批判的因素使得共享变得具有吸引力甚至成为一种必然。不断增长的人口，与此不相匹配的低工资增幅、物质资源的匮乏以及人类对环境可持续性的承诺都在宣示着消费主义和物质所有权这一经济范式的过时。反过来，人口的激增已经足以产生

concepts is perhaps merely a reflection of the ambiguities peculiar to the English language. If all these four concepts are now also rendered in Chinese—namely, cooperating as 合作；participating as 参与；commoning as 共用 and finally sharing as 分享/共享 —then not only do these concepts become clearer, but their respective sphere of usage is also rendered much more distinct. Importantly, this comparative linguistic approach demonstrates that at least for the Chinese language, sharing denotes a dimension of "enjoyment" that is absent from cooperating, participating and commoning even in their respective equivalent meanings in Chinese. In this way, this reading of sharing with the valence of enjoyment and joy is consistent to its approbative definition and social meaning in English: one does not share what is bad (or evil), but only what is good with others—hence, joy.

Following this and secondly, the concept of "sharing" then offers an emotive valence that is radically distinct from all these proximate concepts. In this regard, pundits of the Sharing Economy have exploited this emotive dimension of "sharing" to the fullest: why own things—and all the burdens that come with them—when you can share and make many new friends in the process of sharing? Or why stay in an alienating hotel, when you can get to know local people and their cultures by using Airbnb? In leveraging on the concept of "sharing" in this way, "sharing" has taken on a dimension of positive—emotive—sociality that no other concepts can equal; "sharing", in other words, evokes a sentiment that can be described as "warm and fuzzy" that should not be under-estimated especially from the perspective of designing more convivial and sociable communities and cities. Unlike the more procedural, and "colder" concepts of "cooperating", "participating", or "commoning", the concept of "sharing" can effectively eliminate distance and amplify intimacy between people. Just for this attribute alone, "sharing" becomes an immediately unique and

群聚效应，不仅带来了对共享的需求，同时也提供着可共享的资源（McLaren and Agyeman, 2015）。共享进而自然成为了社会过程以及社交媒体活动的一部分（John, 2017）。其与高科技的数字经济一起形成了推进未来城市发展、塑造未来城市环境的主要动力，并随着未来的城市共同演进。由此我们不禁要问，共享是否可能成为奠定另一种不基于空间的封闭、分离和控制的都市性的基石？而共享作为一个以信任、共同性、公平以及同情等伦理价值为基础的社会实践模式，也似乎正向我们许诺着对于不同的都市性的想象。因此，创造一个能够有效地推动真正的、合伦理的共享实践的城市环境就变得尤为重要。其关键更在于设计一个能够寻回那些已经丢失的以及正在逝去的伦理价值的共享系统，并由此为我们的城市建立一种团结。如果说城市仍旧是人类文明的未来，那么未来的城市恰恰应该建立在团结的基础上，而绝不是当前这种普遍的、支离破碎的、充满无尽冲突与争执的局面。

对于共享的几点初步思考

如果说基于以上原因共享应该被特别重视，那么究竟何为共享，又应该如何为共享而设计呢？基于初步的研究，我们对本文的这个核心问题形成如下几点思考。

第一，在探讨为共享而设计时，我们首先便假定了设计的功效性。事实上，历史上绝大多数共享实践都是社会演进的必然结果，而非设计本身所致。在大多数情况下，不令人愉快的共享往往在社会进程中被淘汰，而有着正面意义的共享，即那些实践着人类社会和人类生活的可贵道德情操的共享，则被反复强化并延续至今。因此，为共享而设计首先必须认识到，共享实践是一个包含着不断变化着的文化、意象以及价值的复杂动态系统，而并非是由社会过程或者人造系统所支撑的单一化的不变的活动。

第二，为共享而设计需要辨清共享究竟是达到某种目的的途径，还是其本身便是目的所在。比如，共享作为利他主义实践的一种形式往往被认为是一种目的而并不需要其他的解释。与此相反，在通过共享去实现更好的合作以及更有收获的参与这两种情况下，共享都是达到其他目标的途径或者手段。因此，为了作为目的

relevant concept for architects, planners and urban designers in conceiving spaces that attempt to diminish distances between people.

Last but not least, presumed by the concept of "sharing" is a set of ethical values, such as trust, communality, equity, and justice in human relations that are pre-requisites for any thriving and sustainable urbanism. However in today's cities, these values have evidentially atrophied—perhaps even completely nullified—in the face of excessive commercialization and consumerism, rapid technological advancement and change, ambiguous identities, and increasing placeless-ness. In short, the ascendancy of globalized capitalism (Sklair, 2017) has reinforced a form of neoliberal urbanism that tends to privilege enclosures, segregation, and the privatization of public spaces (Hodkinson, 2012). Ironically however, it is exactly these same undesirable attributes that render sharing appealing and perhaps even necessary. A continuously growing urban population amid the simultaneous conditions of low wage growth, material scarcities and commitment to environmental sustainability is likely in time, to eschew the paradigm of consumerism and ownership. In turn, this population presents the necessary critical mass that will not only demand, but also come to supply, the shared resources required for a thriving and sustainable urbanism (McLaren and Agyeman, 2015). Sharing then becomes a constitutive activity of society and social media (John, 2017), and together with the high-tech digital economy, will likely form a primary driver of future urban development—shaping the built environment, which in turn will further evolve sharing practices of the future. Could sharing and sharing practices then establish the foundations for an alternative form of urbanism (see Sklair, 2017)—one that is not predicated on the enclosures, segregation and control found in neoliberal urbanism? As a practice that has to assume trust, communality, equity and in many cases, also

的共享而设计与为了作为途径的共享而设计有着本质的不同并且应该区别对待。下文还将就此做详细讨论。

第三，共享常常被简化理解。然而我们的初步研究则提出，共享至少有三种理想类型，并且每一种类型所共享的内容及其背后的逻辑和伦理基础都大不相同。第一种可以被称之为对有限的有形资源的零和共享，即分享的越多则拥有的越少。由于这一类型的共享主要关乎有限资源的分配，其背后的分配原则的公平性尤为重要。与此相反，第二种则可被称为非零和共享，特指对无形资源的非竞争性共享，因而共享的频率和量纲都无限制。分享的越多并不会减少他人的分享，也不会对资源的提供者产生影响。最后一种也是最常见的一种则是前两种类型的结合。此类型的共享同时具有前两者表面上相互矛盾的特质：一方面从资源上看，被分享的越少，分享者们共同拥有的就越多；另一方面从经验上看，分享的越多越频繁，分享者们也就越有可能因为互惠而更多地分享，从而衍生出一系列先前无法预测的积极正面的结果。对共享的理想化分类也同样意味着在不同的情形下，对于不同的共享系统应该采取不同的设计策略。然而，最后一种类型的复杂性更表明，在实践中如何能够整合利用不同的策略才是完成好的设计的关键所在。我们在下文将对此进行深入探讨。

为共享而设计

以上对于共享的初步思考究竟对于共享系统的设计意味着什么呢？在此，我们将尝试给出答案并结合新加坡国立大学的设计课成果进行探讨。

首先，尽管从概念上而言，对于作为途径和目的的共享系统的设计需要区别对待，然而在实践中，共享系统的设计则常常是为某种目标而服务的。反过来，这些目标也同时影响着共享系统的设计过程。这也就意味着，为了设计真正有效而非空有躯壳的共享系统，设计师必须时刻保持着一个对共享系统所承载的共享活动及其与整体目标的关系的清晰理解。然而，需要指出的是，即便一个精心构想的、充满智慧的共享设计也无法全然保证其承载的共享活动的成功和整体目标的实现。这是因为每一个设计自身都不可避免地有着潜在的风险，并且这

compassion, sharing appears promising for re-imagining this alternative urbanism. It is therefore of utmost importance for architects, planners and urban designer to create an environment that can effectively drive and enable authentic and ethical sharing. More specifically, it is crucial to design better sharing systems that can restore these "lost and disappearing" values, which can build solidarity in and for our cities. An urbanism built on solidarity ought to be the future of humanity, rather than the contemporary fragmentations, strife and endless conflicts observed in cities all around the world.

Some preliminary thoughts on sharing

If sharing ought to be prioritized based on the reasons above, then what is sharing and design for shareability? In trying to answer this primary research question, we discovered a few things.

Firstly, design for shareability has to presume the efficacy of design. In contrast, most, if not all time-honoured sharing practices, were likely the outcome of social evolution rather than design. For instance and in all likelihood, bad sharing practices (however "bad" is defined) were most likely weeded out by social evolution while the good sharing practices—practices that most likely exemplify what was, and still is, considered virtuous and praiseworthy for the human society and humane values—were repeatedly reinforced, and come to be retained even in the present. To design for shareability then, it is vital to recognize sharing practices as a complex and dynamic system comprising of changing cultures, intentions, values and artefacts instead of statically or singularly, constituted either by social process or artefacts.

Secondly, to design for shareability implies clarity on sharing within the means-ends spectrum. In other words, it is important to distinguish sharing as a means toward some ends, and sharing as an end in itself. For

些风险无法通过进一步的设计而得以弱化或者回避。反之，进一步的设计则可能带来其他的负面效应（see Beck, 1992）。

其次，就不同理想类型的共享活动而言，为第一种类型设计共享系统对于有经验的设计师或许并不具有太大的难度，只要分配或者分割的原则是清晰的并且能够被恰当地引入到设计中。对于第二种类型，设计一个能够促进人们分享无形资源如信息或者知识的系统也同样并非难事，然而设计本身所无法掌控的正是分享的经历以及可能出现的结果。换言之，仅就传统意义上的设计而言，设计师并无法对非零和共享的道德和伦理的正当性产生影响。为了更好地实现设计的意图，这也就意味着设计师不仅仅需要关注有形的物质或者非物质环境，同时应该对隐形的组织制度体系进行设计，从而推动符合伦理道德的共享实践。对于第三种混合类型，如上文所述，其复杂性本身对于设计师就相当具有挑战性。零和共享与非零和共享之间的动态关联意味着共享者之间的关系会因为其分享的经历而不断变化。也正是因此，有形资源的分配原则在不同的时间或者情形下也可能会大不相同。这就要求共享系统的设计需要满足最大化地灵活可变并且能够被重复地自定义。

我们对于共享的初步研究是由新加坡国立大学和清华大学的联合设计研究课程所促成的。共享系统的设计正是这一设计研究课程的核心所在，而上文总结的关于共享系统设计的要点也在这一课程中得以应用并对有效地指导着设计研究。新加坡国立大学的两个设计课小组分别以不同的视角和方式对共享和设计之间的复杂关系进行了深入探讨。

以城市共同体（Urban Commons）为主题的设计研究小组着重探讨了共享如何能够作为促进居民共创、共建、共营社区共同体的一个有效途径。首先，高层社区及其附属的移动社区活动站的设计旨在为社区共同体的营造创建一个社会基础。这一共享系统让不同背景和身份的居民通过分享公共文化生活和兴趣爱好，建立更加密切的社会交往和对话，并最终实现新老居民的融合以及对公共活动的积极参与。在此基础上，社区共同体的营造则主要通过居民共同进行垃圾的回收再利用而实现，比如将不

instance, sharing as a form of altruism is often seen as an end in itself; there is no further need to justify sharing here. In contradistinction, sharing that was done for greater cooperation, or sharing more in order to participate more fruitfully, is each an instance of sharing as a means towards some other end goals. Clearly, to design for sharing as an end in itself, and to design for sharing as a means toward some other end goals, are two diametrically different tasks and ought to be treated as such in any design process.

Thirdly, sharing is often perceived as a monolithic practice. But our research indicates that there are at least three ideal-types of sharing, and each is predicated not only on different kinds of entities, but also on vastly different sets of logic and ethics. First, zero sum sharing pertaining to limited and tangible goods implies that the more I share, the less I would have. And because this type of sharing deals with the distribution of limited goods, the quality—fairness—of the distributive rules behind sharing is paramount. And following this, non-zero sum sharing that pertains to non-limited, non-rivalrous and intangible goods implies that there is no limit imposed on either the frequency or magnitude of sharing. Consuming more of one's share neither diminishes the amount consumed by another individual nor impacts the originator of sharing. The total value of sharing is also not conserved. Finally, the most common type of sharing is perhaps a hybrid of the first and second type. This type of sharing has both the attribute of the first type of sharing in that the less is being shared, the more each sharer will get to keep; and the attribute of the second type in that the more is being shared, the more likely other sharers will reciprocate with more sharing—leading to a situation where sharing can lead to many desirable outcomes, which however, cannot be predicted in advance. Categorising sharing practice into different ideal types seems to suggest that different strategies should be taken when

可降解的垃圾转化为三维打印租赁自行车的原材料，以及将食物垃圾转化为洁净能源和特别服务于城市农业的肥料（依照前文讨论的共享的定义，共享在此可被理解为参与并分担某项活动）。合作工坊的设计便是为了承载这一系列的共享活动，并服务于创建一个共同的可持续社区的最终目的。最后，艺术流中心的设计则是为了推动大家分享传统应用艺术的创作和推广，其根本目的在于通过建立一个社区共同体来与这些传统艺术因为新媒体而出现的滥用和贬值抗衡。

尽管上述三个共享系统的设计均经过细致推敲，然而依旧无法回避其内在的风险以及实现最终目标的阻力。比如垃圾再利用这一共享实践可能会因个体的过分节俭或者懒惰导致的垃圾总量不足而无法开展，进而影响到社区共同体的营造。又比如保护并延续传统应用艺术的努力可能在经济层面并不可持续，最终仍然会对市场形成妥协。显然，这些潜在的风险和负面的效应很难通过建筑或者城市设计本身得以解决。如果要进一步完善共享系统的设计，那么探究这些系统背后的组织体系并在可能的条件下巧妙地将其转化为对物质环境的设计则是关键所在。

新加坡国立大学的另一个设计研究小组则侧重于探讨以下两个问题：零和共享与非零和共享的动态关联，无形的组织制度体系在共享系统设计中的整合。比如，共享基础设施系统被构想为一个垃圾—能源—动力的循环。在这一系统中，每家每户产生的垃圾被回收并转化为电力以驱动无人驾驶车。无人驾驶车能够实现居民在社区内部以及到公共交通站点的便利出行，并同时承担垃圾收集和电力能源配送的任务。从前文提及的理想化共享类型的角度看，零和共享（即电力能源的生产和使用）与非零和共享（即无人驾驶车服务）相互交织在一起，并且长久的非零和共享依赖于个体充足的贡献，也即一定量的垃圾（在此共享可以被理解为参与活动和分担责任）。这也是此基础设施系统正常运转的关键所在。为了更好地激发居民持续性的贡献，保证足够的垃圾输入，这一设计同时也引入了一个信用体系以强化这一循环：居民可以通过贡献垃圾获取相应的信用值，从而利用这些信用值免费使用共享的无人驾驶车。由此，这一信用体系将个体及公众权

designing sharing systems under different conditions. However, the complexity of the last type indicates that in real practice achieving a concerted application of different strategies is the key to a good design, and which will be elaborated below.

Design for shareability

Then, what does the preliminary findings from our research specifically mean for the design of sharing systems? We will summarise our observations below and illustrate these points based on a few design schemes from the NUS studio.

First, although conceptually design for sharing as a means toward some ends and design for sharing as an end in itself ought to be treated differently, in practice however, the design of sharing systems is often undertaken in the service to some desired (end) goals. In turn, these goals also conditioned how things are shared. This means that to design a truly effective, rather than just an approbative sharing system, designers should always have in mind a clear understanding of the correlations between the sharing activity under consideration and the attainment of these desired goals. However, it has to be pointed out that even a deliberately conceived sharing design, replete with insights, does not guarantee the success of its sharing practice and the attainment of these desired goals. Risks are likely to be inherent in every single attempt of design, and can hardly be prevented or ameliorated by further design interventions, which in return would likely bring about other side-effects (see Beck, 1992).

Second, with respect to different types of sharing, design for the first type is not necessarily challenging for an experienced designer, as long as the division and partition rules are clear and incorporated in the design. For the second type, designing an artefacts or a system that facilitates the sharing of intangible goods such as information and

益整合到设计中，即便或许只具有象征意义上，但也足以使得共享基础设施系统更加持久有效。

同样的设计策略也体现在共享公共设施系统的设计中。这个方案构想了一个完全开放的教育平台，形成对现有的学校体系的补充。基于这一平台，无论是当地的居民还是游客和到访者，只要有所专长并拥有相关认证，都可以面向社区内混合的学生群体开设开放性的课程。反过来，学生们也因此得以有效地拓展接触面并获取多样化的知识和技能。为了以便利的场地和低廉的成本推动这一共享教育平台，社区内闲置的、未充分利用的空间被整合，更新并按时间租用来进行教学。在这个设计中，零和共享（即对教室的分享）和非零和共享（即对知识和经验的分享）同样交织在一起，而同时因为教室的共享则有可能推动知识进一步的传递和分享。因此，就物质空间而言，保证这一共享系统持久运转的关键在于设计一个能够容易改变格局且易于维护的教室环境，从而满足不同课程的需求。这个方案也对此给出了自己的答案，即通过模数化的设计为空间的分隔和重组提供一个便捷的体系。最后，与之前共享基础设施案方案中相类似的信用体系也被整合进入并强化这一系统，即在课后清洁教室的学生可以获取相应的信用值当作其他课程的“学费”。

以上的例子似乎都在不同程度上指出，为共享而设计常常需要应对一个包含着诸多相互关联的子系统的复杂系统。对于不同子系统之间互补性的细致研究可以带来更深入的洞见，从而构思出更好的设计并且使得共享实践更加持久。反之，对于子系统之间的微妙联系和潜在冲突的忽视则可能瓦解所追求的共享活动并导致共享系统设计的失败。总而言之，关于系统方法（Churchman, 1968）的知识以及如何有效地进行系统设计都对为共享而设计大有裨益。

小结：对于共享系统的设计研究?

本文的主要目的在于辨清并阐释设计和共享之间迄今为止仍旧模糊的关系。首先，我们讨论了“共享”作为一个带有诸多语境含义的概念应该如何进行解析，并在此基础上分析了为什么“共享”在当今的设计研究中应该被加以重

knowledge is equally not difficult, but the challenge lies in the fact that the experiences of sharing and the consequences are usually beyond the control of the design itself. This is not to say that there is no room for designers to exercise their discretion over the integrity of non-zero sum sharing. What they should pursue is perhaps to go beyond design of the visible to designing in the "invisible" – the organizational-institutional system that can enable ethical sharing. For the hybrid type, as mentioned above, its complexity is a real design challenge for designers. The dynamic interactions between zero-sum sharing and non-zero sum sharing suggest that the relationship between sharers may change as a result of their experiences of sharing, and in relation to this that the division and/or distribution rules for the sharing of tangible goods are likely to vary at different times. This means that design of sharing system ought to aim for maximal flexibility and repeated customisation.

This research and in particular the abovementioned findings and observations would have been impossible if not with the design research of the NUS-Tsinghua joint studio. The above-summarised implications to design for shareability are the highlights of this studio and guide the design inquiry. There are two design studios within the NUS-Tsinghua joint studio program, and each is structured with a different approach to investigate the varied relationships between sharing and design.

For instance, in the NUS design studio deemed as Urban Commons, sharing is taken as a means to sensitise residents to participate in the co-creation, co-production and co-maintenance of certain commons of a community. Specifically, the sharing of social experience and cultural interests enabled in the design of Mobile Community Tower and its satellite CC hubs aims for encouraging interpersonal communications and civic

视。此后，我们延伸这一问题，探讨了共享的目的，不同的理想类型及其对共享系统设计的启示，并通过新加坡国立大学的两个设计研究小组的成果对此进行了详细的阐述。

此外，本文也阐明了设计可以通过不同的方式与共享相连，并且可以被借以推进共享。反过来，共享也同时可能限制甚至破坏共享系统的设计所希望达成的目标。更重要的是，我们认为为共享而设计并非易事。我们不能将其简化为对创造一个有利于共享的先决环境或者是对某一理想化共享类型的产生条件的单纯关注。反之，有效的设计需要一个广阔的视角，以及对于所设计的共享系统作为一个更大的复杂系统的子系统之一是如何与整体目标相关联的清晰理解。

共享系统设计所面临的这些挑战似乎为我们提出了这样一个亟须回答的问题：在着手共享系统的设计前，我们应该进行怎样的一种设计研究？更准确地说，我们需要什么样的设计知识体系，才更能有效地服务于为共享而设计？

REFERENCES

1. Beck, U. (1992). Risk Society: Towards a New Modernity. London, UK: SAGE.
2. Churchman, C.W. (1968). The Systems Approach. New York, NY: Delta Books.
3. Harari, Y.N. (2015). Sapiens: A Brief History of Humankind. New York, NY: Harper.
4. Hodkinson, S. (2012). The New Urban Enclosures. City, vol.16, no.5, pp. 500-518.
5. John, N.A. (2017). The Age of Sharing. Malden, MA: Polity.
6. McLaren, D. & Agyeman, J. (2015). Sharing Cities. Cambridge, MA: MIT Press.
7. Sklair, L. (2017). The Icon Project: Architecture, Cities, and Capitalist Globalization. New York, NY: Oxford University Press.
8. Stavrides, S. (2016). Common Space: The City as Commons. London, UK: Zed Books.
9. Tomasello, M. (2009). Why We Cooperate. Cambridge, MA: A Boston Review Book.

dialogues between different demographic groups, assimilating newcomers with local residents and cultivating active citizenships; involving all the residents in collecting and up-cycling waste to produce 3D printed rental bicycles for local mobility, biogas as clean energy and compost for urban agriculture (according to the aforementioned dictionary definition, this form of sharing is to have a part in something, especially an activity) serves the objective of building a sustainable community of new professionals; and finally creating an Artstream centre to accommodate and facilitate sharing of the making and distributing of traditional and applied arts is an attempt to address the threats of exploitation and devaluation that new media is likely to pose onto them.

Despite that the three sharing systems are deliberately conceived to serve respective end goals, there also exist risks and counterforces to the attainment of these goals. For instance, the efforts in building a sustained new professional neighbourhood may be undermined by individuals' frugality in daily consumption and laziness in contributing waste; and the endeavour to protect and sustain the value of traditional and applied arts may not remain economically sustainable and have to be subordinated to market power. Clearly, these risks are not easily ameliorated by either architectural or urban design itself. To further improve these schemes, however, extending the design exploration to the underlying organisation systems and, if possible and appropriate, incorporating and materialising them in the design of the physical environment would be key.

In the subsequent NUS design studio however, the design of sharing systems is explored with a special consideration of (i) the dynamics between zero sum and non-zero sum sharing, and (ii) the facilitation of the invisible organisational-institutional system to the visible forms of sharing. For instance in one design project, the sharing infrastructure system is conceived as a waste-energy-mobility loop. In this system, waste produced by every household is collected and turned into electricity, which is then used to power the self-driving cars that are then shared by local residents for traveling to different places within the neighbourhood as well as to public transport nodes. And at the same time, the sharing cars also serve to collect waste and distribute energies across the neighbourhood. Applying the aforementioned ideal-types of sharing to this case, zero sum sharing (sharing of electricity) and non-zero sum sharing (sharing of local self-driving car service) are intertwined, and more sustained non-zero sum sharing is predicated on individuals' contribution of waste (sharing in the sense of involving or participating in an activity with others), which is critical to the functioning of this sharing infrastructure system. Clearly, the tripartite loop itself does not guarantee sustained sharing practice. To incentivise the residents' continuous contribution and then ensure a sufficient amount of waste inputs, a credit system (as an invisible institution) is then introduced to complement the loop that allows residents to pay for the shared self-driving cars using the credits they would earn from their waste contribution. In so doing, this system design incorporates in itself an ostensible form of stakeholdership that is sufficient for the sustainability of the sharing infrastructure.

The same paradigm of design can also be observed in another scheme on the sharing of public amenities. In this design, an open education platform that complements with existing school system is proposed, whereby both local residents and visitors with certain expertise and certifications can offer open courses to a mix of students of different demographics. In turn, students can benefit from being exposed to a wide variety of knowledge fields and skills through their participation. And to physically realise this sharing system and keep the cost to minimal levels, redundant buildings and temporarily underutilised spaces within the neighbourhood are identified, renovated and then used as the venues for teaching and learning. In this case, again zero sum sharing (sharing of classroom) and non-zero sum sharing (sharing of knowledge and experience) are interlocked, and reciprocal sharing of classroom spaces is very likely to lead to enhanced knowledge impartation (sharing of knowledge). The challenge for a sustained sharing practice is therefore firstly the design of the classrooms, which have to permit repeated customisation and renewability for different courses, and secondly, designing for their convenient maintenance. In response, this scheme uses the modular approach as the key strategy to make the classroom easily divisible and (re) configurable. And a similar credit system to that of the abovementioned sharing car is also introduced, where free lessons will be awarded to students who clean the classroom immediately after their class.

The above examples seem to suggest that design for shareablity often has to deal with a complex system comprising a number of interrelated sub-systems. Deliberate consideration of the complementarities between the sub-systems can offer more insights and produce enhanced design that is likely to achieve sustained sharing practice, whereas ignorance of the subtle links and potential conflicts between the sub-systems may undermine the desired sharing practice and even result in failure of the design of shareability. In short, knowledge of the systems approach (Churchman, 1968), and how to design these systems effectively, appear likely to benefit the efficacious design for shareablity.

Conclusion: towards design research for sharing systems?

Our primary aim in this chapter is to clarify, and then explicate, the hitherto ambiguous relations between sharing and design. In doing this, we suggest how "sharing" as a concept should be distinguished despite its many contextual meanings in design, and following this, argue for why "sharing" ought to be prioritized today in design research. We further reinforce these claims by examining the objectives and different ideal-types of sharing and exploring their implications to design for shareability, which are illustrated in details through the design schemes of the two NUS design studios.

Moreover, this chapter clarifies that design can be connected in varied ways to sharing, and furthermore, leveraged for, and enhance, sharing. In turn, we anticipate that sharing practices may constrain, and even undermine, the very objectives that sharing systems were designed to attain. Importantly, we suggest that design for sharing is challenging — it cannot be reductively simplified, or solely focused on establishing the preconditions for sharing, or to encourage the emergence of any one particular ideal-type of sharing. Instead, efficacious design for sharing requires a broader view and understanding of how the design of sharing system can be correlated to the end goals within a broader context where sharing is merely one of the many sub-systems.

In light of these acknowledgements on the challenges of design for shareability, the question then turns to what kind of design research ought to precede any design projects on shareability, and specifically, what form of design knowledge has to exist to effectively design for shareability.

STUDIO

Introduction to studio setup

联合设计教学组织介绍

In the brief setup, Tsinghua University took the the White Pagoda Temple area in historic Beijing as its main intervention area with 8 themes: National University of Singapore has selected two site, a site in the city center where the traditional Nyonya culture originates: Joochiat and a “sub-urban” site where a railway station and residential life intersects: Jurong East. All sites have a similar aim to discuss the idea of sharing basic facilities, sharing infrastructure and service, sharing production and commerce, sharing culture and everyday life, as well as establishment of the basis of urban community, collaborative production of urban community, and other related aspects.

During this joint studio, two design charettes were held in Beijing and Singapore respectively. Students from Tsinghua University and National University of Singapore collaborated and divided into a few groups, then carried out a 1-2 day workshop according to the site in Beijing and Singapore. After the collabration, the students completed their own design work respectively, including the case studies, site analysis, and urban design proposal. Finally, the “Shating City: Sharing Economy and Urban Renewal” forum and final design presentation from both university was held in Tsinghua University School of Architecture.

Through this collaboration, the joint studio has achieved a positive results for further collaboration in the future, students get to form strong relationship and the results from this joint studio participated the 2017 26th UIA exhibition and 2017 Beijing Design Week.

SETUP

在题目设置上，清华大学选取了位于北京旧城的白塔寺历史文化地区，探讨将共享城市分为8个不同的主题进行讨论；新加坡国立大学选取了位于城市核心地区的传统娘惹文化发源地如切和轨道交通站点地区裕廊东，分别讨论了共享基础设施、共享公共设施与服务、共享经济生产与消费、共享文化与日常生活的主题，以及奠定城市共同体的基础、协同生产城市共同体、预测新兴城市共同体形式等方面的内容。

在教学过程中，分别在北京和新加坡举办了两次工作坊，清华大学和新加坡国立大学的同学们联合成为若干小组，针对北京和新加坡的地段进行为期1~2天的工作坊，就共享城市提出针对地段的解决策略。此后两个学校的同学们分别完成自己地段的设计工作，设计工作包括全球范围内共享城市的案例搜集与分析、地段现状分析及发展策略研究、城市设计及建筑设计。最终，在清华大学建筑学院开展了“共享城市：共享经济与城市更新”论坛和双方设计成果的最终评图。来自高校的研究者、共享经济的市场运营者、规划师、建筑师从不同的角度对共享城市的理念、经济模式、实施途径等方面进行探讨。

此次联合教学取得了积极成效，为后续合作奠定了良好的基础，同学们之间建立了深厚的友谊，教学成果参加了2017年第26届UIA展览和2017北京设计周展览。

DESIGN CHARETTE

Conceptualization of theme, creative brainstorm to bridge cultures

设计主题概念化，建立不同文化间的创新头脑风暴

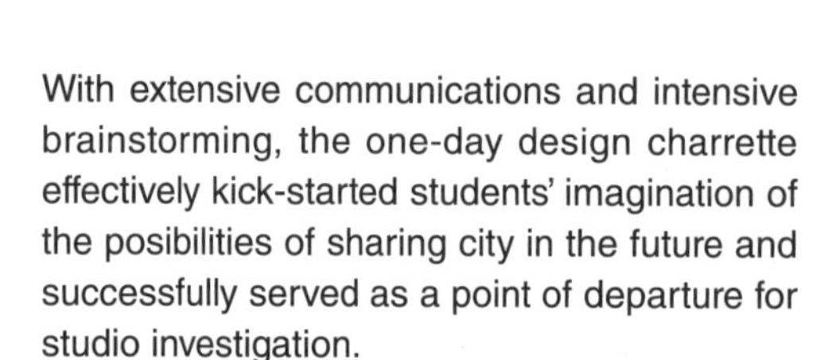

With extensive communications and intensive brainstorming, the one-day design charrette effectively kick-started students' imagination of the posibilities of sharing city in the future and successfully served as a point of departure for studio investigation.

A detailed site survey of Baitasi, a historical neighbourhood in the West of the Old City of Beijing, provided students with a concrete basis to explore how sharing can contribute to regeneration of traditional urban neighbourhoods.

伴随广泛的交流和集中的头脑风暴，一天的设计研讨有效地启发了学生们关于未来共享城市可行性的想象，并成功地作为联合设计调研的始发点。

随后对于位于北京老城内西部地区白塔寺地段的详细调研，给学生们提供了探索共享如何贡献于传统城市邻里更新的具体依据。

BEIJING

北京

DESIGN CHARETTE

Conceptualization of theme, creative brainstorm to bridge cultures

设计主题概念化，建立不同文化间的创新头脑风暴

For the second workshop and charrette, two sites in Singapore, Joo Chiat and Jurong East, were selected to continue the design exploration from the first workshop on the one hand (regenerating traditional urban neighbourhood), and on the other, expand the investigation to a completely new urban condition.

Entensive discussions about different design schemes and underlying ideas among students and tutors served as a fruitful reflection on the joint studio halfway through the collaboration.

第二阶段的工作坊和设计研讨针对新加坡的两个地段，如切和裕廊东。一方面，这是第一个工作营的延续（更新城市的传统邻里），另一方面，也将调研扩展到全新的城市环境之中。

对不同设计方案以及老师同学们内在想法的密集讨论，对过程中的合作设计提供了富有成效的支撑。

SINGAPORE

4

DESIGN STRATEGIES 设计

WHITE PAGODA TEMPLE

北京白塔寺

After the rapid urbanization process, Chinese cities development trend shifted from urban expansion to the built up area regeneration significantly, with the main issue focusing on urban life quality promotion by urban regeneration for better human habitats, in which people and activities inextricably intertwined.

The studio chose the White Pagoda Temple area as the urban regeneration site, which is mainly residential, cultural and commercial functions, with abundant cultural heritage in Beijing old city. In this area, there is a white pagoda temple built in the Yuan Dynasty, in which the pagoda is the oldest and largest one in China. The Fusuijing Building, a socialist style building built in early time of P. R. China, is the only high-rise building in the site. Another two landmarks, Luxun Museum and Flower Market, strengthen the cultural context of the site. The White Pagoda Temple area is surrounded by high-rise financial offices and housings, both on north and south boundary. Meanwhile, the White Pagoda Temple area is

中国城市在经历快速的城市化进程后，城市从往外快速扩张的发展阶段进入了城市存量空间更新发展的阶段。如何通过城市更新提升城市生活的品质，营造更为美好的人居环境，让不同的人群、不同的活动交融其间，成为当前中国城市面临的主要问题。

课程选取了位于北京旧城的白塔寺地段作为更新设计的场地。这里是北京旧城内文化遗产保存丰富的地区，以居住、文化和商业为主要功能，地段内的白塔寺始建于元代，寺内的白塔是中国现存年代最早、规模最大的喇嘛塔。中华人民共和国成立后在此建设的福绥境大楼成为地段内唯一的高层建筑，地段内还有鲁迅博物馆、花鸟鱼市，增添了浓厚的文化氛围。地段周边高楼林立，金融办公区、居住区南北环抱。这里也是不同人群汇聚的地区，本地居民、外来游客、周边商务地区的工作人群，往来其间。因此，白塔寺地段的多样性使得它成为讨论共享城市的良好范本。

图片来源：1. Google map (https://maps.google.com) 2. http://www.sohu.com/a/219767422_493897 3. http://www.sohu.com/a/252659008_428290

Shoubi Street
Elementary School
Former Flower Market
Pushuijing
Socialist Building
Luxun Museum
White Pagoda Temple
Zhaodengyu Street
Metro Station
Fuchengmennei Street
Sharing
Transportation
Sharing
Culture
Sharing
Education
Sharing
Housing
Sharing
Workspace
Sharing
Heritage
Sharing
Commerce
Sharing
Infrastructure

also a gathering ponit for different groups of people, where the local residents, tourists, and working people come across with each other. According to the characteristics of this area, the urban design task is divided into 8 themes, like sharing housing, sharing workspace, sharing transportation, sharing education, sharing culture, sharing heritage, sharing commerce and sharing infrastructure, which are the fundamental aspects of sharing city. The urban design group of each theme is required to propose the overall strategy covering the whole site, followed by the detailed design for subdivided sites. The solution of each theme shapes the outline of sharing city as well as the specific features.

依据地段的特性，设计任务被划分为8个主题：共享居住、共享办公、共享交通、共享教育、共享文化、共享遗产、共享商业和共享设施，这些主题是形成共享城市的基本要素。每个主题均要求在整体地段上提出解决策略和在特定主题的地段上给出详细的解决方案。各个主题的工作交融共存，以期勾勒出城市更新中共享城市营造的总体面貌与分类特点。

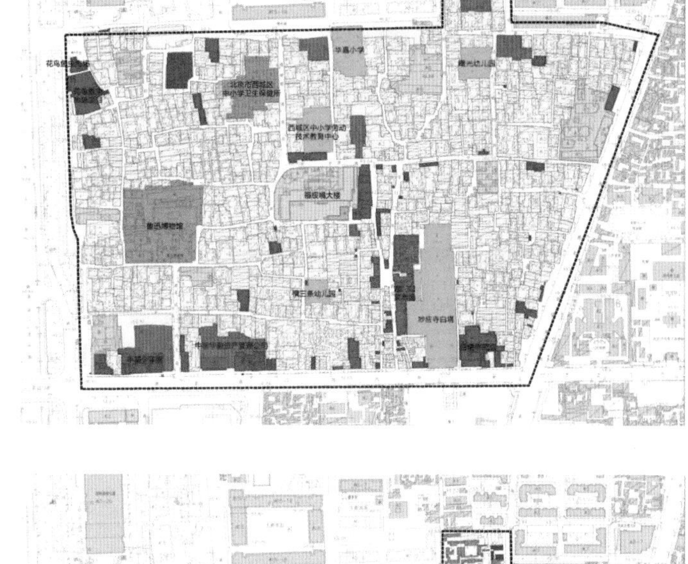

Current Land Use

用地现状

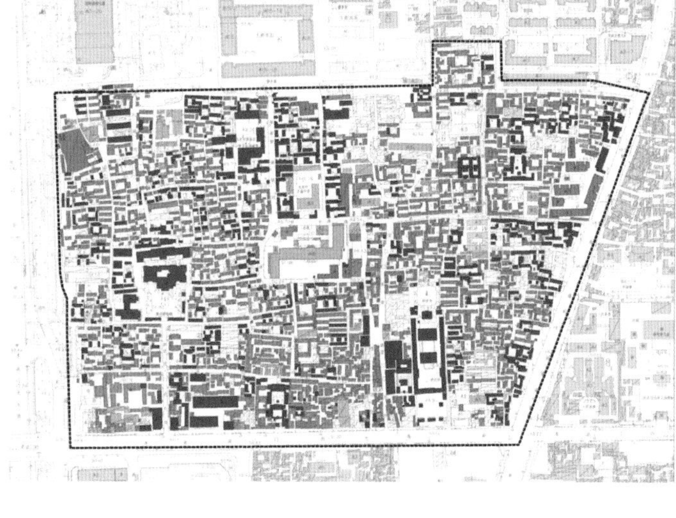

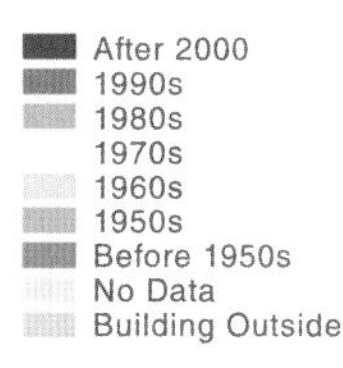

Built Time

建筑年代

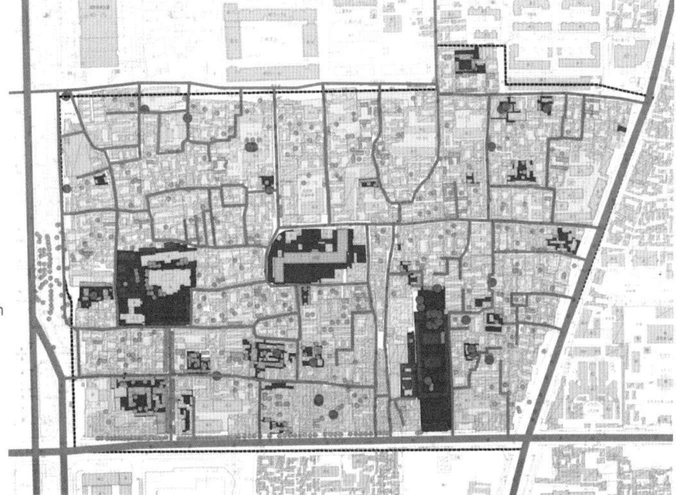

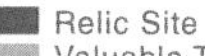

Conservation Elements

保护要素

图片来源：《北京市西城区白塔寺地区规划实施研究》北京市城市规划设计研究院

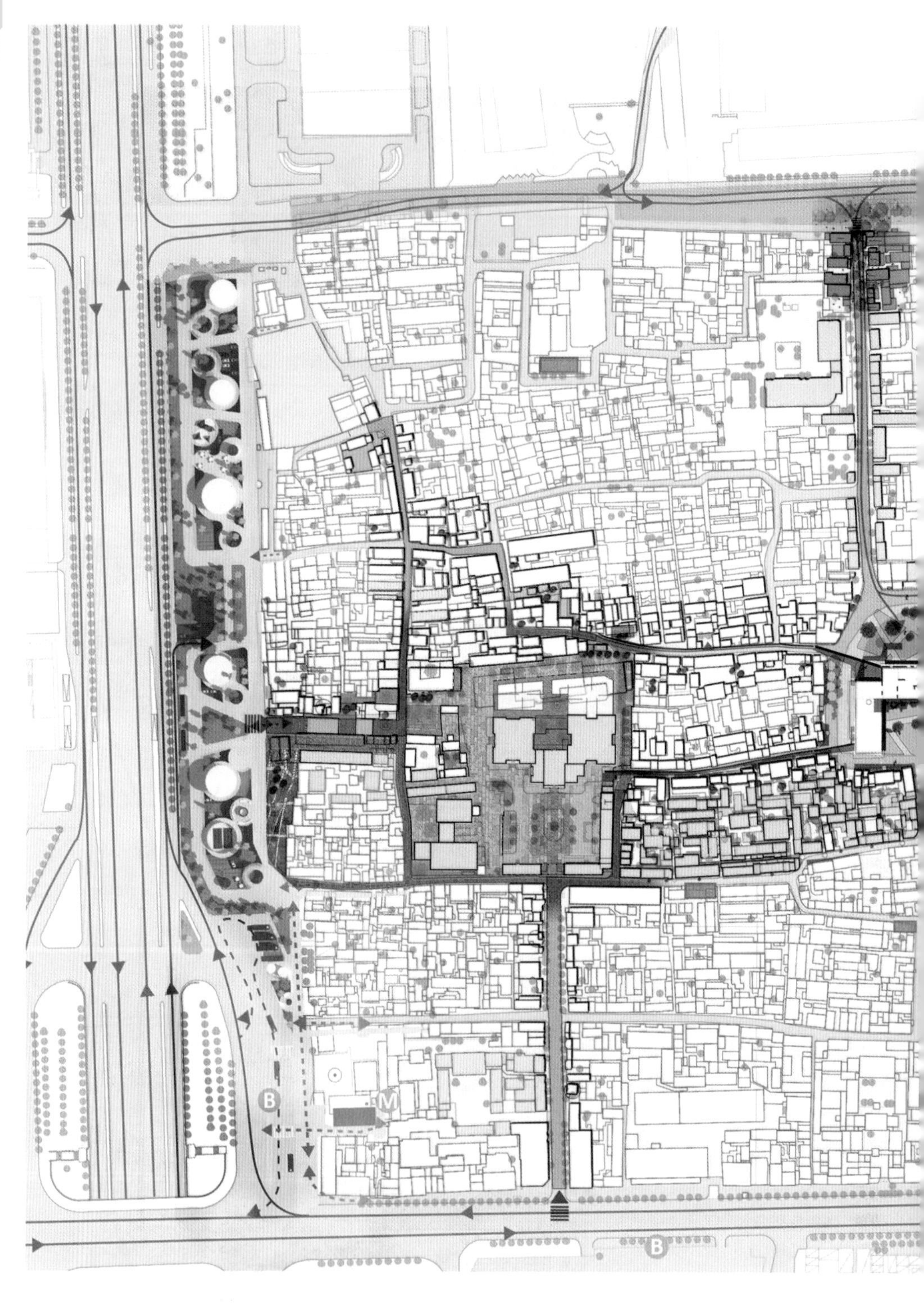
B
M
B

Site plan / 设计总图

01

共享居住

SHARING HOUSING

EXPANDING THE LIVING AREA

BY LAURÈNE XIAOWEN

扩展居住空间

02

共享办公

SHARING WORKSPACE

CREATING A NEW WAY OF WORKING IN BAITASI

BY KATJA CLÉMENTINE

创造新的工作方式

03

共享交通

SHARING TRANSPORTATION

PARK IN PARK

BY GYOUNG_MIN_KO CHIUN_CHENG_FUNG

公园中的停车空间

04

共享教育

SHARING EDUCATION

MAKING MORE VIGOROUS LEARNING ENVIRONMENT

BY BOEY_YING_YAN LOO_HUI_XIN

营造更有活力的学习环境

05

共享文化 SHARING CULTURE

BAITASI SKYWALK

BY DEANDREA VIJAY NEIN

白塔寺空中步道

06

共享遗产 SHARING HERITAGE

SHARING HERITAGE THROUGH PUBLIC SPACES

BY LAURA_HALLER ALESSANDRA_COPPARI

依托公共空间的共享遗产

07

共享商业 SHARING COMMERCE

FROM OWNERSHIP TO ACCESS

BY JAMAR_ROCK DAVID_VARGAS

从拥有到可获取的资源

08

共享设施 SHARING INFRASTRUCTUE

BAITASI PLUS

BY AHMED RICARDO

白塔寺基础设施 +

EXPANDING THE LIVING AREA

By Laurène, XiaoWen

扩展居住空间

The idea is to create a shared public space to expand the living area, and to provide street life and comfortable housing.

设计立意在创造共享的公共空间以拓展居住空间，并提供街道生活和舒适的住宅。

Open house vertically to provide private courtyard and more living space.

竖向敞开住宅，提供私密庭院和更多的居住空间

Open house horizontally to form a sharing living room and kitchen in courtyard.

水平敞开住宅，在院落中形成共享的起居室和厨房

SHARING HOUSING

共享居住

After studying the White Pagoda area, we concluded that, despite the low-standard housing conditions, the current way of shared courtyard living is a cherished aspect of local inhabitants' lifestyle. The present compound courtyard has, in fact, evolved from the traditional courtyard houses, affected by densification; and the contemporary shared courtyard living has represented the way of life of the past generation. Nowadays, the inhabitants are mostly unemployed adults from 30 to 60 years old who are living in poor conditions with less than 20 square meters per person.

在研究白塔寺胡同之后，我们认为居住方式对于居民非常重要。现有的大杂院实际上是原有传统院落不断搭建而成。大杂院代表了以往人们的居住方式。如今，原有居民大多为30~60岁的非就业人群，居住在人均不到20平方米的较差环境中。

Outside everyday life, with inappropriate public space

发生在不合适的公共空间中的日常户外生活

Compound courtyard life, with inaccessible corridors

大杂院生活和可达性差的走道

Residential area, with poor living condition

条件较差的居住环境

总体策略

In our design, we focus on a minimal-intervention strategy, in which we improve local public facilities, we aim to maintain a shared living community yard, and upgrade particular low-quality individual houses.

URBAN INTERVENTION

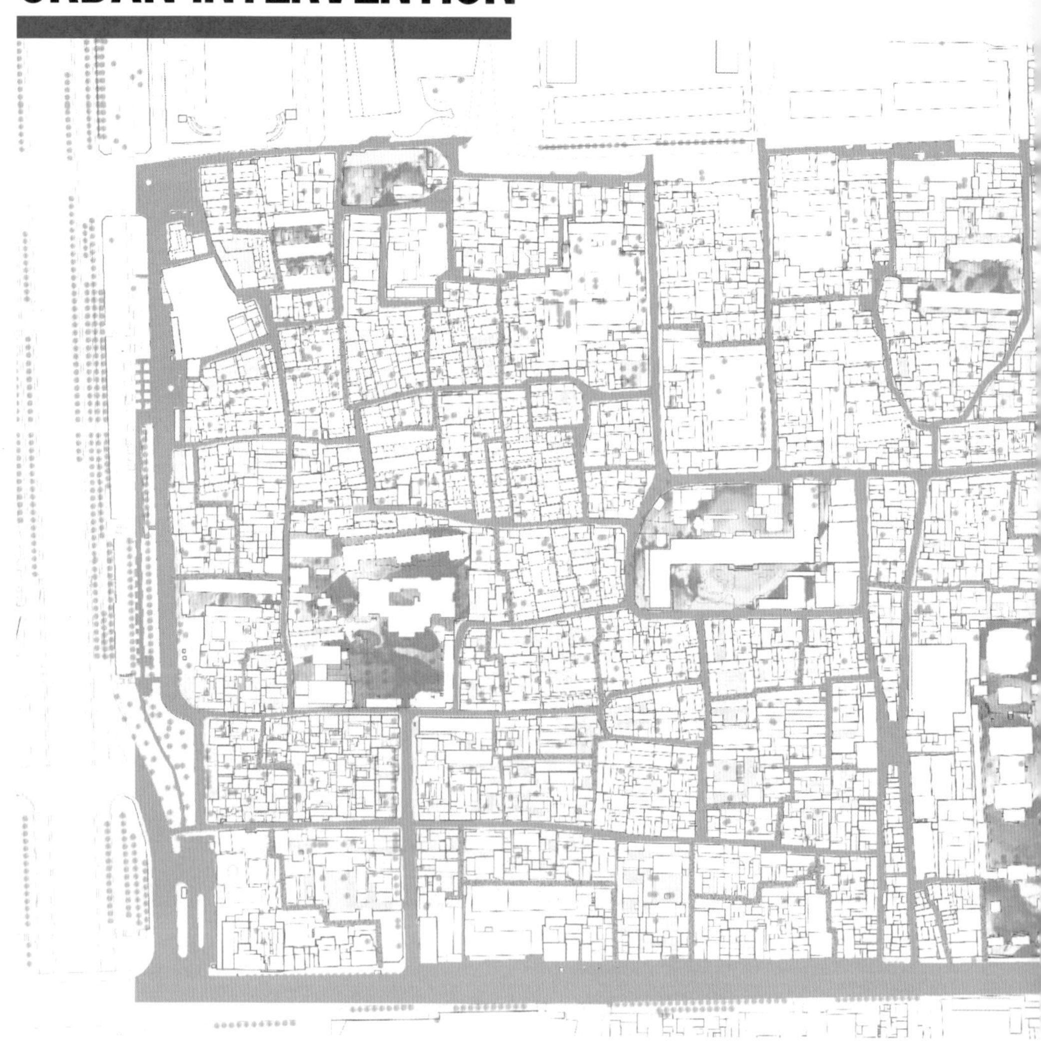

对居住环境的改善，首先分析区域内的房屋产权状况，拆除违章建筑，增加部分通道，并适当打开一些主要道路边的院落。

Master Plan With Detailed Design Site

包含设计基地的总平面图

街道系统

The entrance of the courtyard is more open to the street and the extension of the courtyard roof provide a space in between public and private to allow people to enjoy the neighborhood. Some plazas scattered in the hutong provide the district life with playground, space to dance, meet, drink... Therefore, from inside to outside, by adding more hierarchy within the area, the living space increases.

STREET SYSTEM

Street System

街道系统

通向院落的入口更加开放，并且院落屋顶的拓展为邻里之间半公共半私密的活动提供了空间。一些分散在胡同中的小广场可提供跳舞、集会、用餐等活动的空间。因此，增加由私密到公共的空间层次使得居住空间面积有所增加。

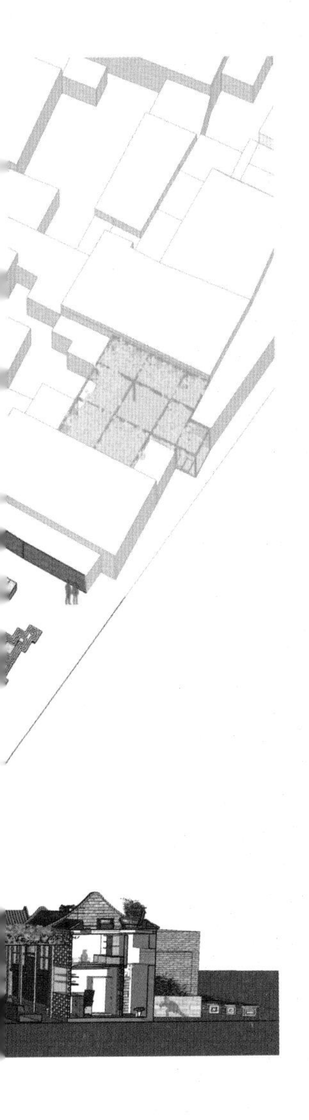

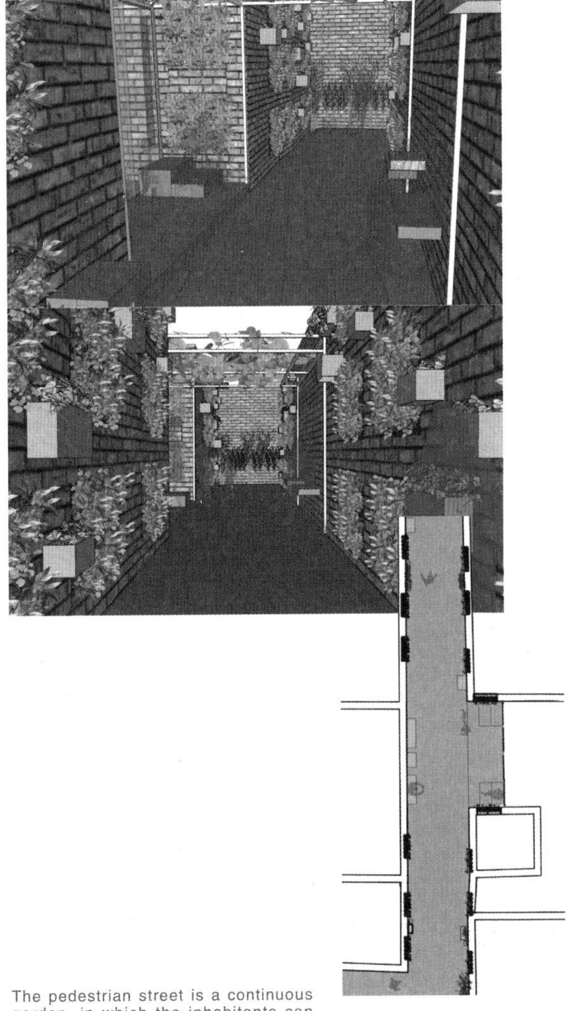

The pedestrian street is a continuous garden, in which the inhabitants can expand the courtyard.

人行道是一个连续的花园使居民的后院得以拓展。

院落系统

COURTYARD SYSTEM

The principle is to open the house vertically and horizontally:
Vertically: by opening the house's flat roof, a private "courtyard" is provided and by plugging inside mezzanine platform with furniture, more space is created.
Horizontally: the brick facades are partially opened, towards a shared courtyard. Partially covered, the courtyard provides a shared living room and kitchen.

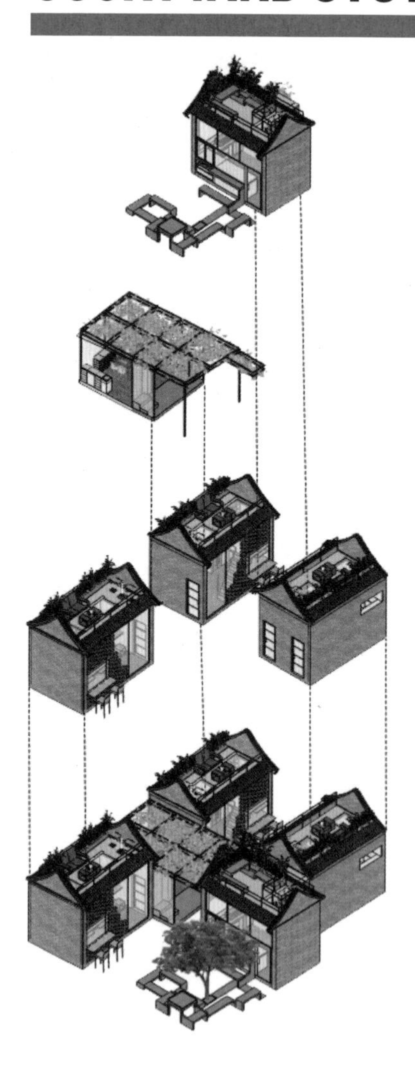

Courtyard System A

院落系统 A

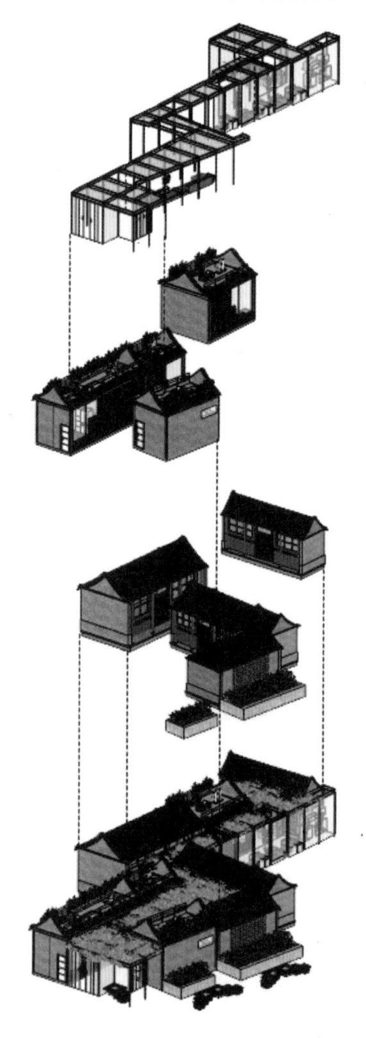

Courtyard System B

院落系统 B

设计原则是在竖直方向和水平方向打开建筑界面：
竖直方向：通过建筑屋顶平台形成私密的庭院，通过家具夹层增加室内空间。
水平方向：将建筑面向共享院落的一部分立面打开，在院落中形成共享的起居室和厨房。

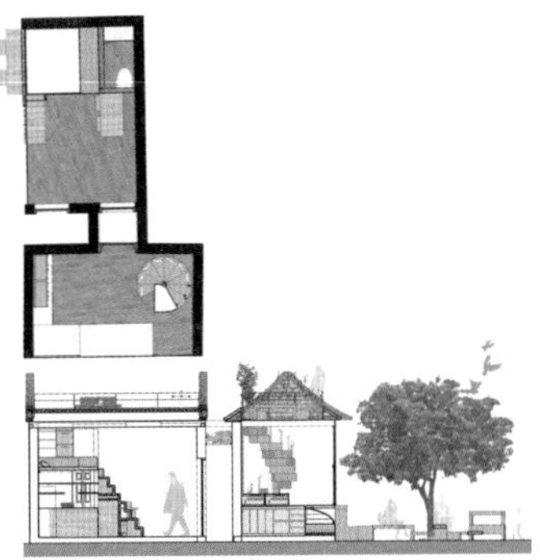

Coffee Shop

咖啡店

Courtyard System A

院落系统 A

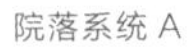

Courtyard System B

院落系统 B

44 m² floor
50 m² used
20 m² courtyard

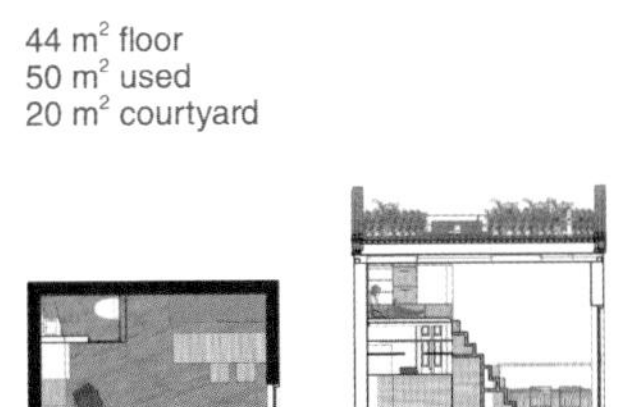

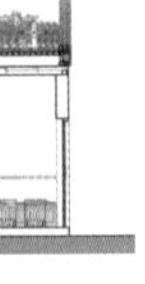

Single Room

单人间

58 m² floor
58 m² used
20 m² courtyard

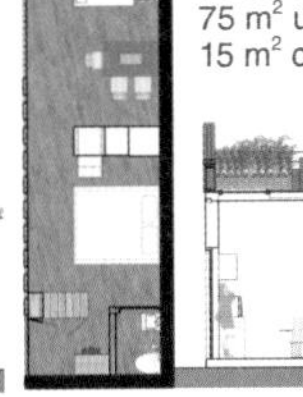

Double Room

双人间

70 m² floor
75 m² used
15 m² courtyard

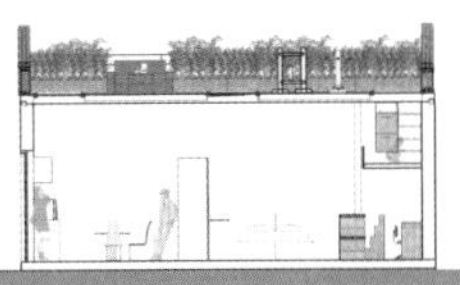

Triple Room

三人间

CREATING A NEW WAY OF WORKING IN BAITASI

By Katja, Clémentine

创造新的工作方式

Sharing work means everyday commuters from the outside, locals living in the Hutong, as well as temporary visitors. To create a space that benefits all these different users, we wanted our spaces to have good connections to the outside, as well as a central location in the BaiTaSi.

Our objective is to create new ways of working in the Baitasi hutong, that engages the locals as well as outside users, and to create spaces where the users can share and exchange work and ideas, in order to strengthen the community and identity of the area. Our proposal focuses on two specific spaces in the BaiTaSi hutong: The Fusuijing Socialist Building, and the hutong block on the south west side of the FuSuiJing.

共享办公涉及地区外的通勤人群、居住在胡同中到人群和旅游人群。为服务这些人群，设计者希望设计的空间与地区外、与白塔寺都有很好的关联。为强化地区的特色，设计者在白塔寺胡同中创造涉及地区内外人群的全新工作方式，创造分享工作和灵感的空间。设计聚焦在两个特定的空间，即福绥境大楼及其西南侧的胡同街区。

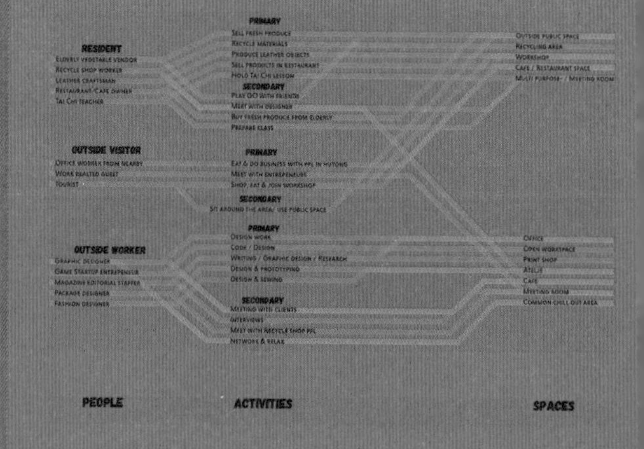

SHARING WORKSPACE

We want to provide spaces where users can share and exchange work and ideas, in order to strengthen the community and identity of the area.

我们希望提供使用者能分享和交流工作内容和创意的场所，以此强化社区和地区特征。

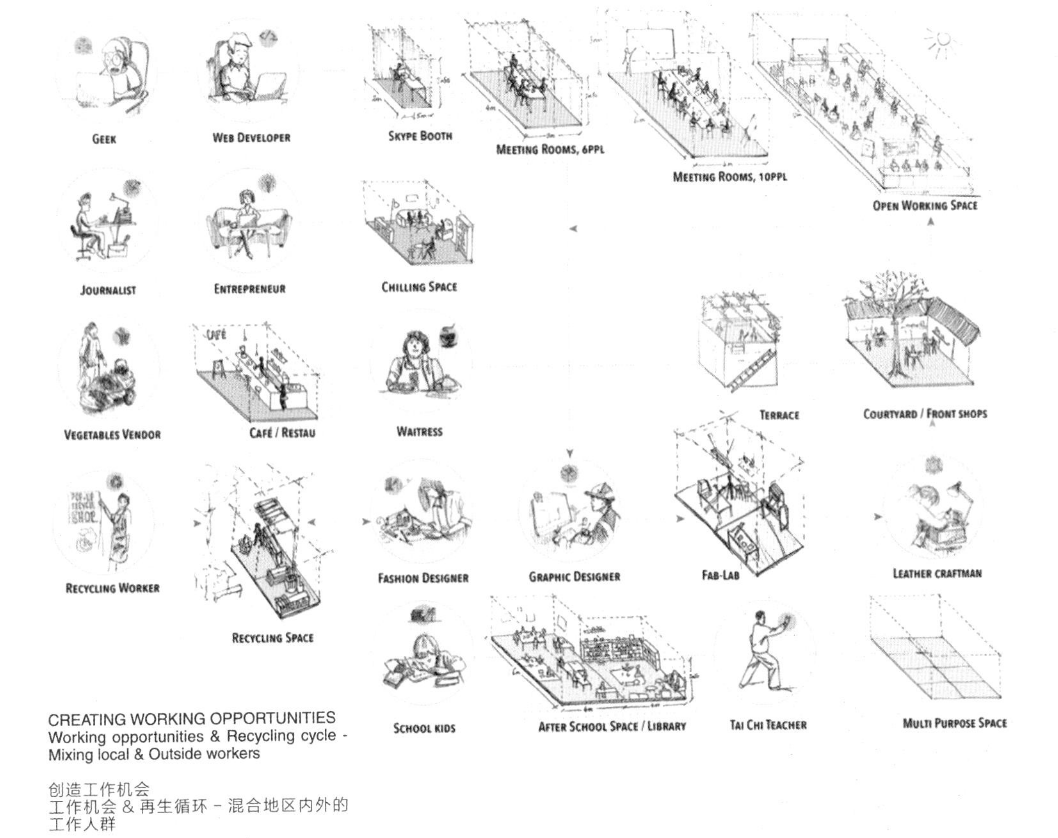

CREATING WORKING OPPORTUNITIES
Working opportunities & Recycling cycle - Mixing local & Outside workers

创造工作机会
工作机会 & 再生循环 – 混合地区内外的工作人群

分析

Following the diverse character of the two sites, our proposal includes two different typologies, which contrast and complement each other. First, the Fusuijing Building, which will work as a flexible hub, where working and collaboration with others is easy.

The former residential building represents

与两个场地的不同属性相关，我们的方案包含两种截然不同又互为补充的建筑类型。首先，福绥境大楼可用作灵活的便于协作与沟通的办公场所。

原住宅楼代表了19世纪50年代后期苏式建筑的风格，这也是为什么我们希望

ANALYSIS

CURRENT PROBLEMS
-The landuse of the area is not diverse enough
-The Fusuijing building is empty and deteriorating
-Lack of public space
-The line between public private is blurred
-The Hutong courtyards and alleys are cluttered and have plenty of informal additions

LANDUSE

土地利用

the Soviet inspired socialist architecture of the late 1950s, and we want to preserve the building as an icon of the area, by making it a hub for work, meeting and events. The building is connected to a new open public space, that is created by removing car traffic and walls from the surrounding site, and it creates a connection between the central Fusuijing building and the surrounding hutong areas.

通过将原有建筑保留成为地区的标志，并将其改造成工作、集会的中心。通过移除车行交通和基地周边的围墙营造出一个开放的公共空间，使建筑与其连通。

概念

CONCEPT

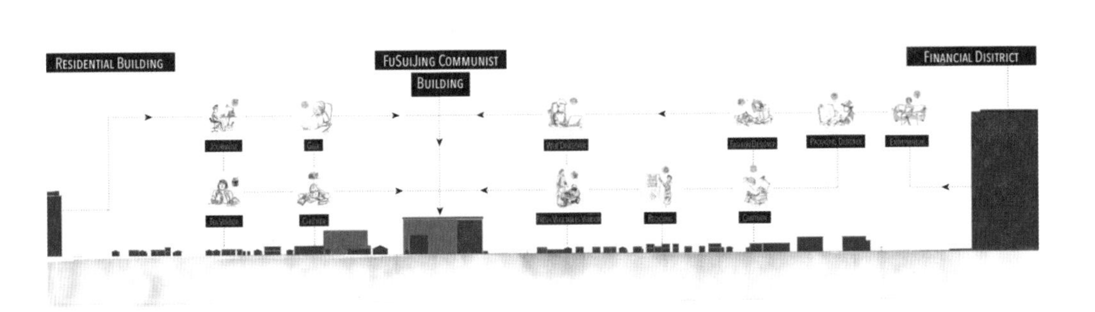

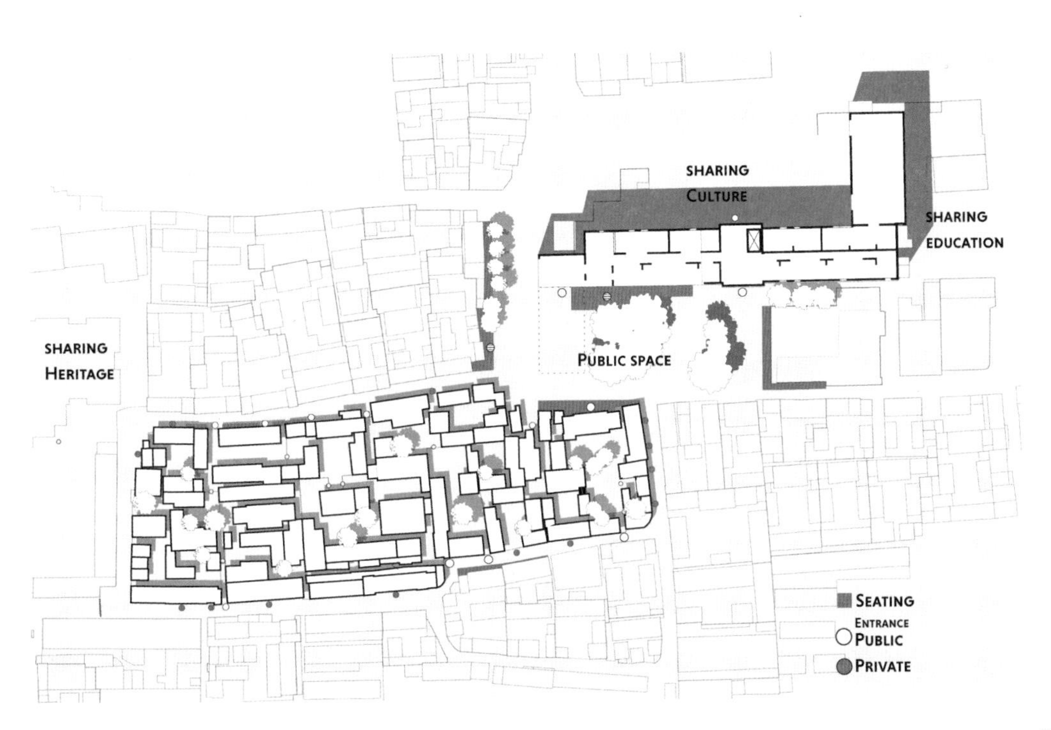

设计方案

Our proposal for the small scale, historic courtyard typology of the hutongs is focused more on permanent small offices and workshops, through a combination of shared and public courtyards. The courtyards provide opportunities for the entrepreneurs to bring in visitors from the streets for business, and the vernacular

对于小尺度的历史性的胡同四合院，我们设想方案更加关注于永久性的小型办公室和商铺，通过共享的公共合院相联系。通过将游人从街道引入，四合院为企业提供商业机会。相对于福绥境大楼的开敞性，胡同的地区性特点提供了亲密和私密的空间。

PROPOSAL
THE FUSUIJING

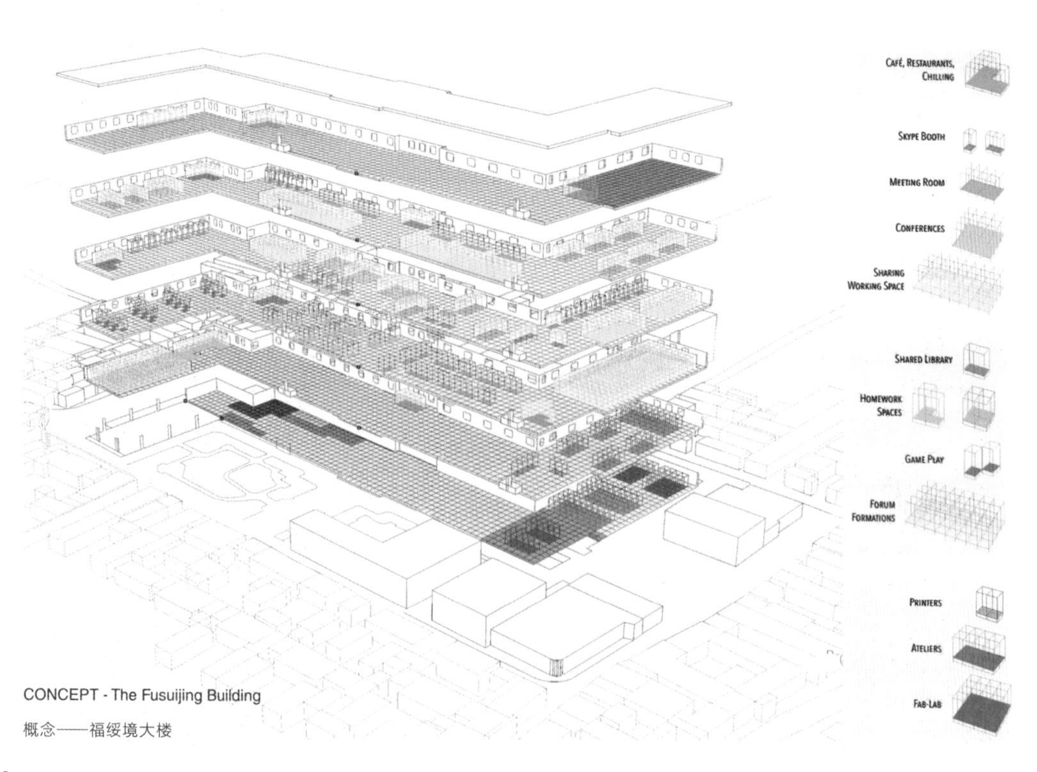

CONCEPT - The Fusuijing Building

概念——福绥境大楼

pattern of the Hutong create intimate and private spaces, that contrast the openness of the FuSuiJing.

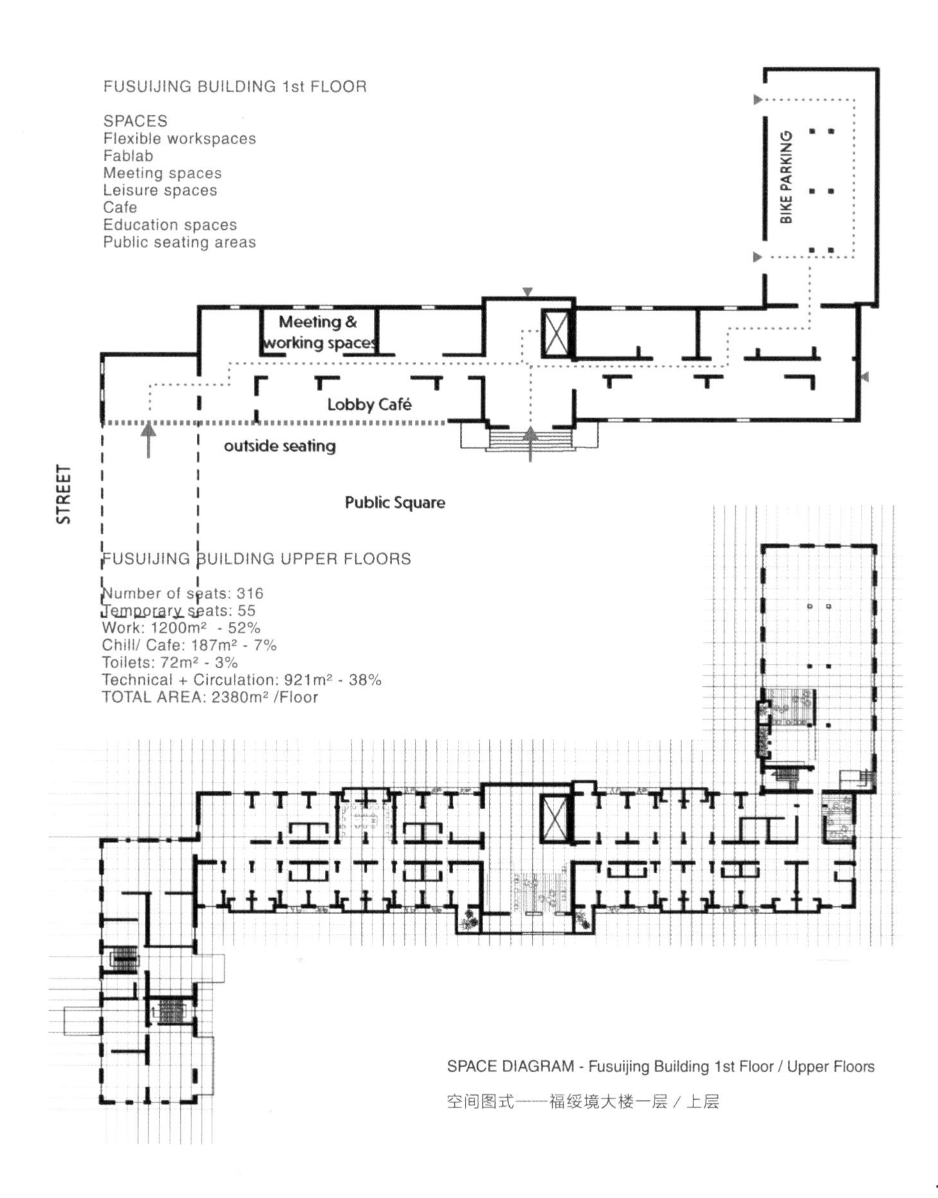

SPACE DIAGRAM - Fusuijing Building 1st Floor / Upper Floors

空间图式——福绥境大楼一层 / 上层

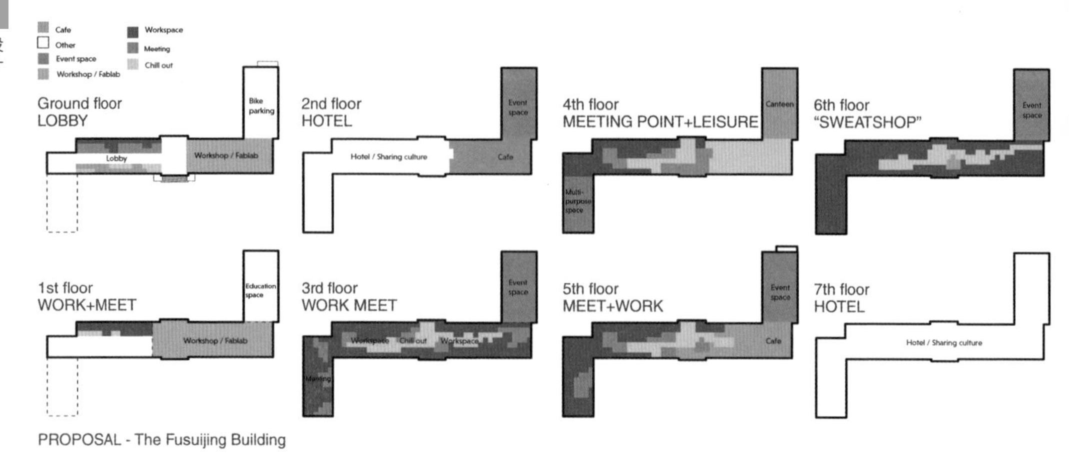

PROPOSAL - The Fusuijing Building

设计方案——福绥境大楼

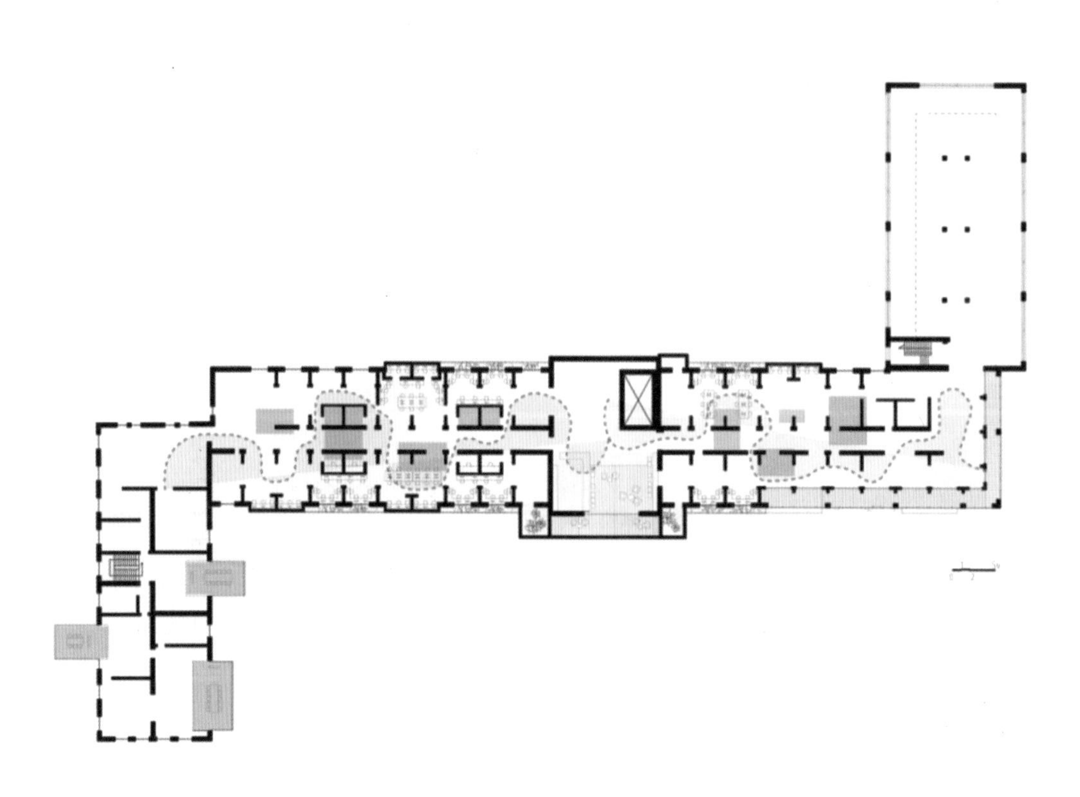

Fusuijing Building - Upper Floor Plan

福绥境大楼——上层平面

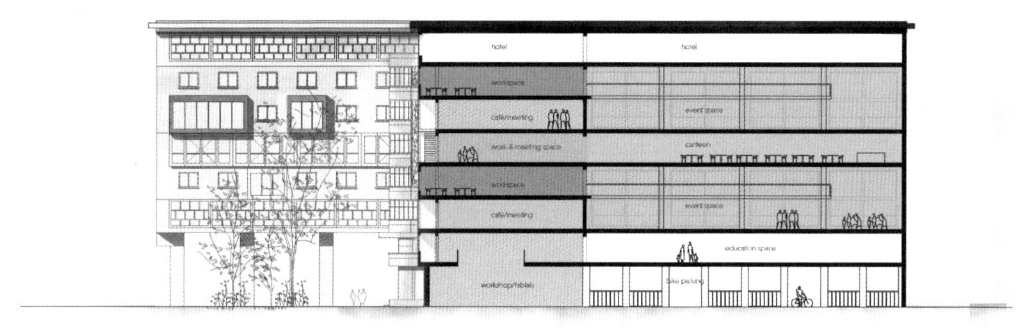

Fusuijing Section

福绥境大楼剖面

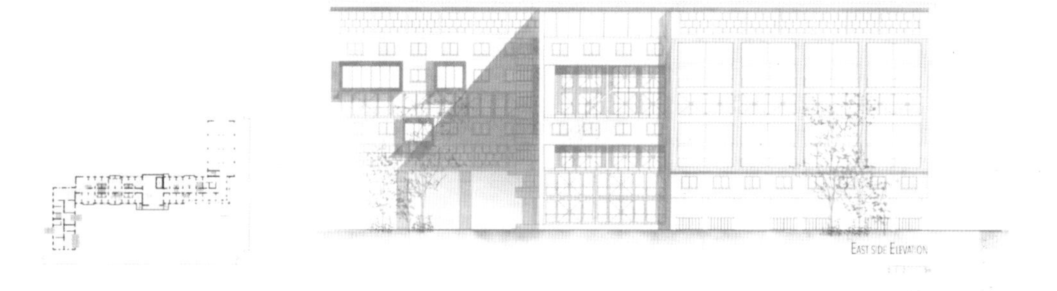

Fusuijing - South Side / East Side Elevation

福绥境大楼——南立面／东立面

设计方案

Lastly, to create an area that benefits all users, we designed our spaces to have good connections to the outside, as well as a central location in the area. Our site connects busy nodes and central parts of the road network with ample public space.

PROPOSAL

THE HUTONG

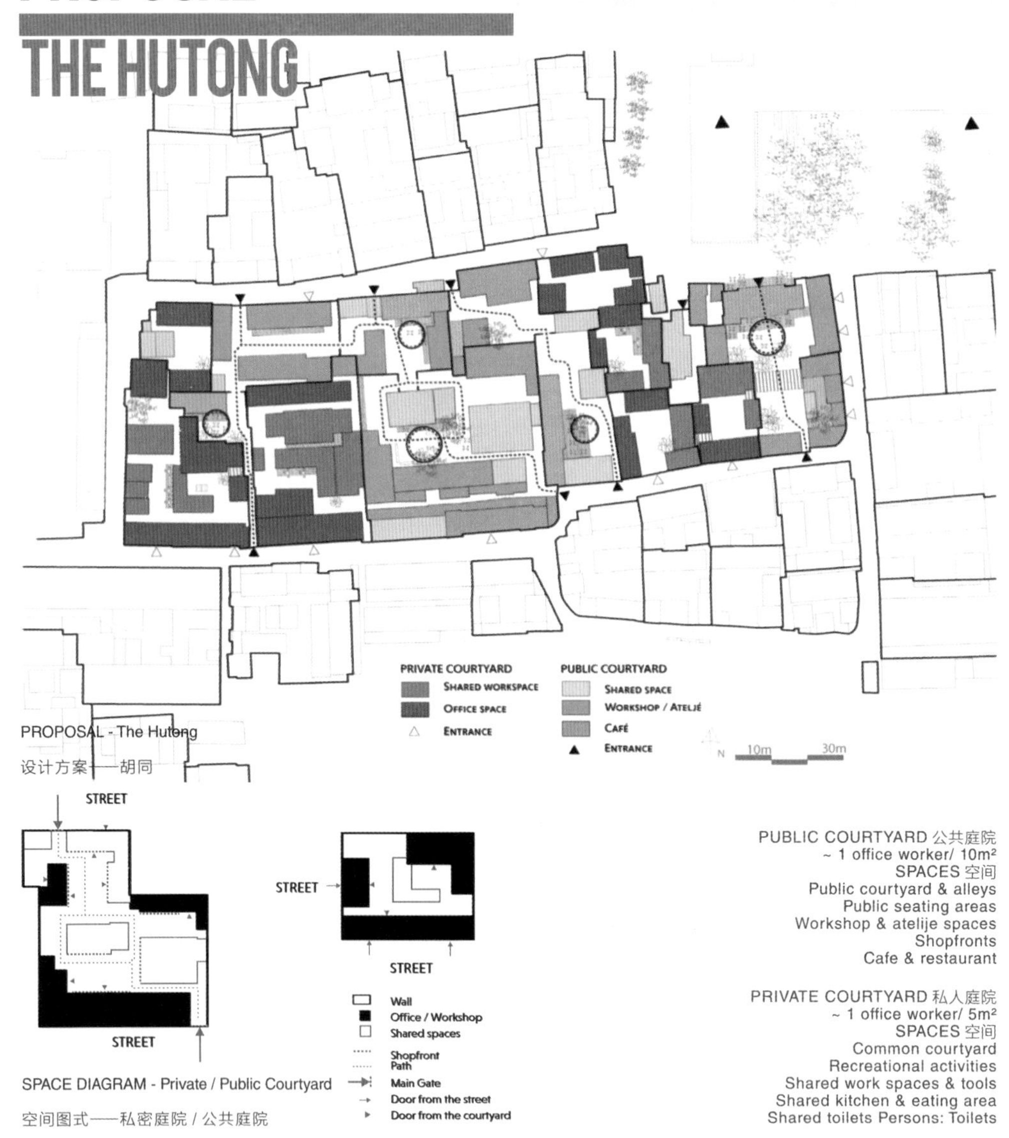

PROPOSAL - The Hutong

设计方案——胡同

SPACE DIAGRAM - Private / Public Courtyard

空间图式——私密庭院 / 公共庭院

PUBLIC COURTYARD 公共庭院
~ 1 office worker/ 10m²
SPACES 空间
Public courtyard & alleys
Public seating areas
Workshop & atelije spaces
Shopfronts
Cafe & restaurant

PRIVATE COURTYARD 私人庭院
~ 1 office worker/ 5m²
SPACES 空间
Common courtyard
Recreational activities
Shared work spaces & tools
Shared kitchen & eating area
Shared toilets Persons: Toilets

最后，为营造使所有使用者受益的环境，我们将设计的区域与外部环境良好联系，同时也位于地区的核心区域、我们的场地以丰富的公共空间将繁忙的节点和地区道路网络的中心地区联系起来。

Hutong - Section

胡同——剖面

PARK IN PARK

By Gyoung Min Ko, Chiun Cheng Fung

公园中的停车空间

PARK IN PARK is a public infrastructure which is proposed to resolve the transportation and social issues in the historic hutong district near Baitasi, Beijing.

The three targets of the proposal are: Resolving parking chaos, Pedestrian Friendly Streets, and Provide suitable spaces for social activities. Advancing the popular idea of sharing bicycle, our proposal applies the idea of sharing not only bicycles, but also motor bikes with different additions and sharing cars.

公园中的停车空间（PARK IN PARK）是一项位于北京白塔寺附近历史胡同地区的、旨在解决交通与社会问题的公共基础设施。

方案的三个目标是：解决混乱的停车、营造宜人的步行道以及提供适宜的社会活动空间。在共享单车概念的基础上，方案更进一步提出了共享摩托车与共享汽车的想法。

SWITCH STATION

SWITCH

CONNECT

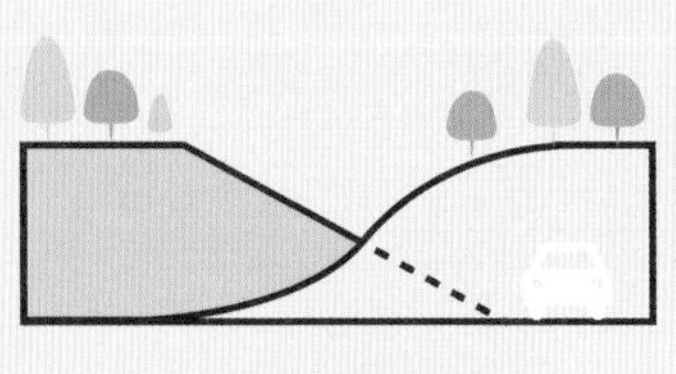

INTEGRATE

SHARING TRANSPORTATION

The condition of the site is highly densified with the hutong buildings with relatively narrow paths with uncontrolled mixture of vehicles which results in the dangerous environment for many residents. In addition, the site lacks spaces for safe social activities of various users groups such as children, students, families, and elders.

基地所面临的现状是：密集的胡同建筑、狭窄的道路、缺乏控制的交通工具令居民感到危险。另外，缺乏满足不同人群社会活动的安全空间。

ROAD TYPE 1
5m<WIDTH<12m

ROAD TYPE 2
3m<WIDTH<5m

ROAD TYPE 3
1.5m<WIDTH<3m

ROAD TYPE 4
WIDTH<1.5m

Road Type Analysis

道路类型分析

PUBLIC TRANSPORTATION
BUS(A) / SUBWAY(B) / TAXI(C)

PRIVATE VEHIVLE
CAR(D)

MOTOR BIKE
PRIVATE(E) / COMMERCIAL(F)

THREE WHEEL BIKE
CARGO TYPE(G) / CAR TYPE(H) / CART TYPE(I)

NON MOTOR BIKE
BICYCLE(J) / WHEELCHAIR(K) / KICK BOARD(L) / SHARED BICYCLE(M)

Transportation Vehicle Analysis

交通工具分析

	[illegible] SHARED TRANSPORTATION				商业 COMMERCIAL		政府 GOVERNMENT	私家交通 PTIVATE TRANSPORTATION					
	A	B	C	M	F	G	D	E	H	I	J	K	L
OFF SITE 地段外	●	●	●	●	●	●	●	●	●	●	●		
ROAD TYPE 1 道路类型 1				●	●	●	●	●	●	●	●	●	●
ROAD TYPE 2 道路类型 2				●	●	●	●	●	●	●	●	●	●
ROAD TYPE 3 道路类型 3				●	●			●			●	●	●
LOW SPEED (30KM>) 低速				●						●	●	●	●
HIGH SPEED (30KM<) 快速	●	●	●		●	●	●	●	●				
REQUIRE PARKING ON SITE 场地内必要的停车				●			●	●	●	●	●		
MUTIPLE PASSEN-GERS 多样化乘客	●	●	●				●	●	●				

Current Transportation Vehicle Analysis

交通工具现状分析

设计方案

PROPOSAL

In order to reach the targeted goals, our project took three strategies of switching of transportation models, connecting various parking/ charging stations throughout the site, and integration of park and parking. The main design proposal sits on the west edge of Baitasi hutong district which used to be a ground level park with streets full of

为了达到目标，我们的方案采用三个策略：转换交通模型、连接基地内各种停车场或充电站，以及整合公园与停车区域。设计方案主体位于白塔寺胡同地区的西侧，地段内曾是一个地面层的公园，公园内的街道停满了车辆，流线也不顺畅。新的方案通过连接不同交通工具停放站和胡同入口提供了更有效的交

cars with unsmooth flow of circulation. The new proposals provides more efficient flow of circulation through connecting different vehicle stations and hutong entrances from people's perspectives. PARK IN PARK also integrates parking, park and community serving spaces in a playful manner to avoid simply layering different functions.

通流线。PARK IN PARK 同时整合了公园、停车、社区服务空间，并避免将不同功能简单地分层排布。

Solve the parking chaos

解决停车混乱

Provide pedestrian friendly zones

提供友好的人行区域

Provide encouraging environment for social activities

提供良好的社会活动环境

Site Plan

总平面

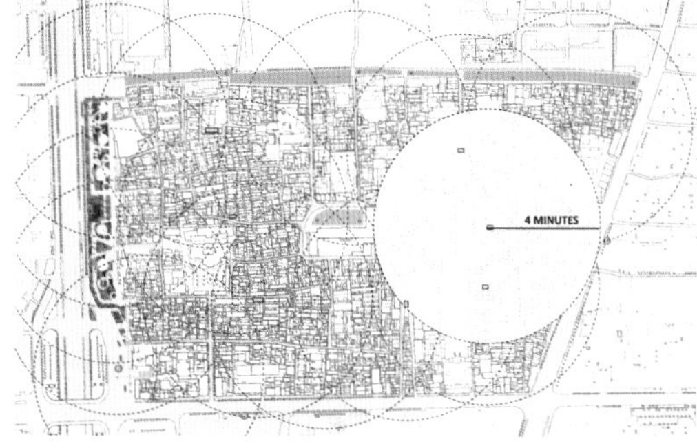

设计方案

PARK IN PARK would serve the people in Baitasi closely in daily basis for daily commute, daily activities and relaxation in green environments.

“公园中的停车空间”为白塔寺地区人群日常通勤、活动、休闲提供了绿色的环境。

PROPOSAL

WEST PARK

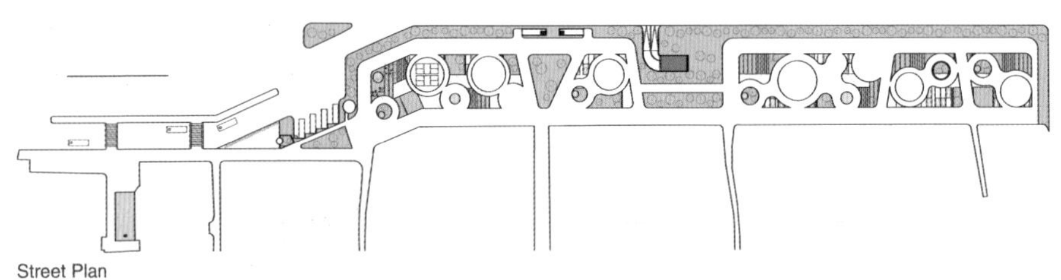

Street Plan

地上平面图

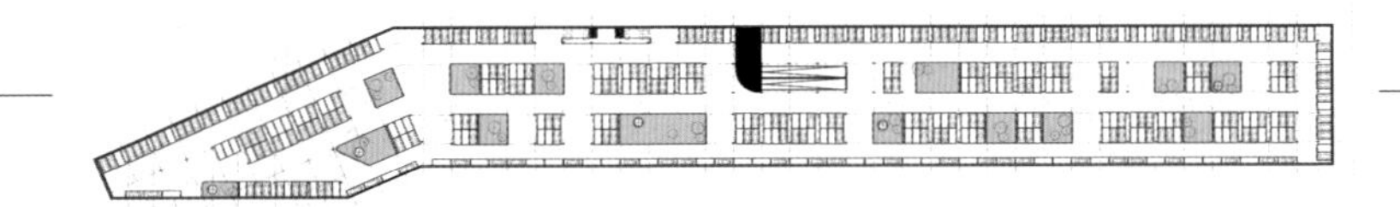

Basement Plan

地下平面图

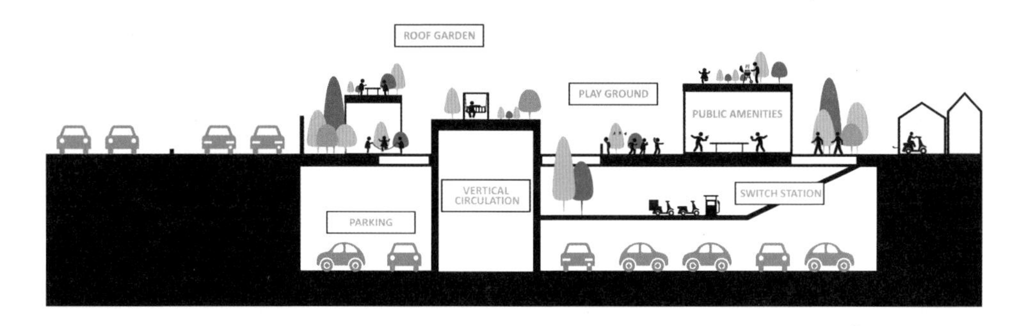

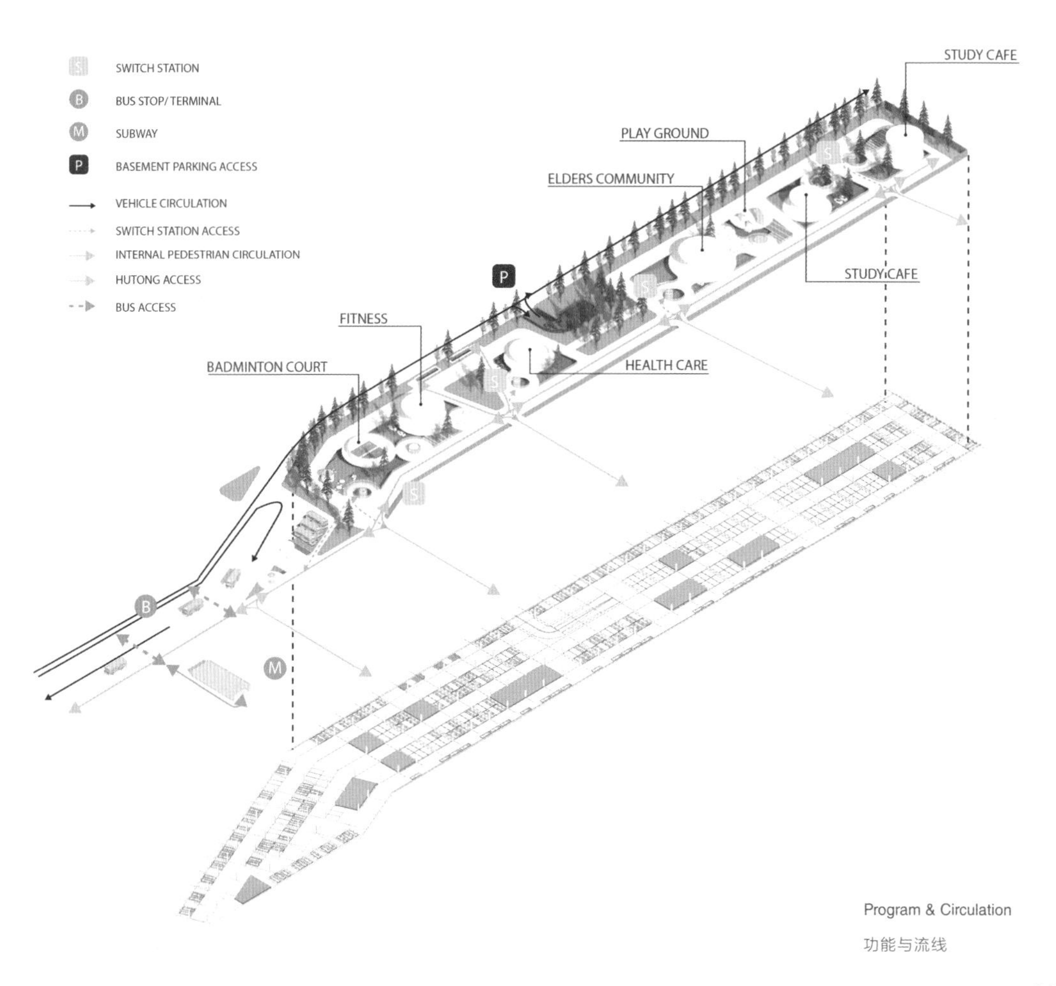

Program & Circulation

功能与流线

MAKING MORE VIGOROUS LEARNING ENVIRONMENT

By Boey Ying Yan, Loo Hui Xin

营造更有活力的学习环境

By breaking the conservative and traditional learning environment to a more vigorous learning environment, a new wall systems connecting the schools to the street, and continues all the way to the hutong residential courtyards that creates a journey of learning that goes beyond the boundary of the schools. Lifelong learning among communities are introduced through the way of sharing and giving in the proposal of school and skill learning centre designs.

通过打破保守的、传统的学习环境，创造一个更有活力的学习环境、一个连接学校和街道的新的外墙系统通向胡同居民的庭院，创造了一个打破学校边界的学习行程。社区中终身学习的概念通过共享的方式加以传达，在学校和技能学习中心的设计中得到体现。

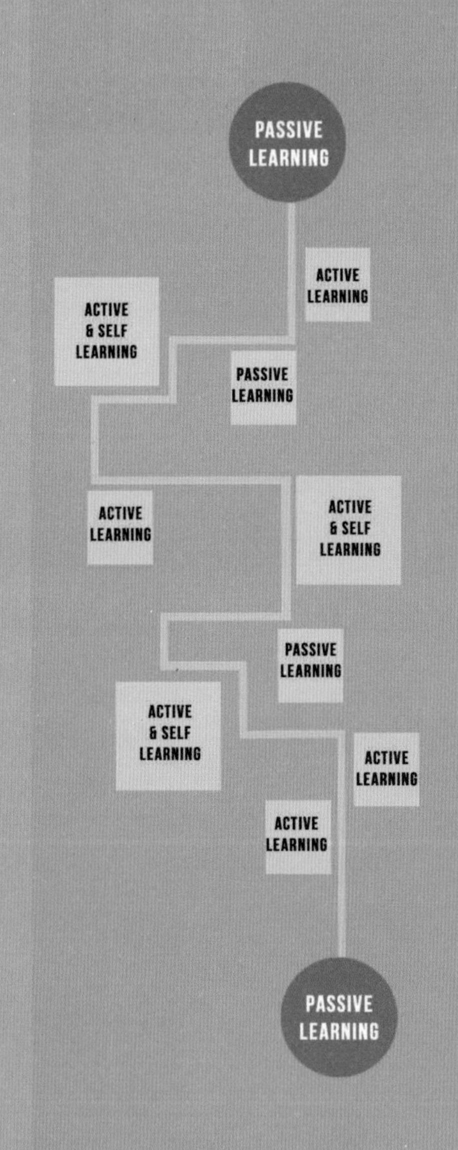

SHARINGEDUCATION

In our analysis of the area, we realized that all educational program is located behind strongly gated walls in which kids have to sit still and listen in uninspiring concrete boxes. With our proposal we have tried to design an alternate educational typology. We redefine the purpose of a school, as a place not merely to absorb knowledge, but a platform that promotes social interaction, develops personal communication skills and character while the students interact among each other. In addition, since we realized that education is not just for children, a skill learning centre and residential learning space are implemented in the site to incorporate more age groups for the idea of life-long learning.

在对该地区的分析中，我们意识到所有的教育项目都位于围墙内那些没有启发意味的混凝土盒子中，小孩子们不得不端坐其中。因此我们试图设计可替代的教育模式，重新定义了学校的目的不仅是在学生互动中吸收知识，更是提升社会技能和发展个人沟通技能。我们进一步认识到教育不仅仅是针对儿童，针对本地居民的学习中心在场地中设立起来以鼓励不同年龄的人投身于终生学习之中。

(A) (B) (C) | (D) (E) (F) (G) (H) | (I) (J) (K) (L) (M)

Current Issue Analysis

现状分析

Types of Learning

Passive Learning:

It is a traditional learning which the lecturing instructor verbalizing information to passive note-taking students. Students are usually passive "tape recorder". There is not much interactions or information sharing in the class and it is usually one way information passing, from the instructor to the students.

Active Learning:

The instructor strives to create "a learning environment in which the student can learn to restructure the new information and their prior knowledge into new knowledge about the content and to practice using it." Students are expected to look up information before and after class. The instructor explains concept, principles and

分析

学习的类别

主动学习

指导者努力创造一个学生能够学习将新信息和原有知识重构成为新知识并加以练习使用的学习环境。学生应在课前课后查阅信息。指导者解释概念、原则，学生练习将这些技能加以应用。

被动学习

一种传统的学习方式，发布演讲指导者向被动记笔记的学生用语言传递信息。学生通常是被动的“录音机”。课堂中并没有很多的互动或信息共享，通常是由指导者向学生的单向的信息传递。

自我学习

是一种自我发展，学生为自我学习课程或一些学校课程表之外的阅读检索信息。在检索合适的阅读材料的过程中，学生易于掌握更多在学校课程表之外的信息，并因此通过阅读完善知识结构。

Passive learning

被动学习

Active learning

主动学习

Self learning

自我学习

Boring Journey

无聊的街道

Gated School

大门紧锁的学校

Used Walls

未被利用的墙面

Underutilized Space

未充分使用的空间

methods for geological interpretation, while the students practice applying these skills to geological interpretation.

Self Learning:
It is more to a self development where students search information for the active learning class or some readings beyond the school syllabus. In the process of searching suitable reading materials, the students tend to know more beyond the school syllabus and thus improve the knowledge via reading.

Current Issue Analysis

现状分析

设计策略

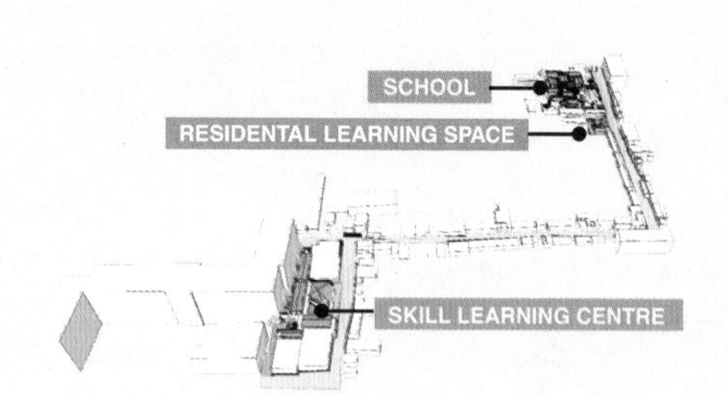

DESIGN STRATEGIES

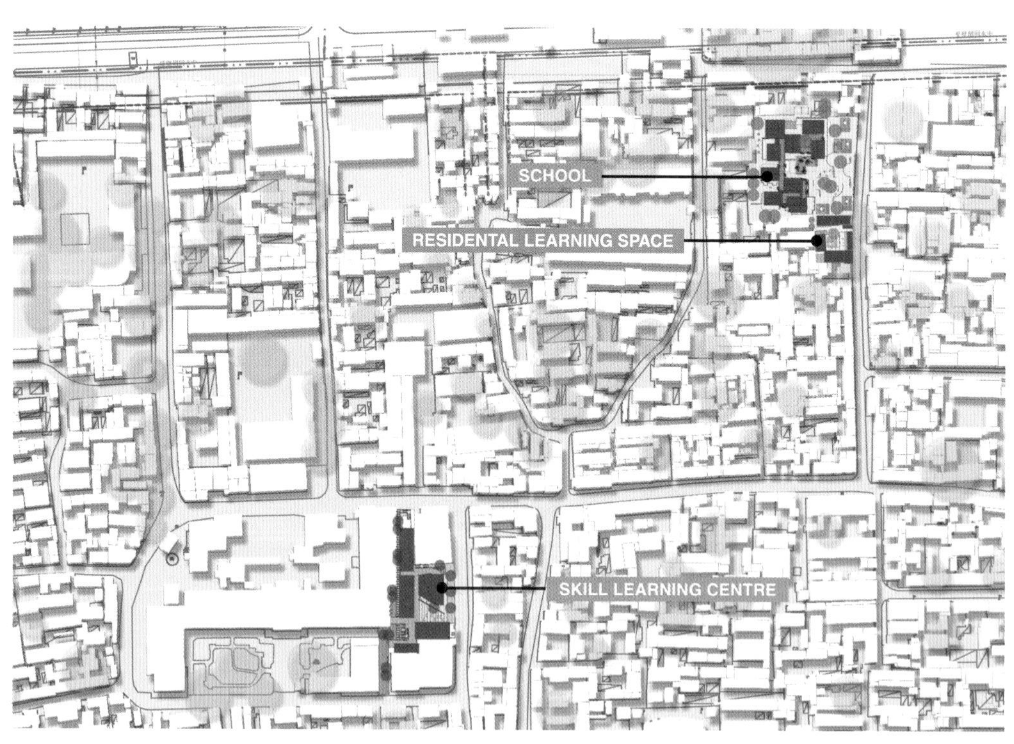

Distribution of Education Facilities 教育设施分布

Existing solid wall

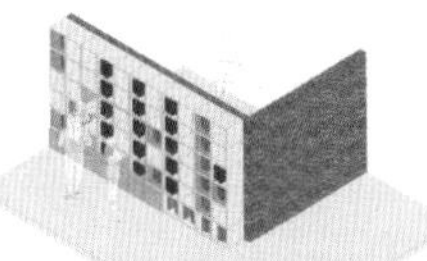

book shelves introduced to the wall

replaced the wall by the book shelves and seatings

create interactive opening along the wall and introduced book shelves and seatings in certain locations on site

Wall System Dedign Development/ 外墙系统设计改进

PROPOSAL

THE SCHOOL

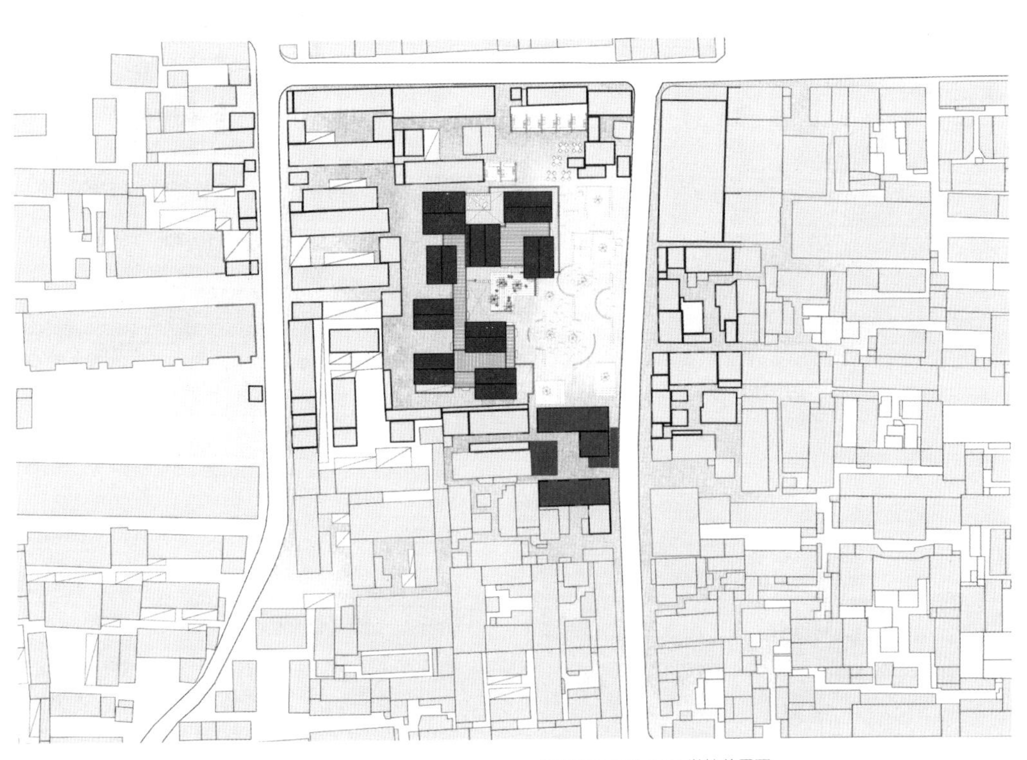

Siteplan of School 学校总平面

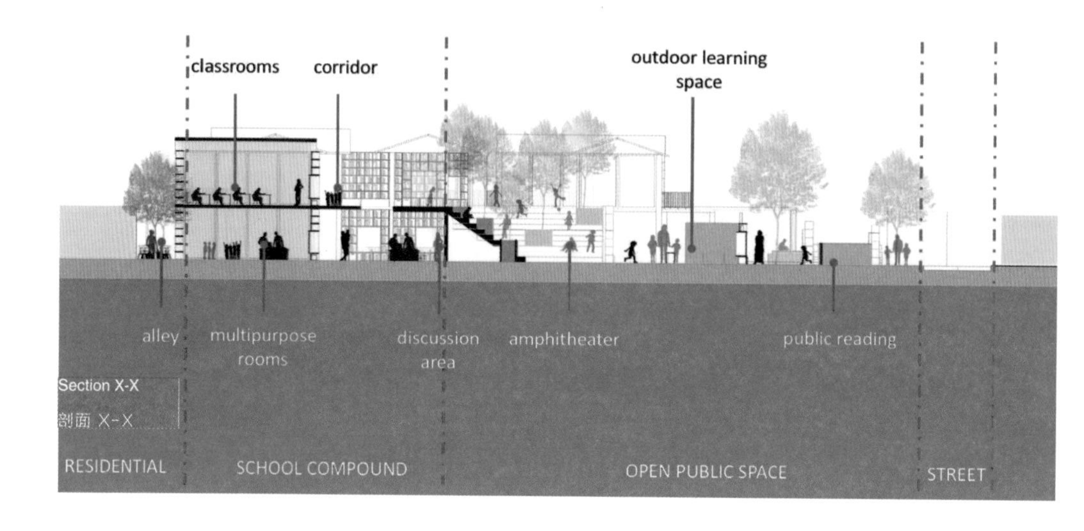
classrooms
corridor
outdoor learning space
alley
multipurpose rooms
discussion area
amphitheater
public reading
Section X-X
剖面 X-X
RESIDENTIAL
SCHOOL COMPOUND
OPEN PUBLIC SPACE
STREET

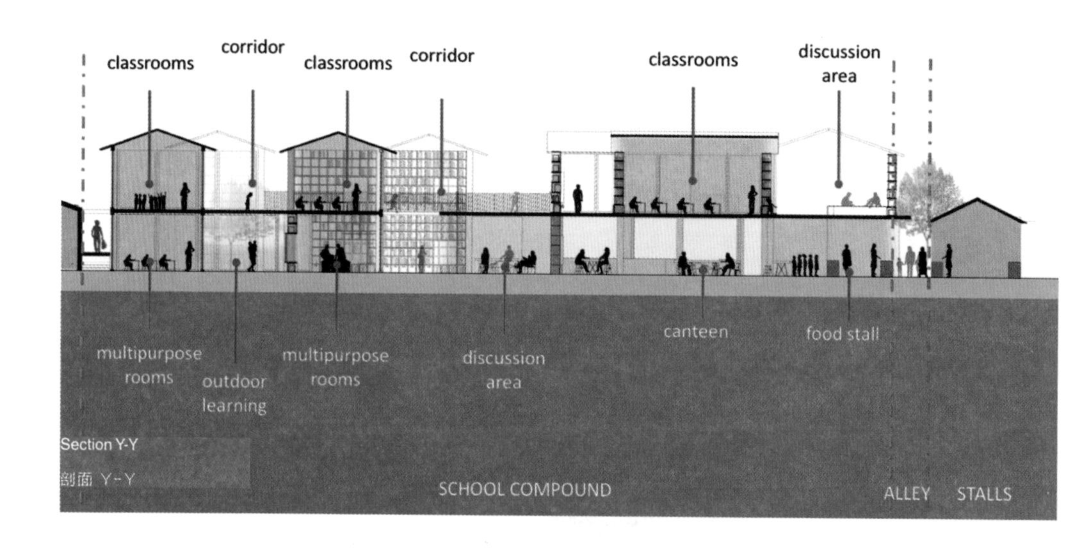
classrooms
corridor
classrooms
corridor
classrooms
discussion area
multipurpose rooms
outdoor learning
multipurpose rooms
discussion area
canteen
food stall
Section Y-Y
剖面 Y-Y
SCHOOL COMPOUND
ALLEY
STALLS

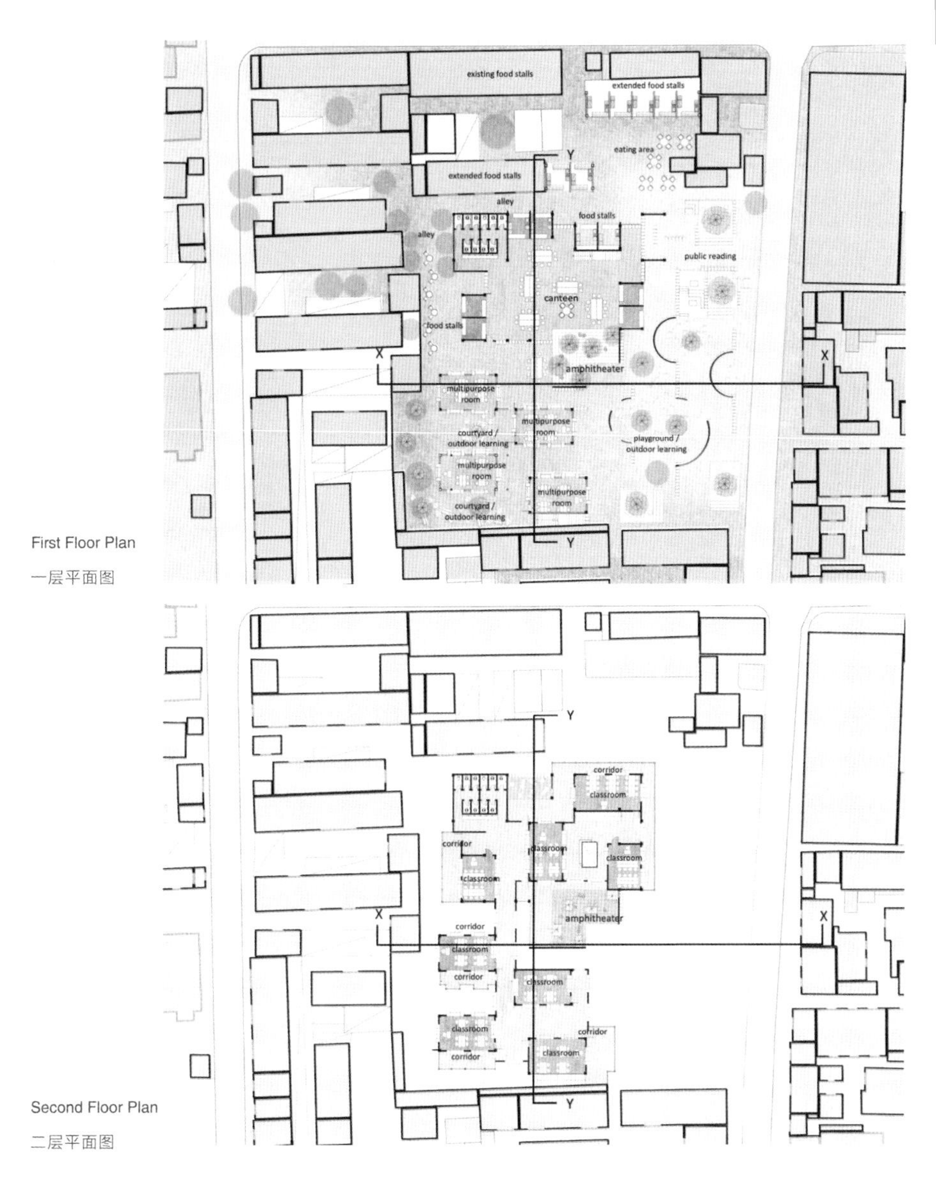

First Floor Plan

一层平面图

Second Floor Plan

二层平面图

设计方案

PROPOSAL

THE SKILL LEARNING CENTRE

First Floor Plan

一层平面图

Second Floor Plan

二层平面图

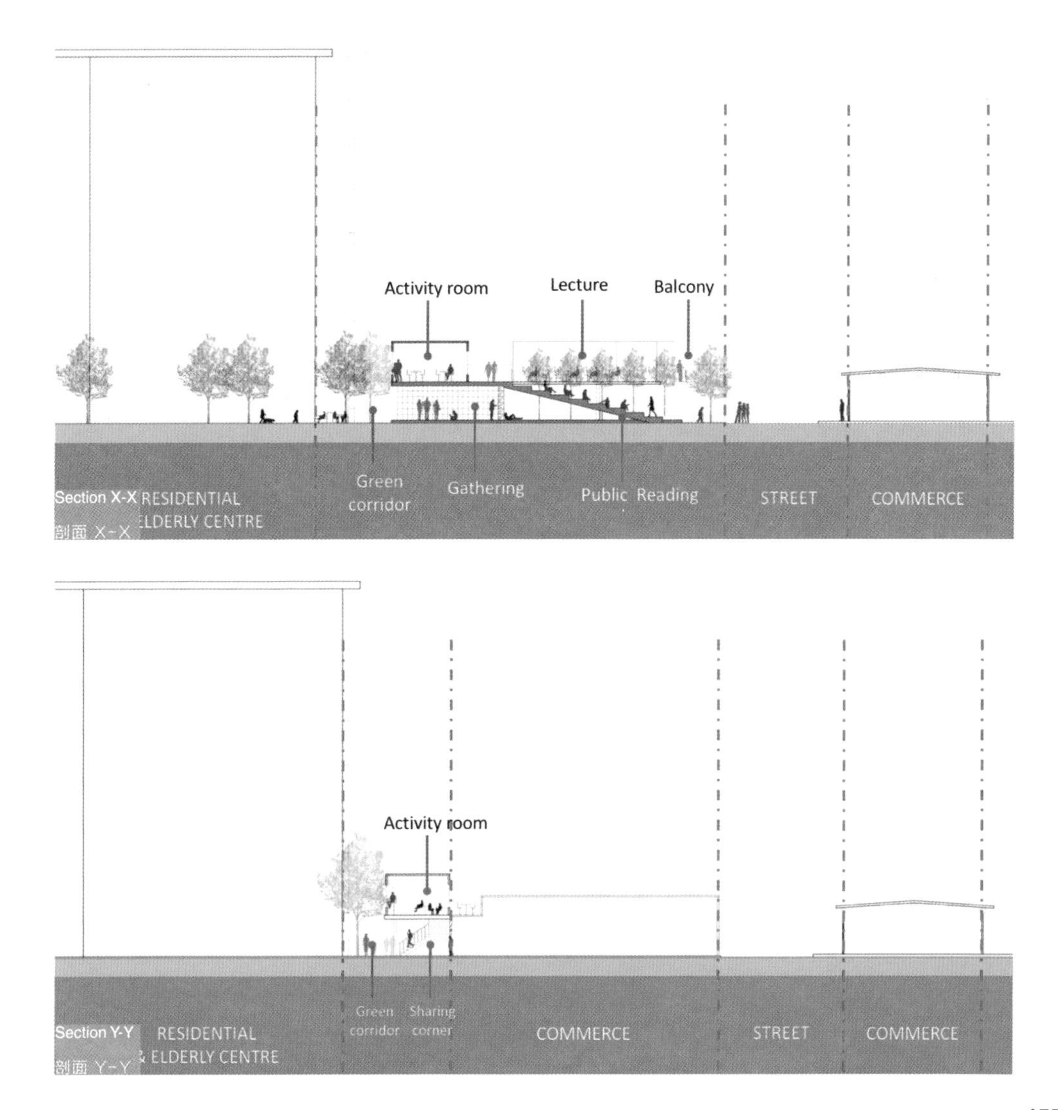
Activity room
Lecture
Balcony
Green corridor
Gathering
Public Reading
Section X-X
剖面 X-X
RESIDENTIAL
ELDERLY CENTRE
STREET
COMMERCE
Activity room
Green corridor
Sharing corner
Section Y-Y
剖面 Y-Y
RESIDENTIAL
& ELDERLY CENTRE
COMMERCE
STREET
COMMERCE

设计方案

PROPOSAL

THE RESIDENTAL LEARNING SPACE

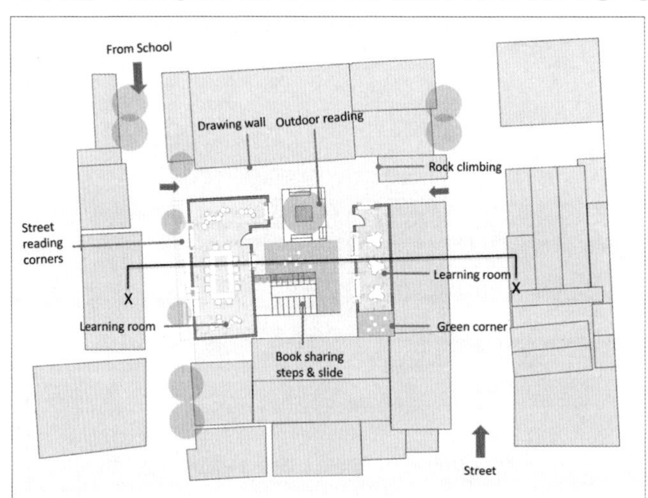

First Floor Plan

一层平面图

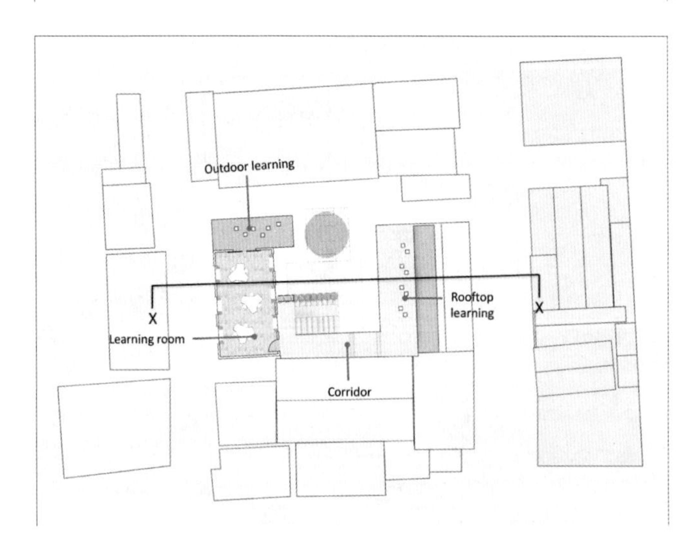

Second Floor Plan

二层平面图

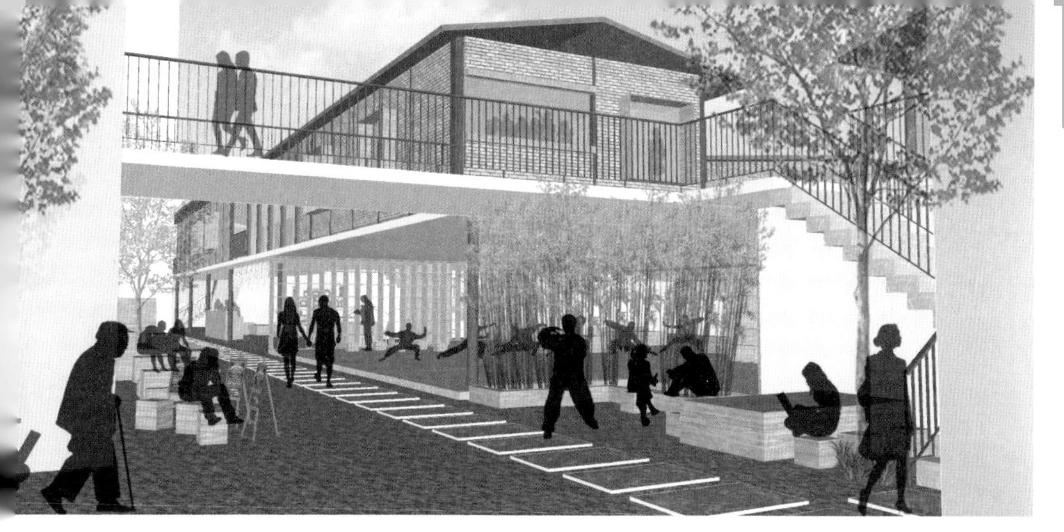

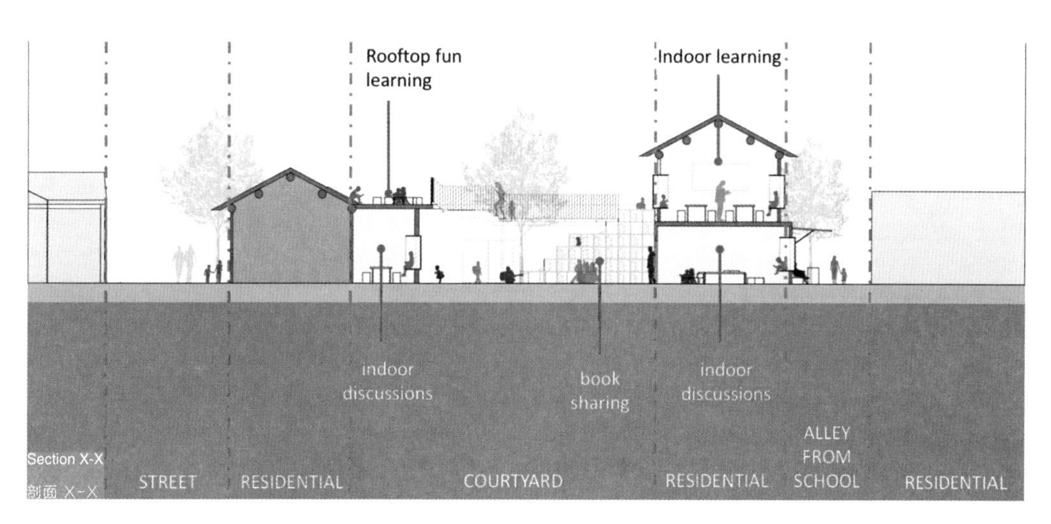
Rooftop fun learning
Indoor learning
indoor discussions
book sharing
indoor discussions
Section X-X
剖面 X-X
STREET
RESIDENTIAL
COURTYARD
RESIDENTIAL
ALLEY FROM SCHOOL
RESIDENTIAL

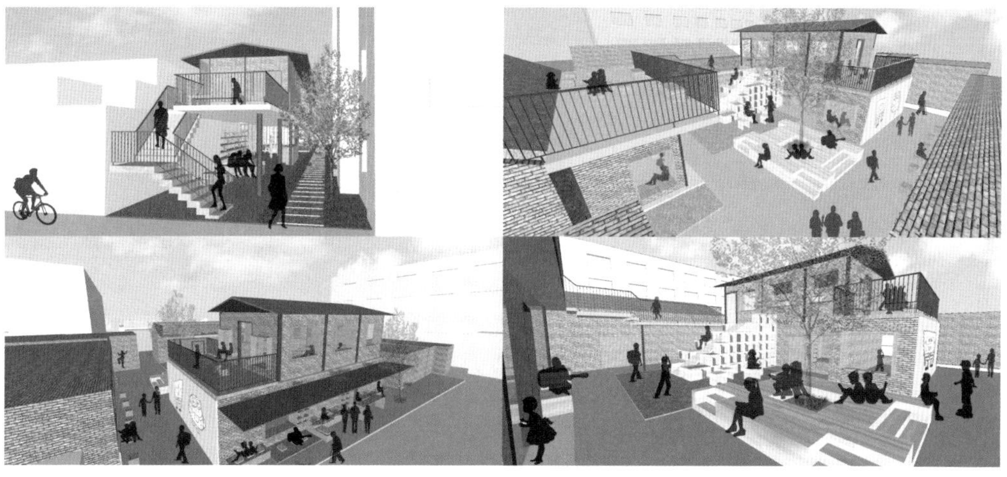

BAITASI SKYWALK

By Deandrea, Vijay, Nein

白塔寺空中步道

The Baitasi Skywalk is a path that connects to the major cultural nodes of the Baitasi hutong. The system starts from the central point, formally known as the Fusuijing building. The building will be redev*eloped into a cultural hub to introduce the new and existing culture of the Baitaisi hutong.

白塔寺空中廊道是一个连接白塔寺胡同主要文化节点的路径。这一系统起始于中心点，即福绥境大楼。大楼将被发展成为一个展现白塔寺胡同新的和既有文化的核心。

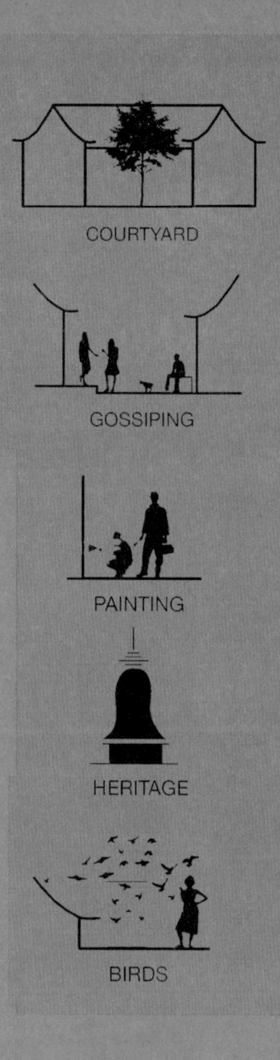

SHARING CULTURE

Culture talks about the intangible aspects that make us who we are. We tried to distill what culture means for the White Pagoda Area, and how local inhabitants experience this. In our research we found that historically there used to be more living space inside the courtyards, filled with green areas where people would gather to gossip or play board games. Raising singing birds, or training pigeons was another strong cultural element that survived throughout the years. We propose that Baitasi presents itself as a cultural oasis, a peaceful enclave among newly developed business areas.

文化是塑造我们是谁的软性要素。我们提取白塔寺地区的文化要素，观察当地居民如何体验这样的文化。我们发现，长久以来在四合院内有更多的生活空间，有着绿色空间供人们聚集聊天或下棋，遛鸟和养鸽是另一长久保留下来的文化传统。我们设想在一片新开发的商业地区包围之中，白塔寺代表着一个文化绿洲。

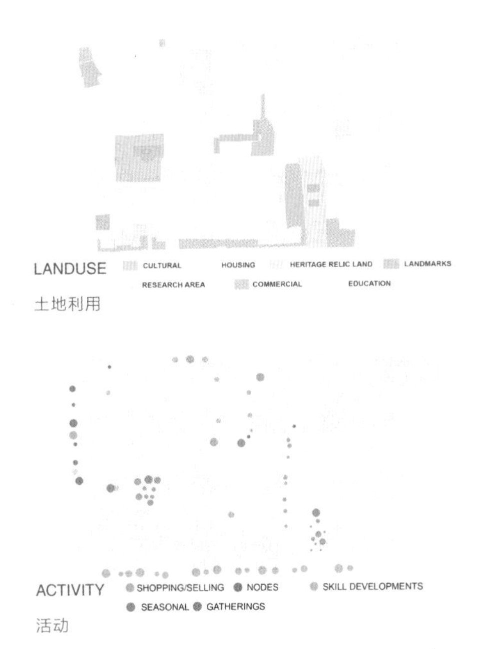

Unseen unique spaces

被忽视的特色空间

Narrow spaces

拥挤的空间

Forgotten landmark

被遗忘的地标

Insufficient space for games

游戏空间不足

Insufficient space for new talents

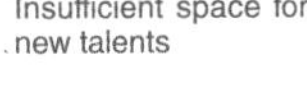

新兴艺术空间不足

Insufficient space for community gathering

社区集会空间不足

空间策略

1. CONNECT THE NODES
Connecting the major landmarks on site. Take people to the areas that are easily missed from while visiting Baitasi.

2. INTEGRATE
Carefully integrating the design within the spaces available above and within the partially abandon building.

1.连接节点
连接基地内主要的地标。带领人们来到参观白塔寺时容易忽略的区域。

2.整合
小心整合可利用的半废弃建筑物之上或之中的设计。

SPACIAL STRATEGIES

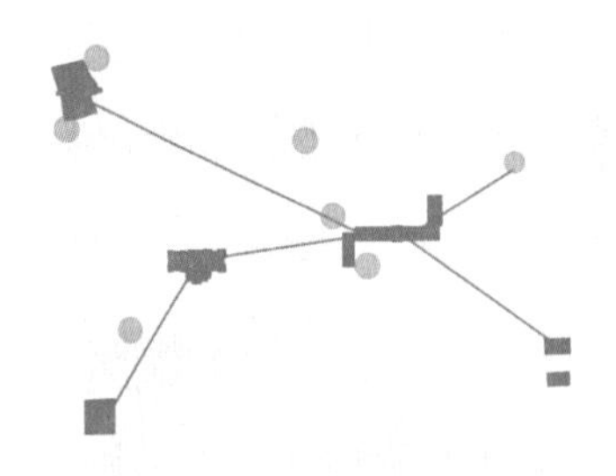

Connect the nodes

连接节点

Integrate

整合

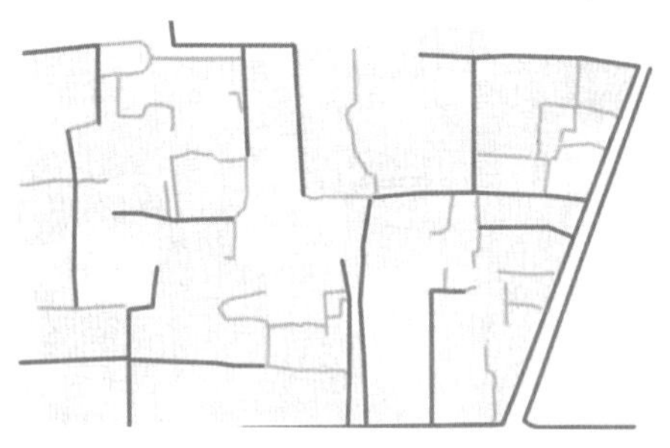

Follow the circulation

遵循流线

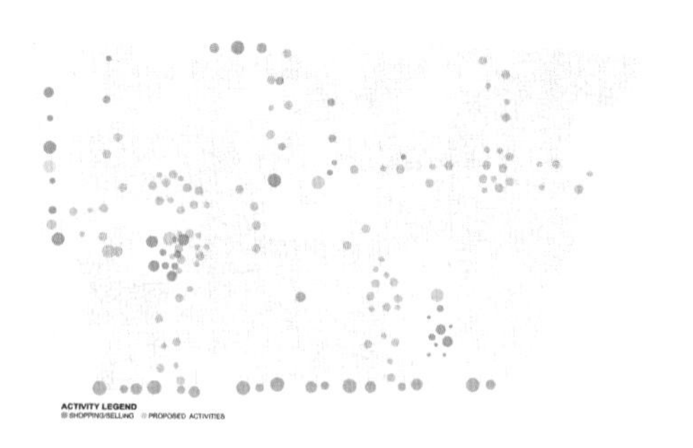

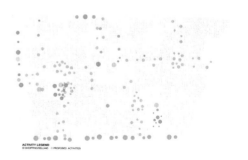

Generate more activities

生成更多活动

3. FOLLOW THE CIRCULATION

Based on the circulation analysis, the circulation of outsiders and locals intertwine. So the skywalk is to separate public and private spaces. Provide the locals with the spaces and direct outsiders mostly away from invading the local privacy.

4. GENERATE MORE ACTIVITIES

The design is focused on also creating more life to the Baitasi hutong with the skywalk path as the main attraction.

3.遵循流线

由于外来者与当地居民的流线相互纠缠，所以天桥将公共和私密的空间分开。为本地居民提供空间并引导外来者避免侵入本地居民的隐私。

4.生成更多的活动

通过将天桥路径作为主要吸引点，为白塔寺胡同创造更多的生活。

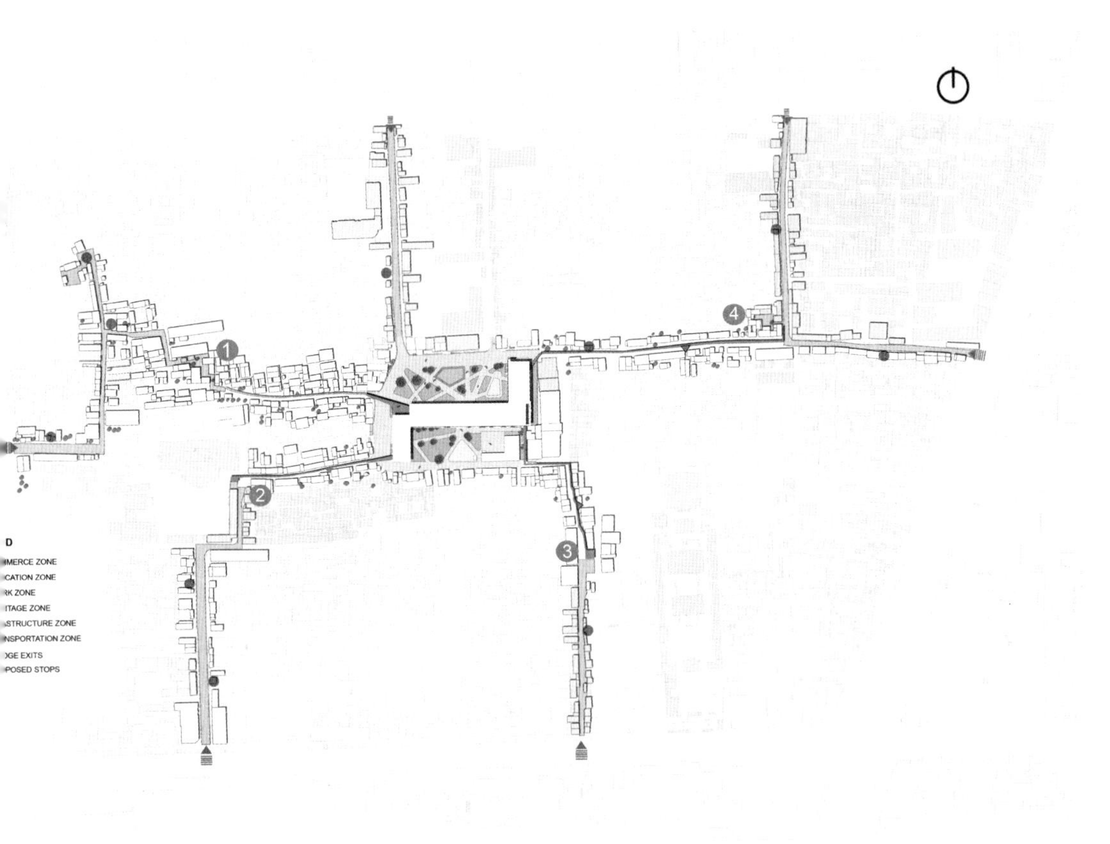

设计方案

The Baitasi skywalk was developed to recreate these lost spaces for the local activities and to share it using a parallel path for outsiders to appreciate their cultural development without disturbing the local living.

PROPOSAL

CULTURAL HUB

Perspective

透视图

白塔寺天桥的开发为当地重塑失落的空间，并通过一个平行的路径在避免打扰本地人生活的前提下，与外来者共享。

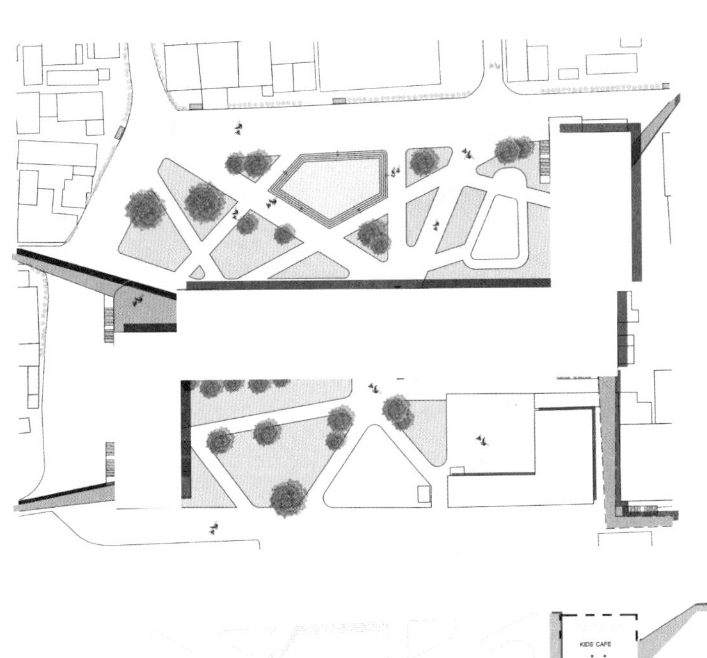

Site Plan

场地平面图

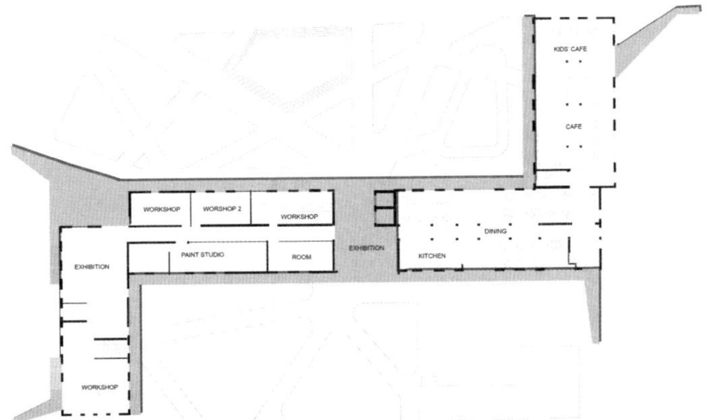

Second Floor Plan

二层平面图

设计方案

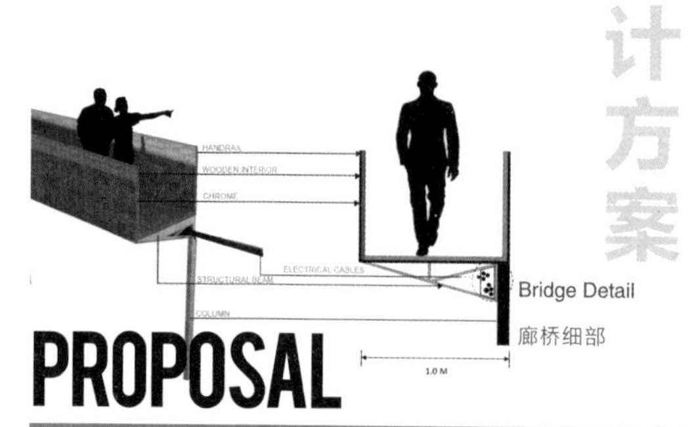

Bridge Detail

廊桥细部

PROPOSAL

SECTORS

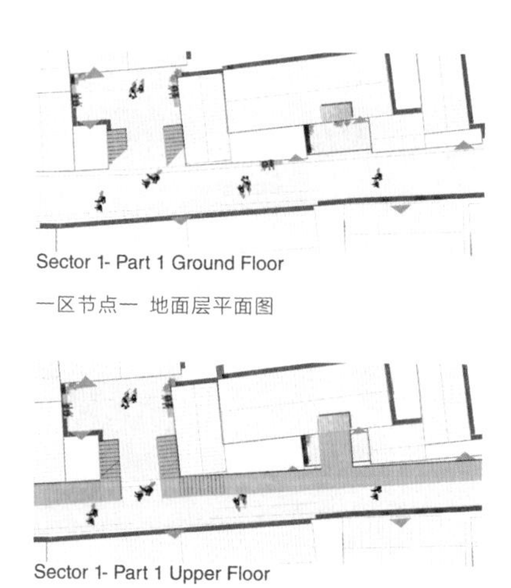

Sector 1- Part 1 Ground Floor

一区节点一 地面层平面图

Sector 1- Part 1 Upper Floor

一区节点一 屋顶层平面图

Sector 1

一区节

Sector 2 Section

二区 剖面

Sector 2

二区 剖

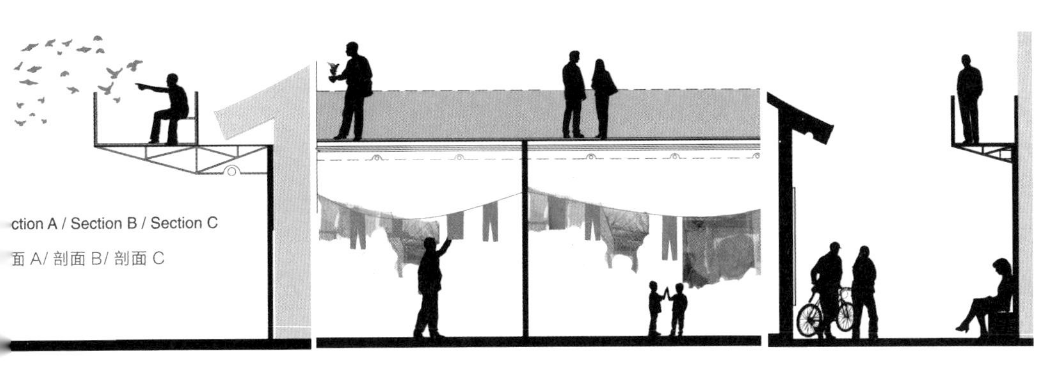

ction A / Section B / Section C

面 A/ 剖面 B/ 剖面 C

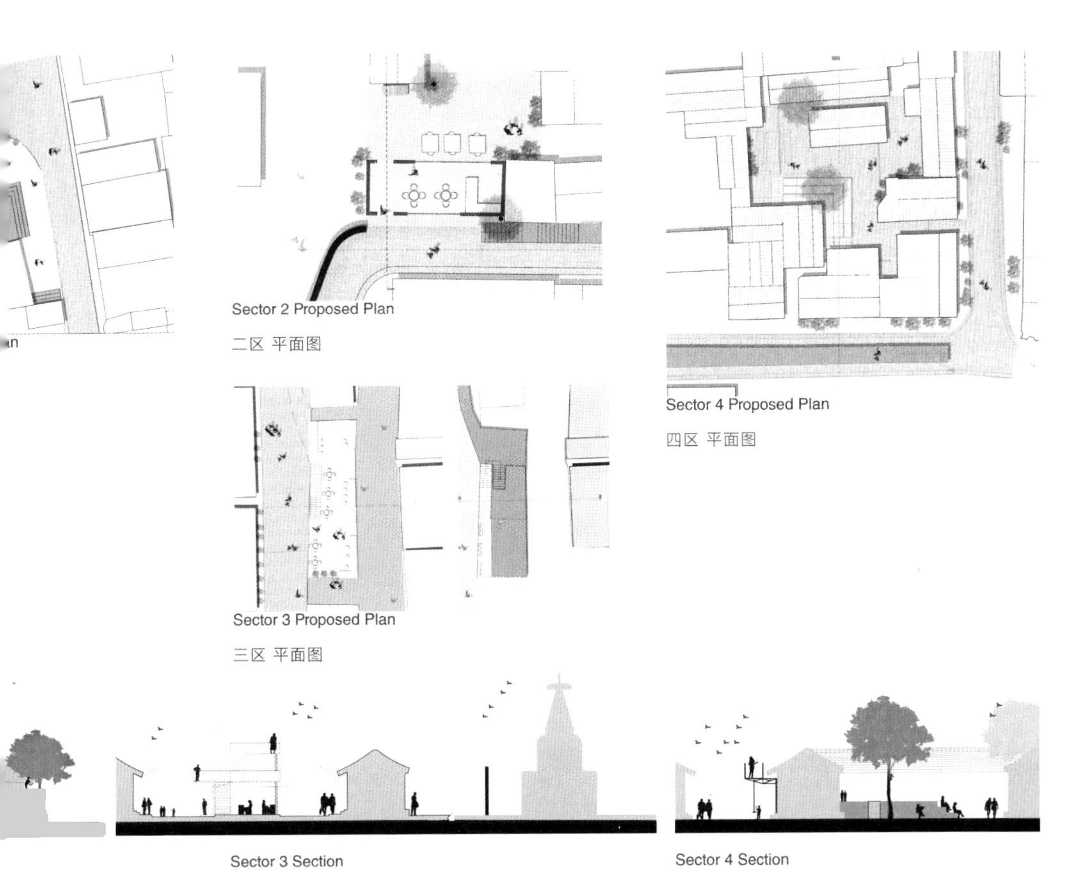

n

Sector 2 Proposed Plan

二区 平面图

Sector 4 Proposed Plan

四区 平面图

Sector 3 Proposed Plan

三区 平面图

Sector 3 Section

三区 剖面

Sector 4 Section

四区 剖面

SHARING HERITAGE THROUGH PUBLIC SPACES

By Laura Haller, Alessandra Coppari

依托公共空间的共享遗产

This project proposes to share underutilized heritage spaces, located in the district of Baitasi, with the local community, visitors and tourists. It aims to bring different types of people together and generate a positive diversity, creating a tangible and intangible heritage route in the district of Baitasi.

项目旨在与当地居民、来访者和旅游者分享坐落于白塔寺地区未充分利用的遗产空间。目标是把不同类型的人群聚集在一起以形成多样性，生成白塔寺地区物质和非物质文化遗产路线。

SHARING HERITAGE

共享遗产

Potential public spaces used as parking lots

潜在的公共空间被用作停车场

Heritage sites are mostly unaccesible

遗产场地大多不可达

设计策略

STRATEGIES

Heritage is not limited to a particular time in history. In fact, the White Pagoda Area is remarkable in its layered heritage. It consists of old temples and historic courtyard houses from the imperial era, but also has some key heritage from the early days of the Republic of China, and an icon of Soviet architecture, with the FuSuiJing building. In our proposal we aim to integrate these three main components into on spatial heritage narrative.

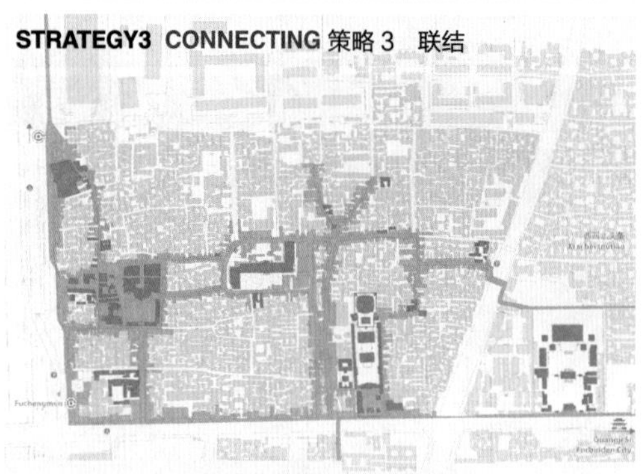

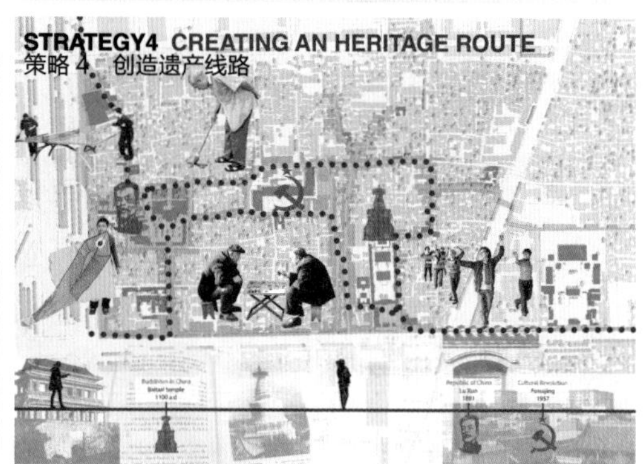

Spatial Strategies

空间策略

遗产并非局限在历史上的某一特定时间，实际上白塔寺地区在不同年代的历史发展中均有值得标注之处。它包含了帝国时代古老的寺庙和传统四合院，也包含了中华人民共和国建立初期的重要遗产和作为标志性建筑的福绥境社会主义大楼。在我们的方案中，我们希望将上述三类遗产通过空间手段整合起来。

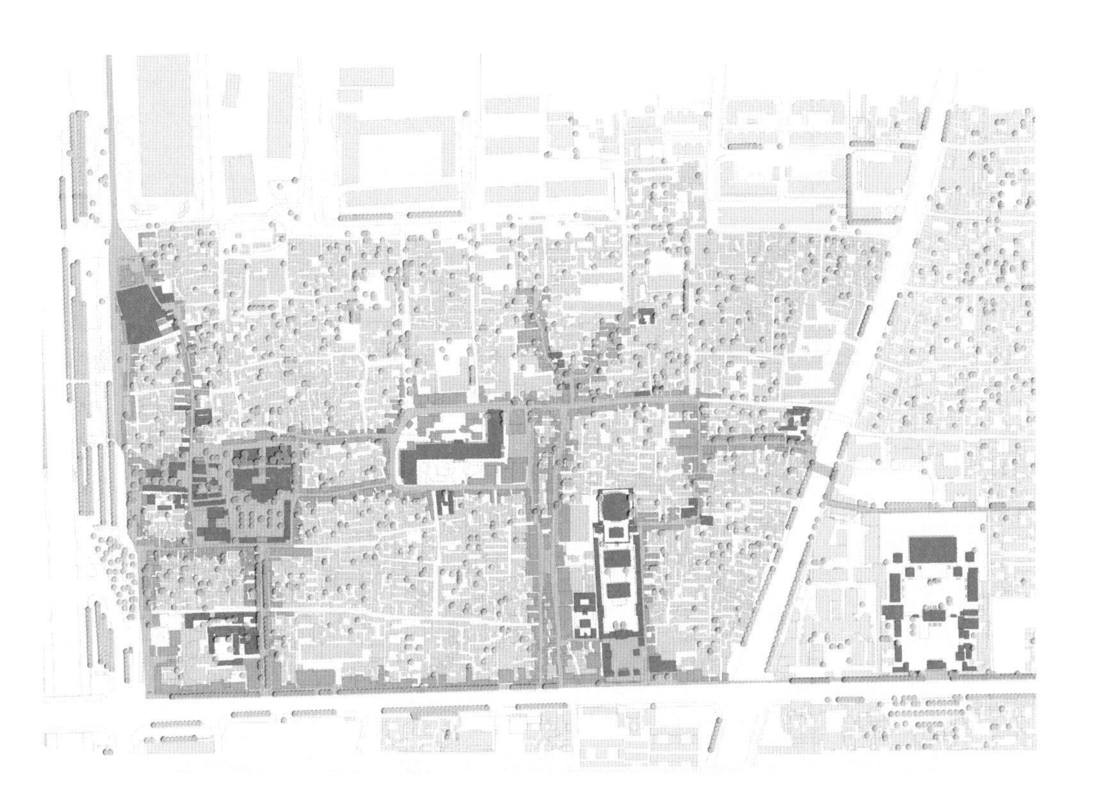

Master Plan

总平面图

设计方案

PROPOSAL BAITASI

Our proposal aims to revitalize the current parking area of Baitasi temple into a car-free public square. A multifunctional room is proposed on the west side of the square while on the east side people can play chess under a pergola and drink tea in a tea house. The center of the square becomes a multi-functional public space that enhances the view of the Baitasi temple.

Perspective

透视图

针对白塔寺现有停车区的设计方案是将其改造成一个无车辆停放的公共广场。广场西侧设有一个多功能的房间；在广场东侧，人们可在藤架下下棋、在茶室品茶。广场中心是一个开放的区域以强化白塔寺的景观。

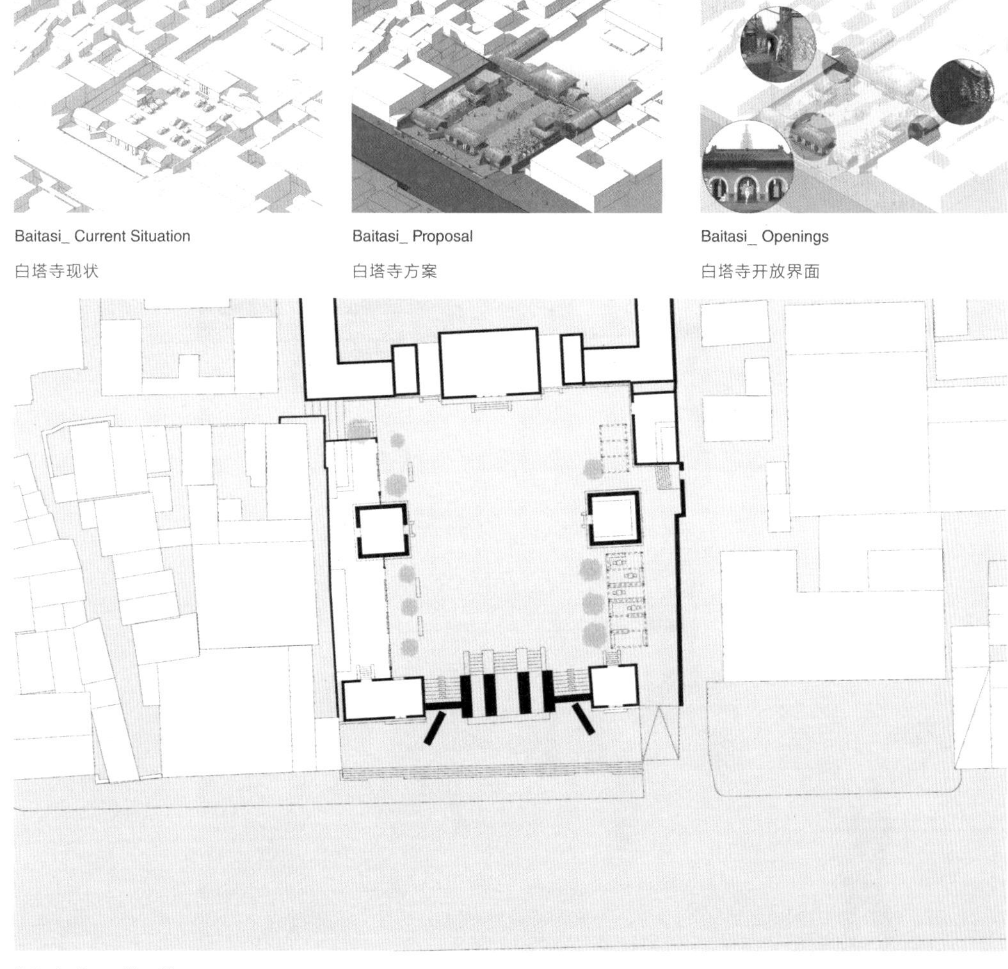

Baitasi_ Current Situation

白塔寺现状

Baitasi_ Proposal

白塔寺方案

Baitasi_ Openings

白塔寺开放界面

Baitasi_ Groundfloor Plan

白塔寺平面

设计方案

PROPOSAL

LUXUN MUSEUM AREA

LU XUN MUSEUM

We propose to transform the closed-off, uninviting "museum", into of is an open, public "cluster museum park", connected with the Second Ring Road linear park. A new museum of intangible heritage is proposed as an addition to the existing museum and connected through the existing courtyard. Workshop spaces, graffiti and historical maps walls, chess tables, a new Lu Xun library are other elements included in the revitalization of this area.

Perspective

透视图

鲁迅博物馆

鲁迅博物馆的复兴方案是将其与二环沿线的公园相连接，形成一个“博物馆公园集群”。将在现有博物馆的基础上增加一个关于非物质文化遗产的新博物馆，两者通过现有的庭院相连接。场地复兴的方案中还包含作坊、涂鸦和历史地图墙、棋牌桌以及一个新的鲁迅博物馆等元素。

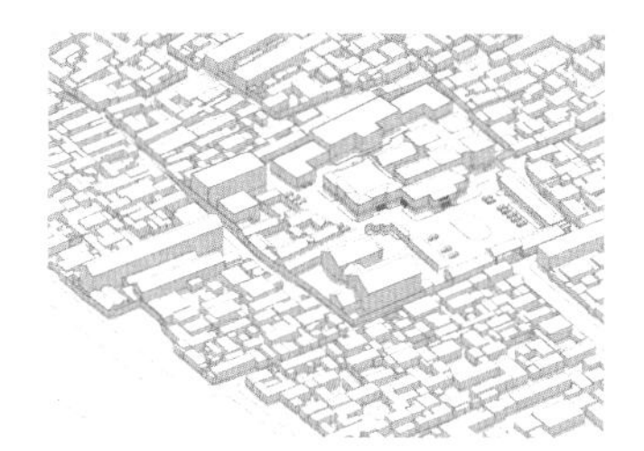

Luxun Museum Area_ Current Situation

鲁迅博物馆地区现状

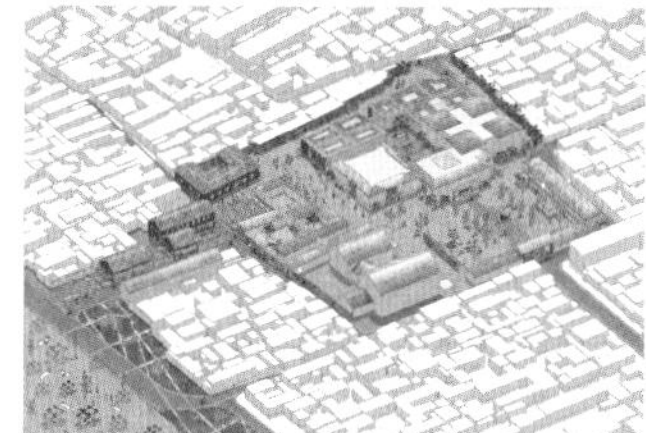

Luxun Museum Area_ Proposal

鲁迅博物馆地区方案

Luxun Museum Area_ Openings

鲁迅博物馆地区开放界面

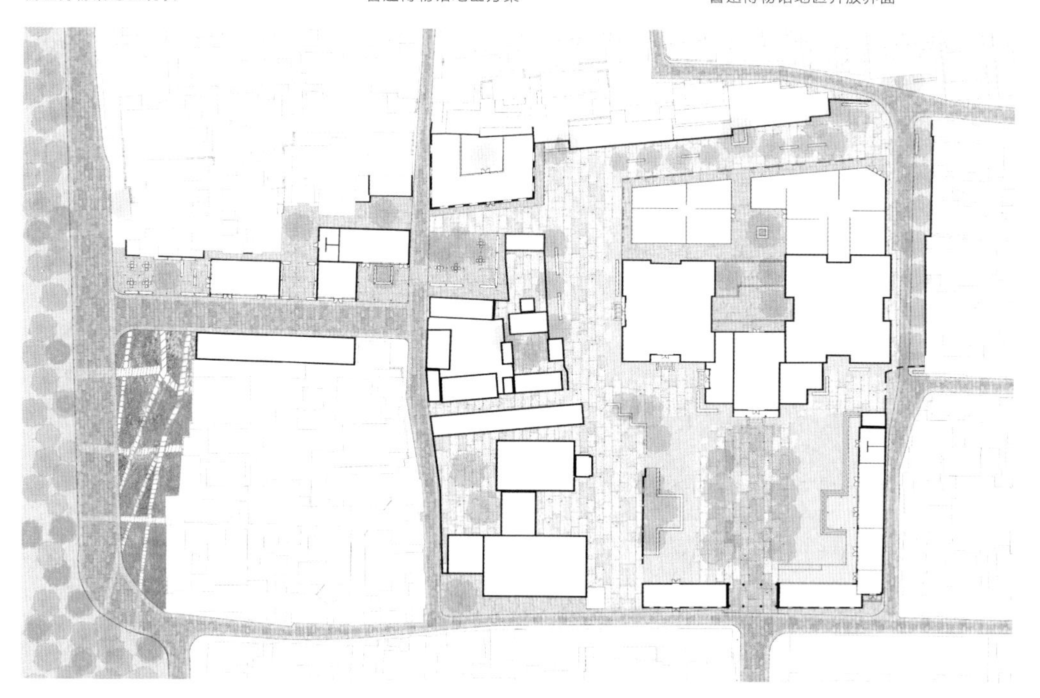

Luxun Museum Area_ Groundfloor Plan

鲁迅博物馆地区平面

FROM OWNERSHIP TO ACCESS

By Jamar Rock, David Vargas

从拥有到可获取的资源

This project proposes to re-energizing current commercial dymamics of Baitasi hutong by implementing strategies in its main linear axis. The strategies include creating a central commerce hub, active facade, green pockets, re-propose spaces & temporary transformation.

通过在原有的主要商业轴线上实施设计策略，激活白塔寺的商业活力，营造混合多元的活力氛围。设计策略包括培育商业节点、塑造积极的建筑立面、设置绿色口袋公园和重新定义空间及多样化用途。

Commerce for us:

"The buying and selling of goods and services"
"From Ownership to Access"

SHARING COMMERCE

After interviewing dozens of local business owners, we realized that the Baitasi area offers a variety of goods and services from private-owned and public- owned businesses. But, despite this variety the present commercial activity is on a steady decline. In order for the existing Baitasi commercial areas to compete with the current and growing economy of China, re-energizing the current commercial dynamics of the Baitasi Hutong is essential.

经过对当地商业运营者的调研，我们认识到白塔寺地区提供从私人商业到国有商业不同类型的多样化商品和服务，但是这样的多样化商品和服务却在持续衰退。为保存既有的白塔寺商业地区以应对增长的中国经济，白塔寺胡同地区商业的复兴是必要的。

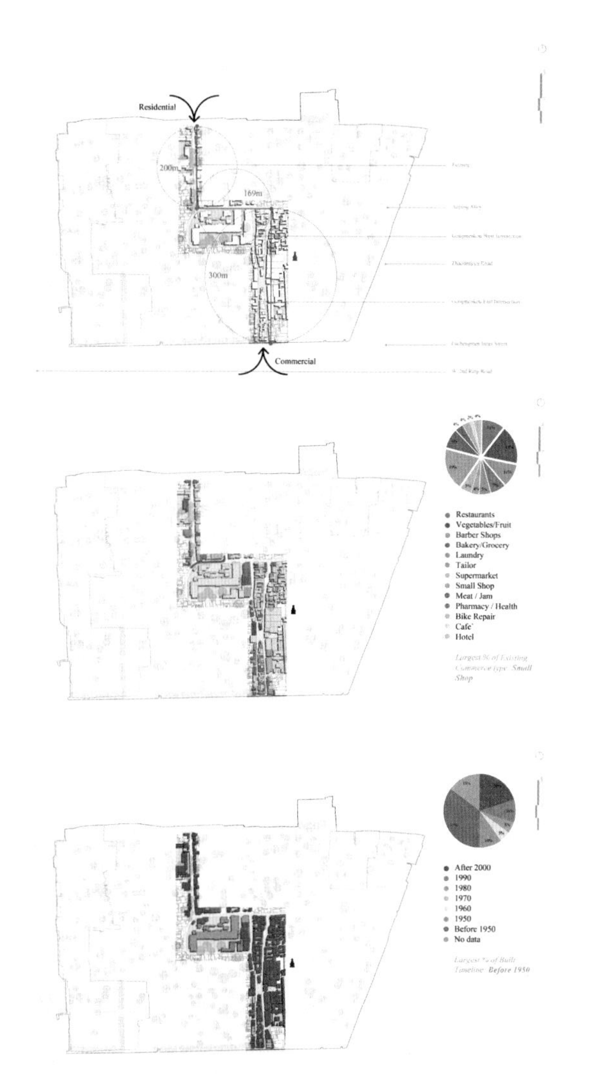

Site

基地

Existing Commerce

现有商业

Built Timeline

建成年代

分析

ANALYSIS

Tsinghua University/ Architecture School
Interview Batsi Sharing Commerce
清华大学建筑学院调问调查

Age 年龄： 33
Country of birth 出生国家： China
Male/female 性别：男 / 女

(Circle the answer) 圈一下答案
Do you live here? Yes or no
你住在这儿吗？是 / 否

Do you work here? Yes or no
你在这里工作吗？是 / 否

If you work here only, where do you live? (Inside the community or outside)
如果你不住这里，你住在哪儿？

Where do you spend your money?
你有什么消费？
For my daughter, education

In what you spend your money?
你大部分的钱是花到哪里的？

When do you spend?
你多久消费一次？
Daily / Weekly / Monthly 每天 / 每周 / 每个月
Morning / Afternoon / Evening 早上 / 中午 / 下午

Tsinghua University/ Architecture School
Interview Batsi Sharing Commerce
清华大学建筑学院调问调查

Age 年龄： 28
Country of birth 出生国家： 中国
Male/female 性别：男 / 女

(Circle the answer) 圈一下答案
Do you live here only, work here only, live and work here?
你只住这里 / 只在这工作 / 还是两者都是
Who are your costumers? Local/Chinese Outsiders/Foreigners
谁是你的顾客？
当地中国居民 / 外地中国居民 / 外国人
Do you rent or do you own this building?
你租还是拥有这间店？
How old is your business? (1-2years, 3-5years, 6 years-above)
你营业多久了？
1-2 年 / 3-5 年 / 六年或以上
Daily work hours?
Opening time............... / Closing Time.............
你每天的营业时间是几点到几点？
开店时间：AM10.......... / 关店时间：PM 12
How many days do you open?
Monday/Tuesday/Wednesday/Thursday/Friday/Saturday/Sunday
一个星期营业几天？
星期一 / 星期二 / 星期三 / 星期四 / 星期五 / 星期六 / 星期日

Research Methods

调查方法

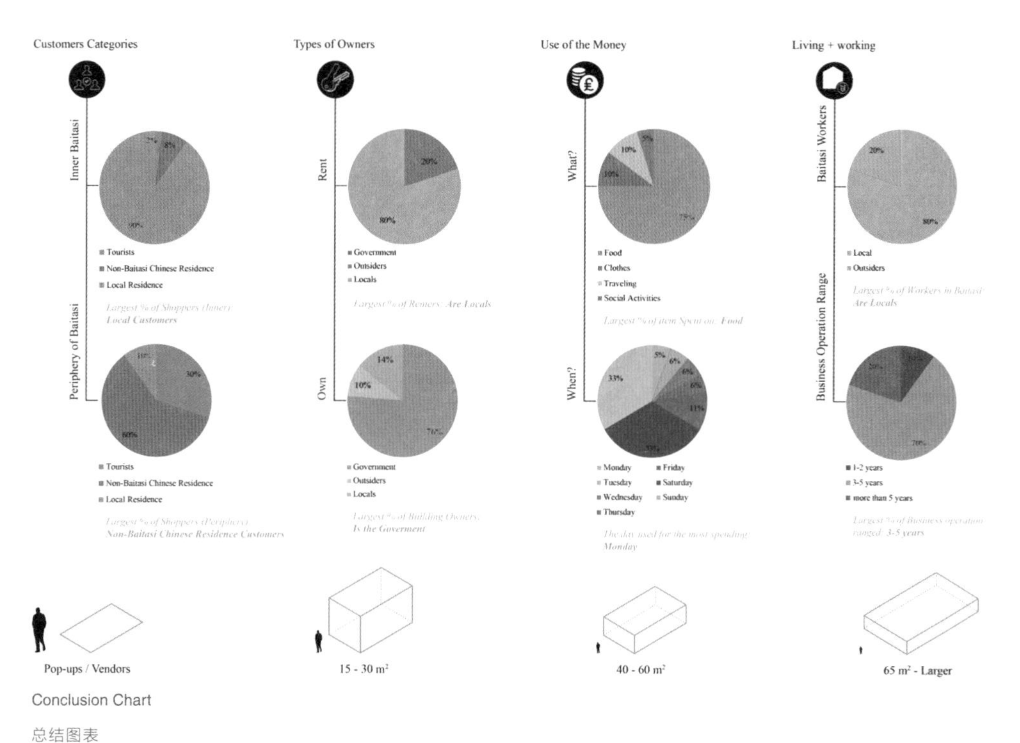

Conclusion Chart

总结图表

The White Pagoda area is home to many cultural treasures. The area is surrounded by the Xizhimen business district to the north, the Fuchengmen commercial area to the south, the Xidan and Xisi shopping districts on the east side.

We aim to utilize local qualities and business owners as a starting point to develop a shared business platform

白塔寺胡同是许多文化瑰宝的所在地。西直门商务区在其北侧，阜成门商业区在其南侧，西单和西四购物区在其东侧。

我们致力于利用当地特色和商业运营者作为发展共享商业平台的起点，通过一条培育而成的、联系周边商业地区的商业走廊得以实施。

spatially expressed through a curated, connected commercial corridor that links to surrounding commercial areas.

To attract customers from different districts, five specific nodes are designed for local business promotion and recreative space revitalization. This linar business corridor will attract potential customers to Baitasi business center if the proposal works.

为了吸引来自不同地区的消费者，方案设计了五个各有特色的中心，促进本地贸易，复兴娱乐空间。如果成功，这个线性商业带将吸引潜在的消费者来到白塔寺商业中心。

策略与原则 STRATEGIES & RULES

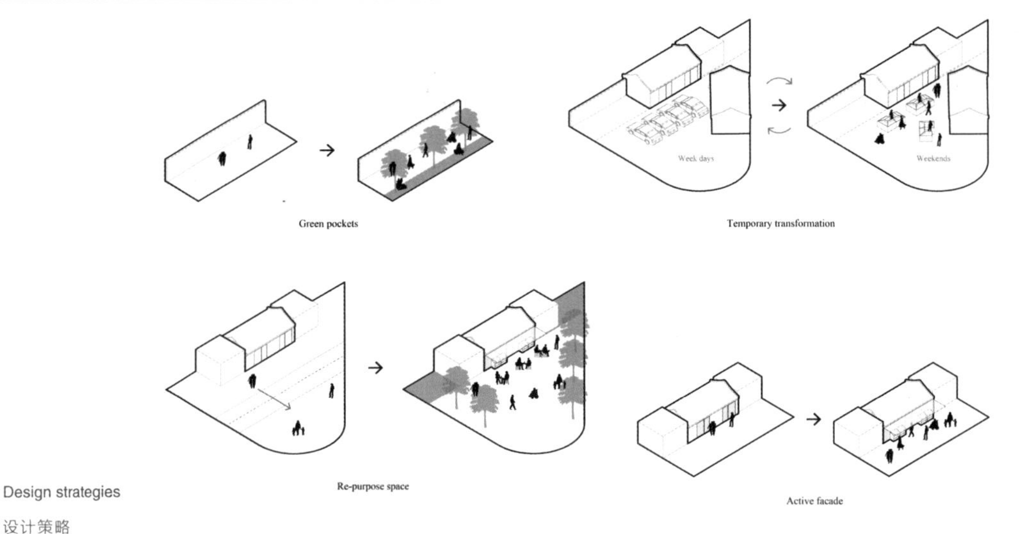

Design strategies

设计策略

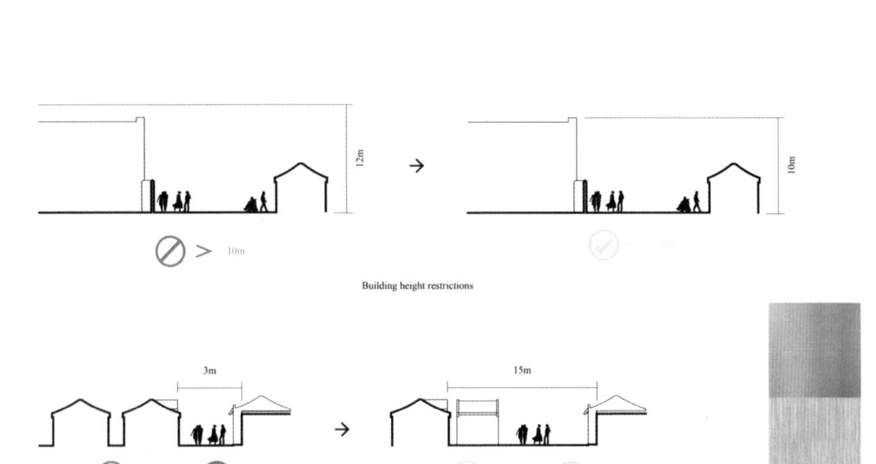

Design Rules

设计原则

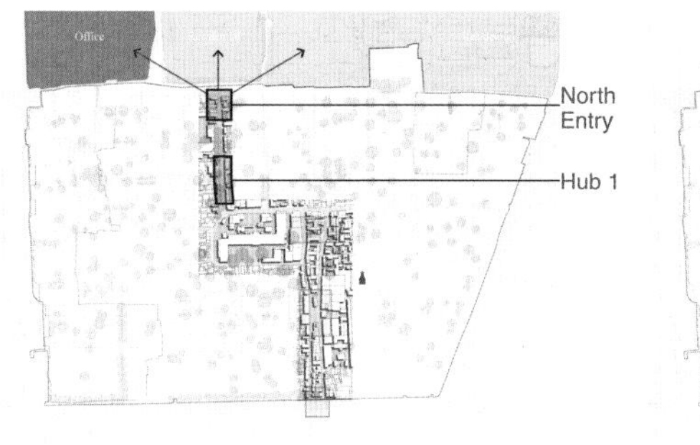

North Entry & Hub 1

北入口及节点一

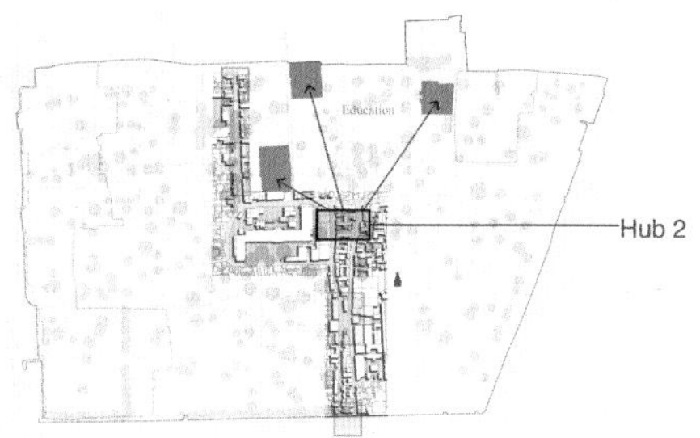

Hub 2

节点二

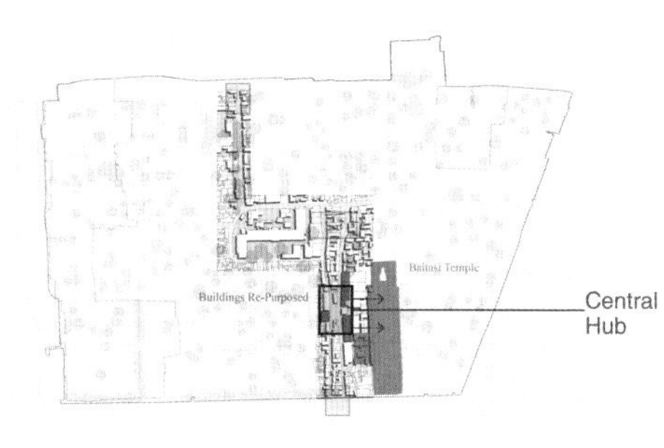

Central Hub

中心节点

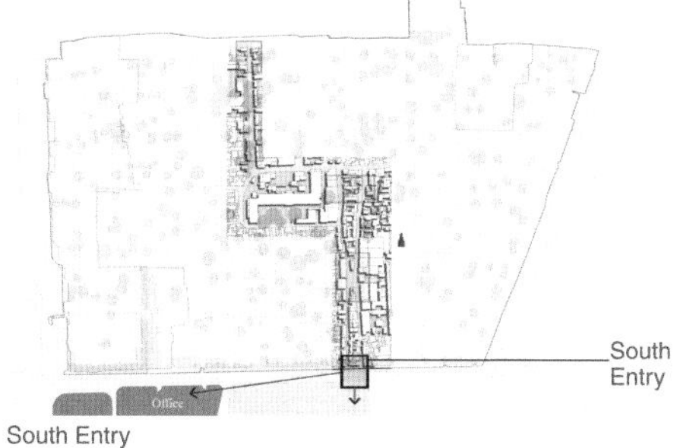

South Entry

南入口

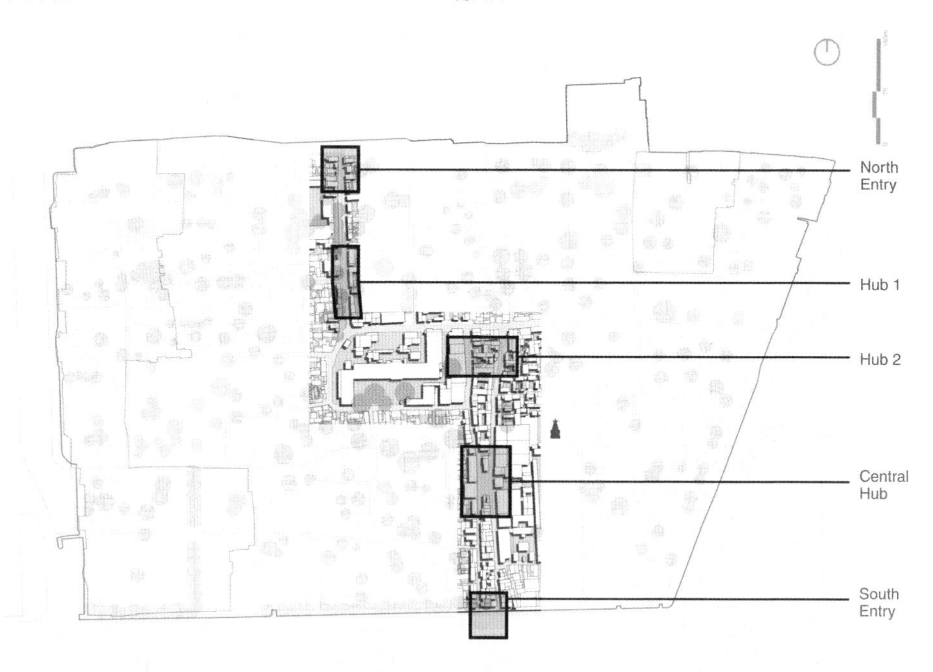

5 Focused Zones

五个重点关注的区域

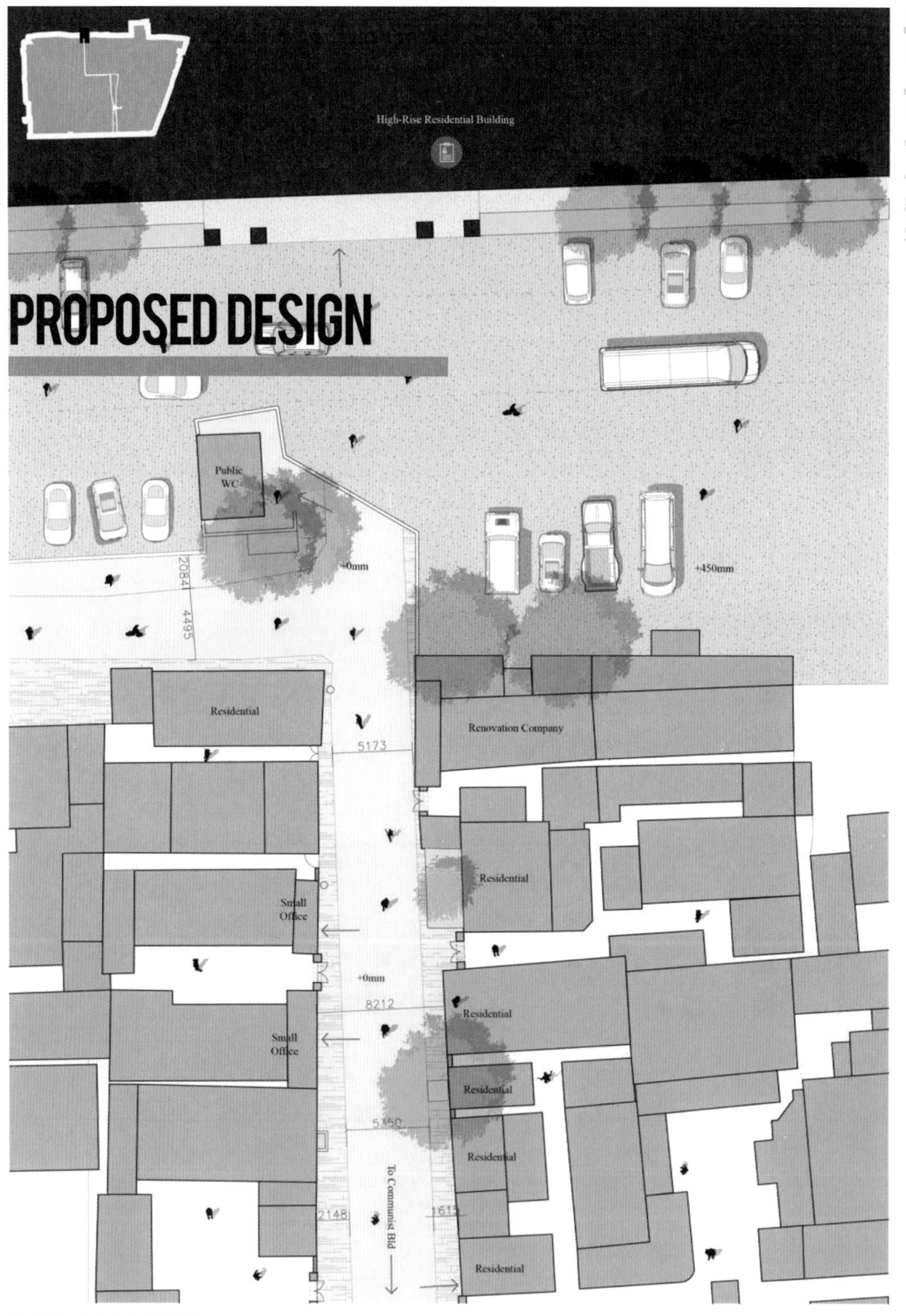

North Entry & Hub 1 - BEFORE

北入口及节点——设计前

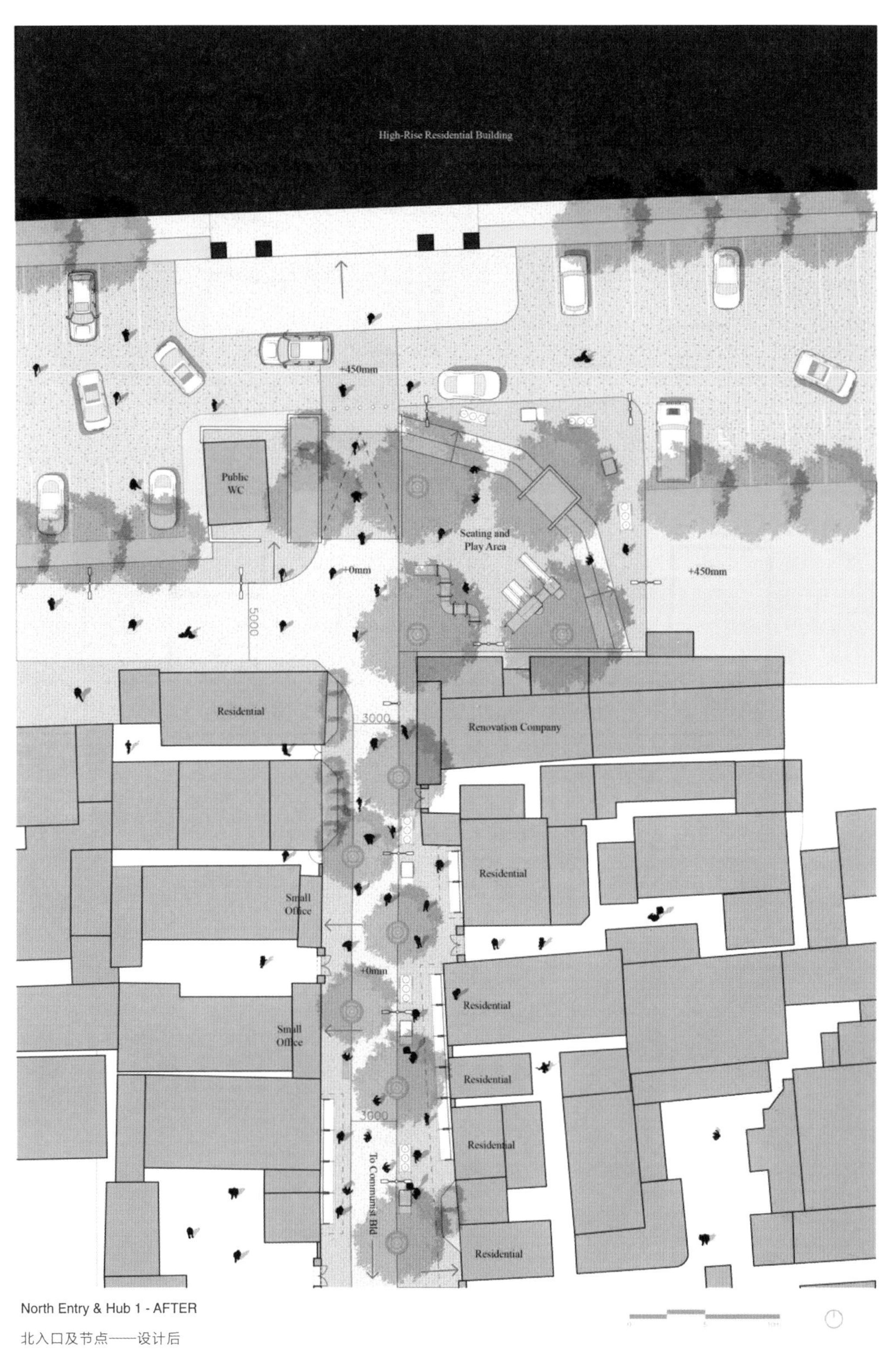

North Entry & Hub 1 - AFTER

北入口及节点——设计后

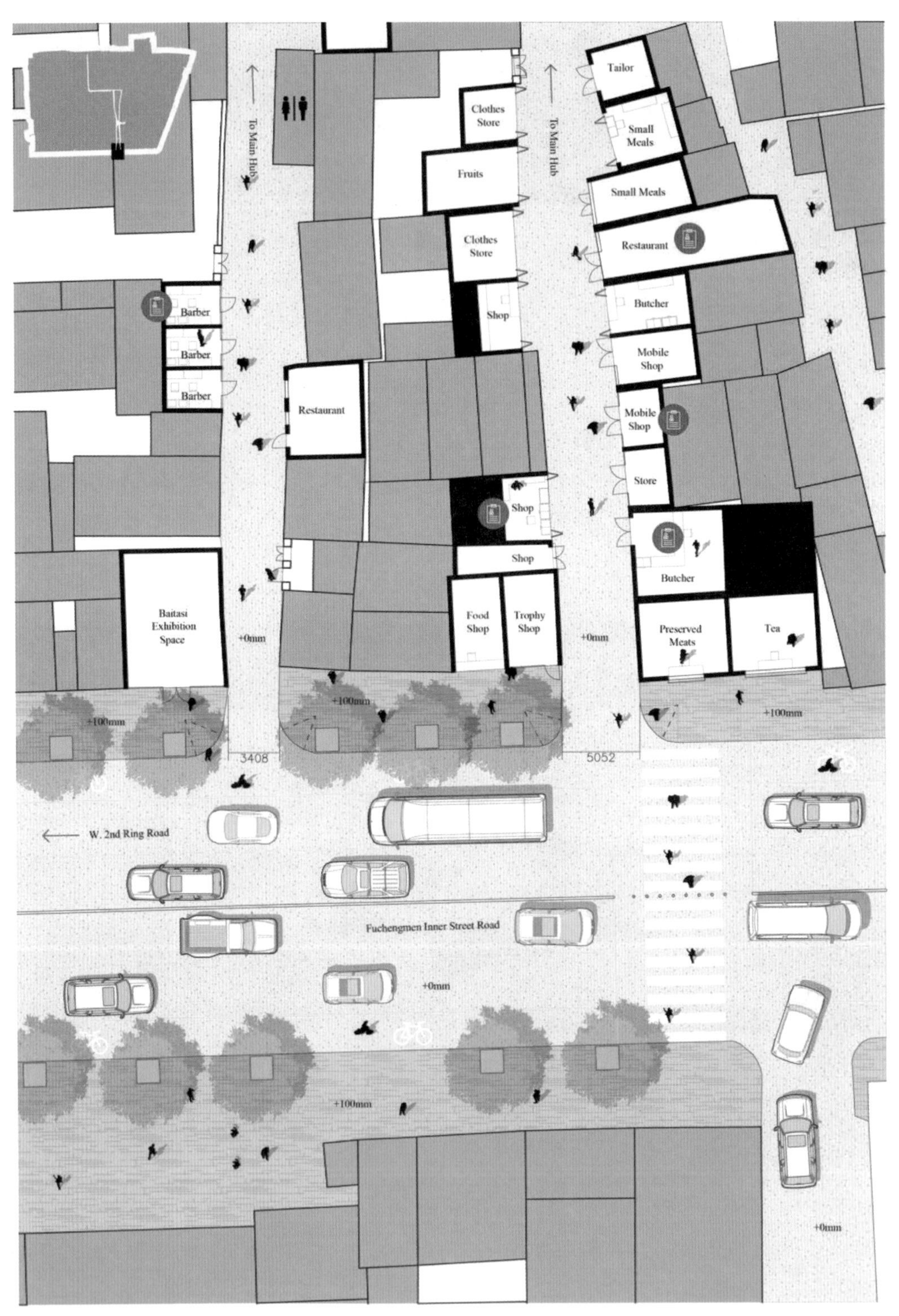

Central Hub - BEFORE

中心节点——设计前

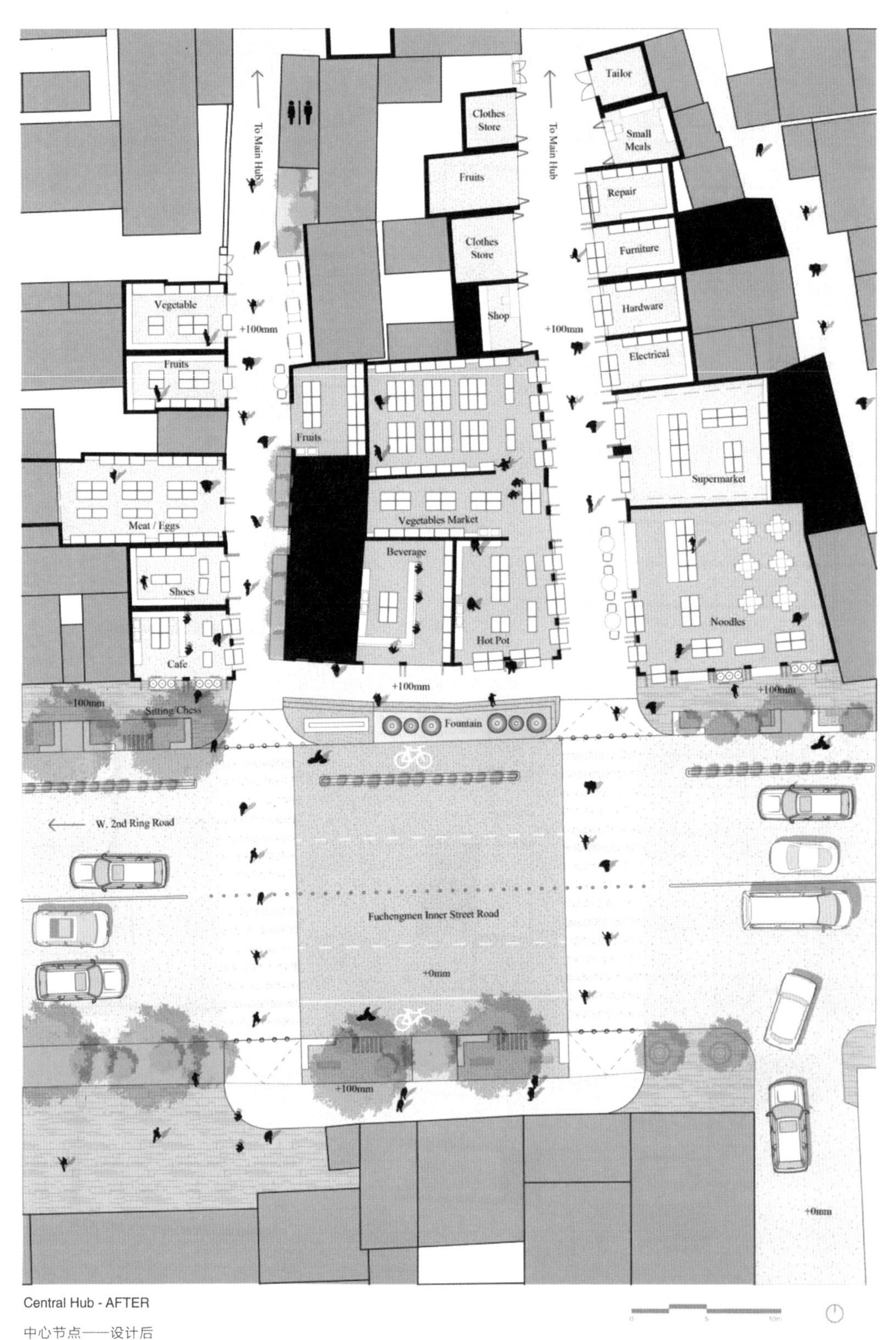

Central Hub - AFTER

中心节点——设计后

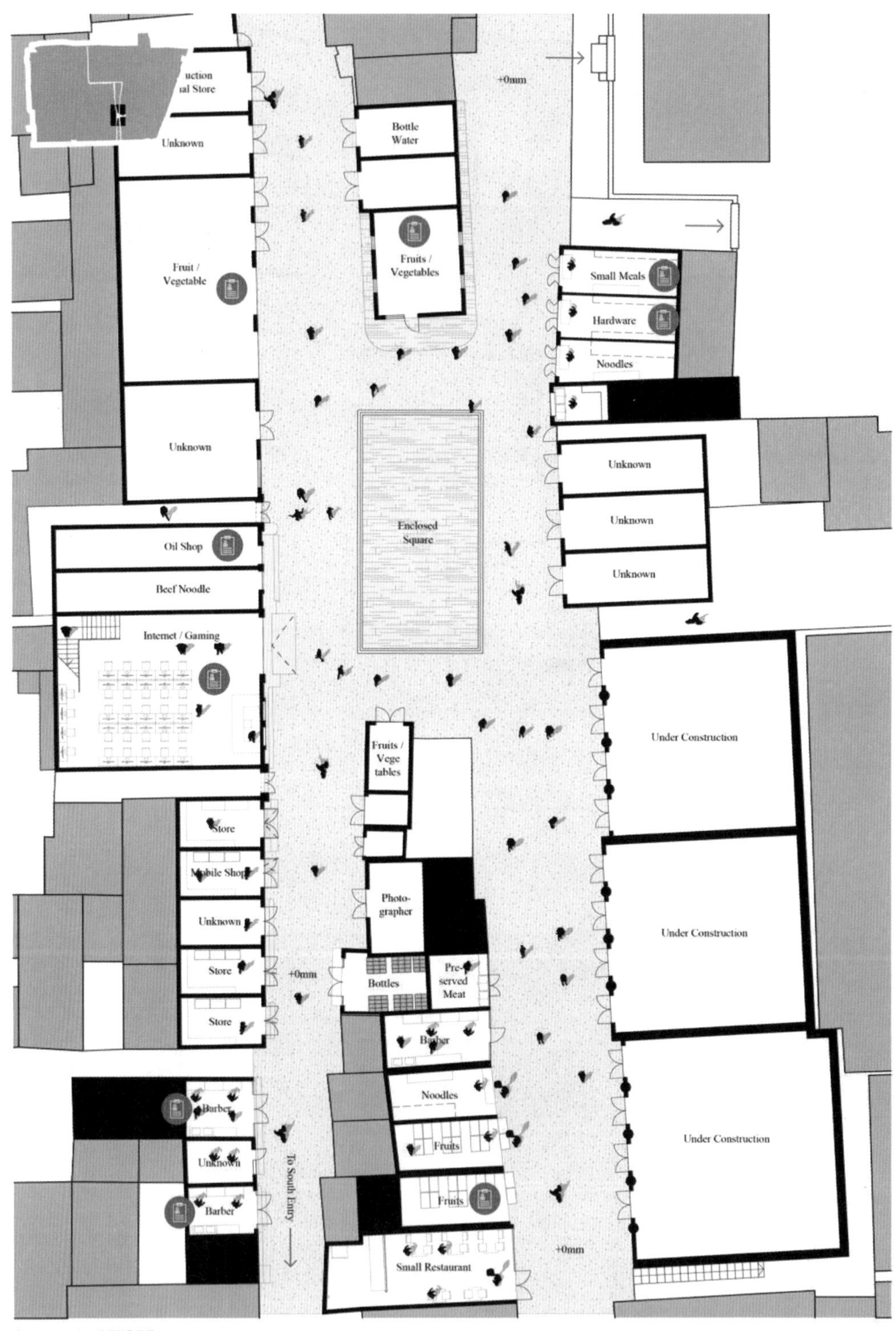

Central Hub - BEFORE

中心节点——设计前

Central Hub - AFTER

中心节点——设计后

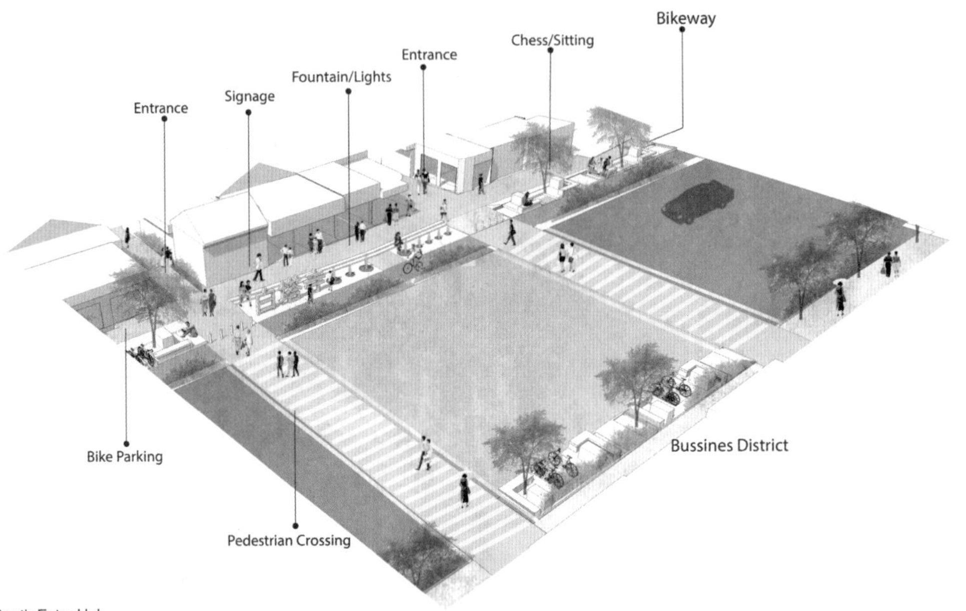

South Entry Hub

南入口

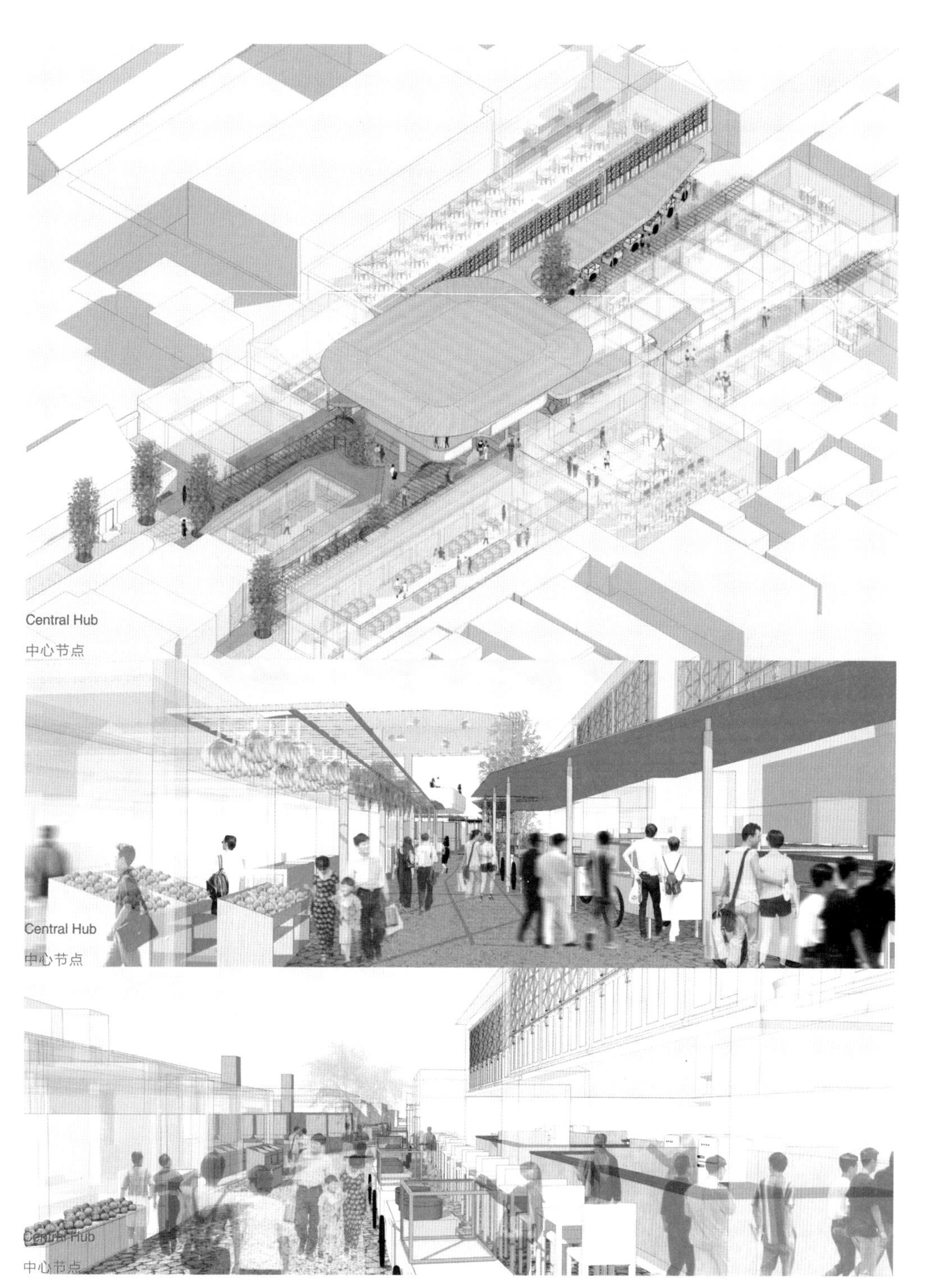
Central Hub
中心节点
Central Hub
中心节点
Central Hub
中心节点

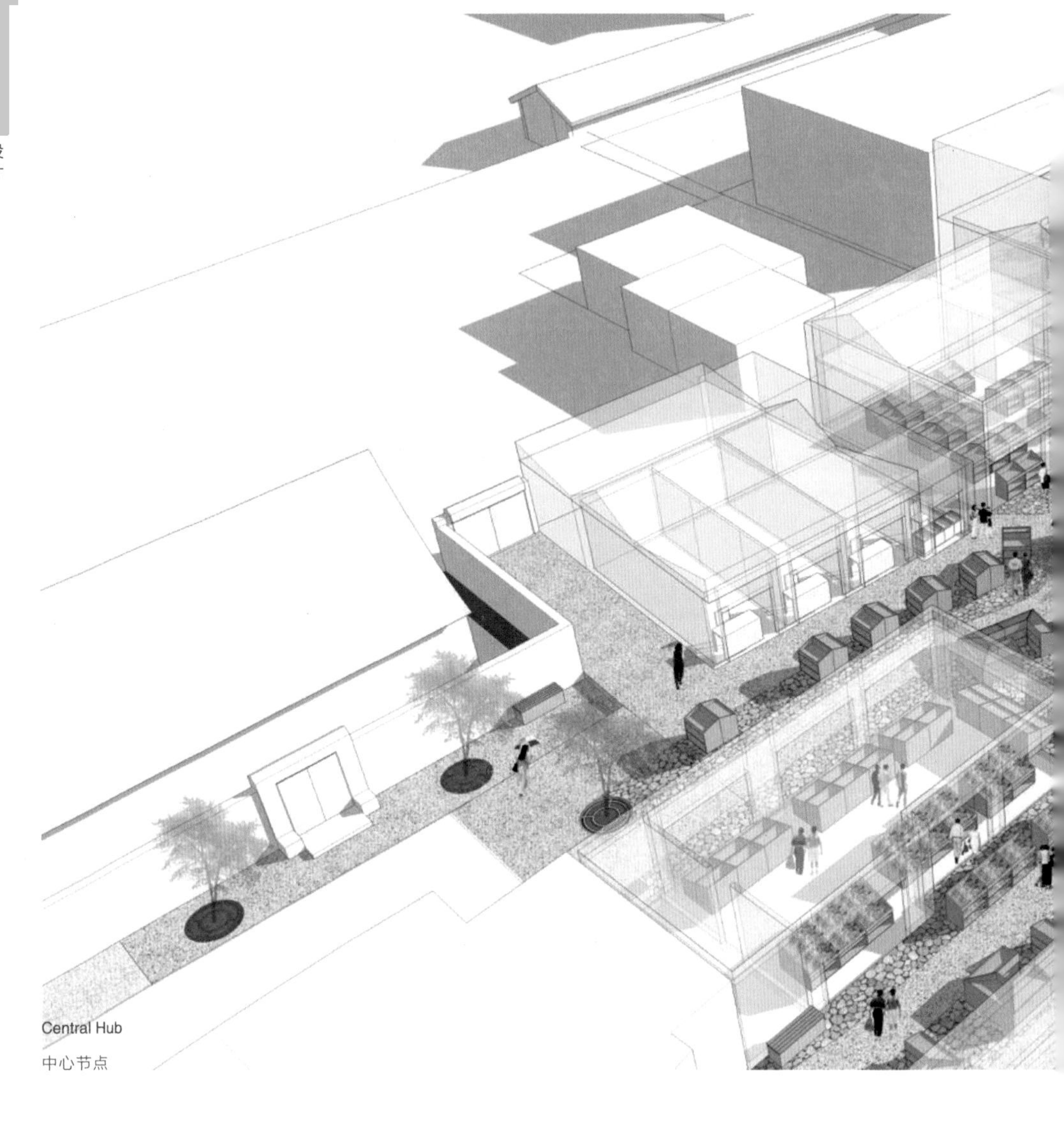

Central Hub

中心节点

Central Hub

中心节点

Central Hub

中心节点

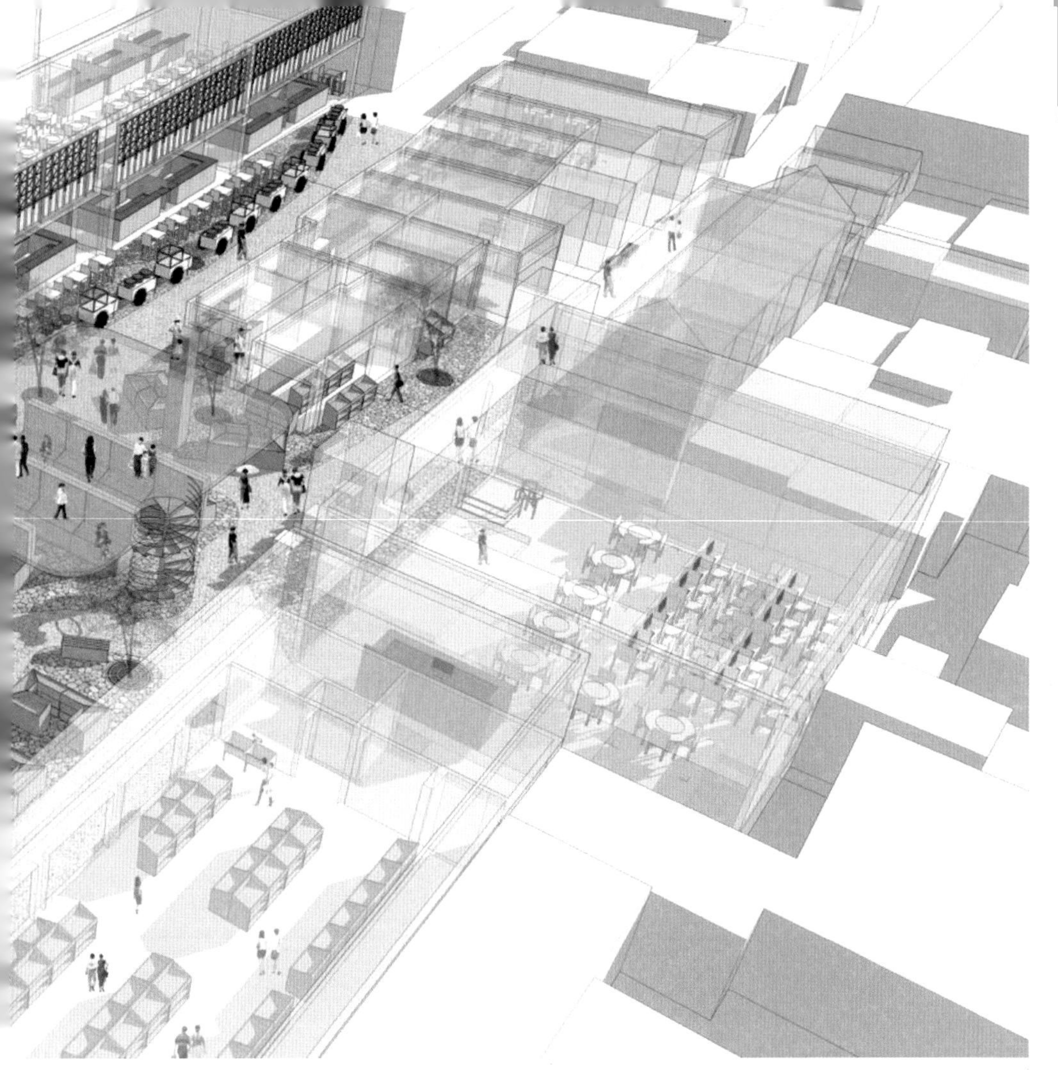

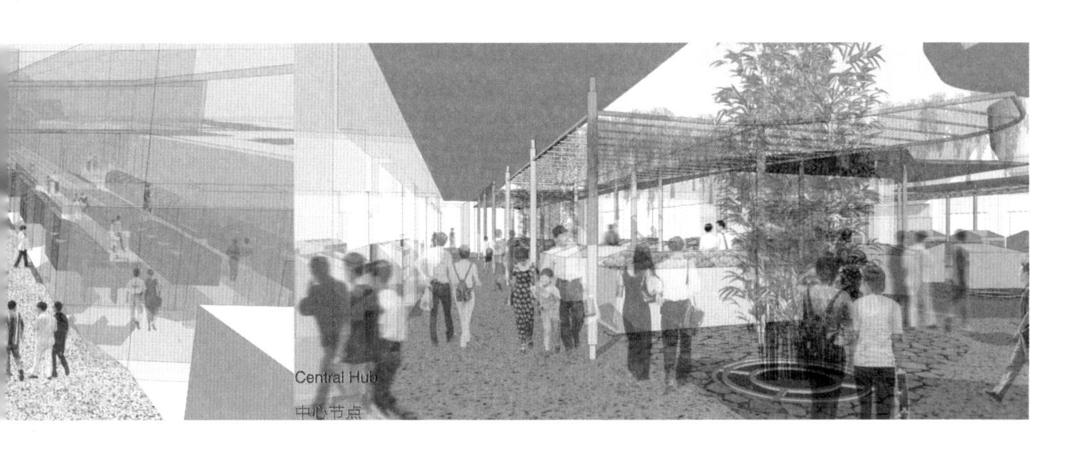

Central Hub

中心节点

BAITASI PLUS

By Ahmed, Ricardo

白塔寺基础设施 +

Baitasi Plus is an infrastructural enhancement initiative for the Baitasi community, aimed at improving the conditions in which the community conducts daily activities.

白塔寺基础设施 +（Baitasi Plus）是一项针对白塔寺社区的基础设施增强计划，旨在改善社区内的日常活动。

SHARING INFRASTRUCTURE

Currently the community lacks critical hygienic functions such as wash sinks in public washrooms and evenly distributed showering facilities. Baitasi Plus consists of three phases which address these issues and add new functions where appropriate.

目前社区内缺乏严格的卫生设施，如公共卫生间内的洗手池，甚至分散式的淋浴设施。方案分三个阶段解决这些问题，并在适宜之处添加新的功能。

Existing Infrastructure Issues
基础设施现状

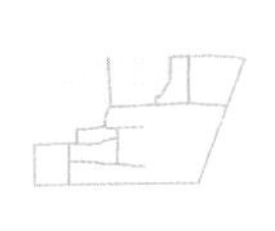

Telecommunications
电讯设施

Electricity
电力设施

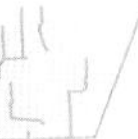

Water Supply
供水设施

Rainwater Drainage System
排水设施

Public Bathrooms
公共卫生间

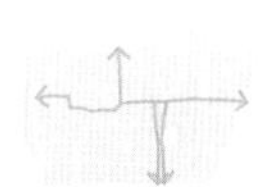

Commercial Spine
商业轴线

(1)Wired infrastructure does not cover entire Baitasi area.
(2)Area is well connected but population density results in large power consumption and overpowering physical infrastructure.
(3)Water is supplied to residents' courtyards but distribution method could be improved.
(4)Rainwater drain used for removal of kitchen Waste and light sewage
(5)No source of water for washing hands / Semi-Private partitions
(6)The commercial spine is important to Baitasi and the supporting role proper hygiene can play is crucial.

(1) 有线通信并未覆盖整个白塔寺地区。
(2) 电力设施覆盖整个区域但较大的人口密度造成的消耗使基础设施严重过载。
(3) 供水配给到居民到后院，但分配方式需要改进。
(4) 雨水排水设施被用于排污。
(5) 缺乏洗手功能 / 厕位半私密。
(6) 商业轴线对白塔寺十分重要，其中适当对供水设施起到关键对辅助角色。

Different social activities of hutong residents
胡同居民的不同社会活动

Barbering / Socializing
理发 / 社交

Eating / Air-drying laundry
用餐 / 晾衣

Hot water adjacent to roadway
沿街的热水器

Narrow roads and alleys
拥挤的道路和走廊

Electrical Poles
电线杆

Crowded courtyards
拥挤的后院

Traditional Methods of Infrastructural Upgrade
传统基础设施升级方法的局限

(7)Roadways are too narrow to implement large sewage or other infrastructure pipes.
(8)An abundance of electrical poles prevent any attempt to utilize sidewalk space.
(9)Illegal constructions and tight courtyards hamper any practical means of sewer integration.

(7) 过于狭窄的道路不足以铺设大型排水设施或基础设施管道。
(8) 过多的电线杆阻碍了人行道空间的利用。
(9) 非法建筑和拥挤的后院妨碍了管线的集成。

设计策略

STRATEGIES

Phase 1: Decentralize and Declutter
Existing electrical infrastructure will be replaced with a solar energy system which consists of roof (Hutong) mounted photovoltaic panels and solar powered street lamps. This system harvests solar energy through photovoltaic panels which is then converted and stored in batteries to supply lighting and appliances. Additionally, wireless telecommunications system will be implemented to supply unconnected areas using wireless routers.

Phase 1 阶段 1

Decentralize
& Declutter 分散与梳理

Phase 2 阶段 2

Remodel 改造

Phase 3 阶段 3

Integration 集成

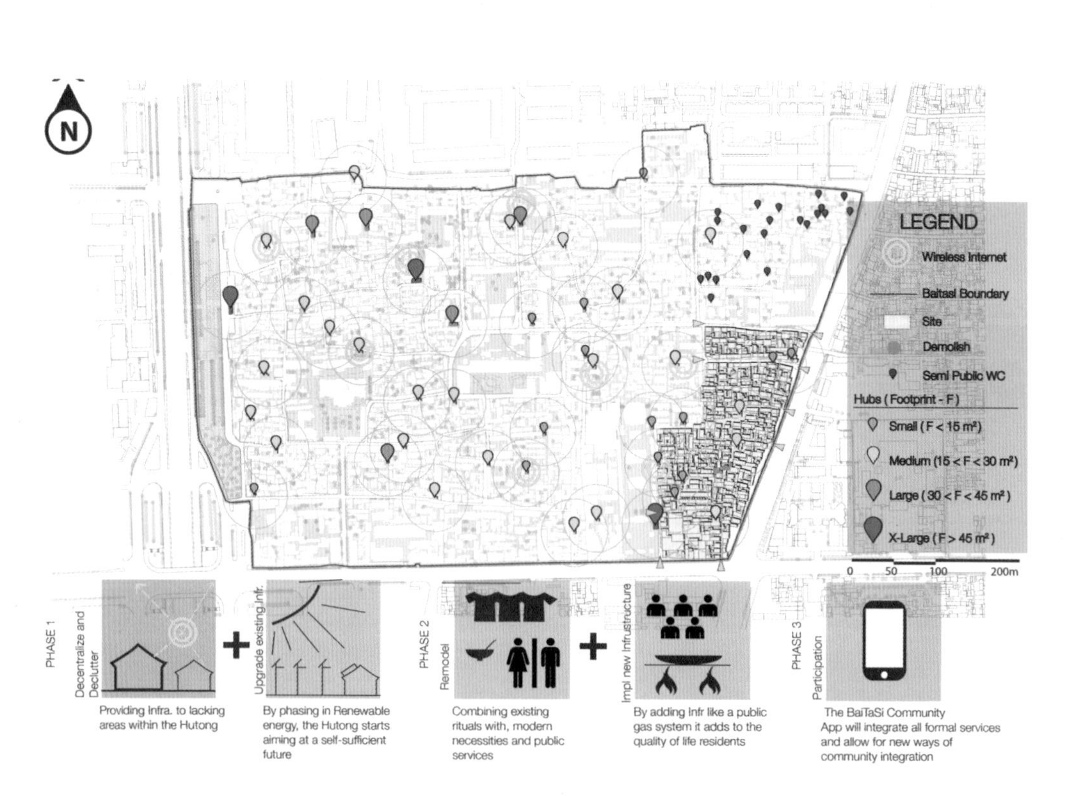

阶段一：分散与梳理
现有的电力基础设施被太阳能系统所替代，系统内包含屋顶太阳能光伏板和太阳能路灯。这个系统依靠光伏太阳能板获取太阳能，将其转换成电能储存在电池里，供给照明和其他电器。另外，无线通信系统将用无线路由器覆盖没有通信的区域。

PROPOSAL

PHASE 1: SOLAR ENERGY

ase 1: Solar Energy

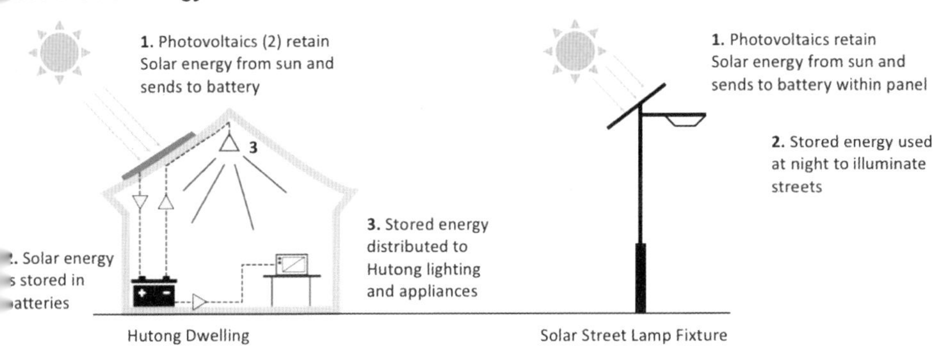

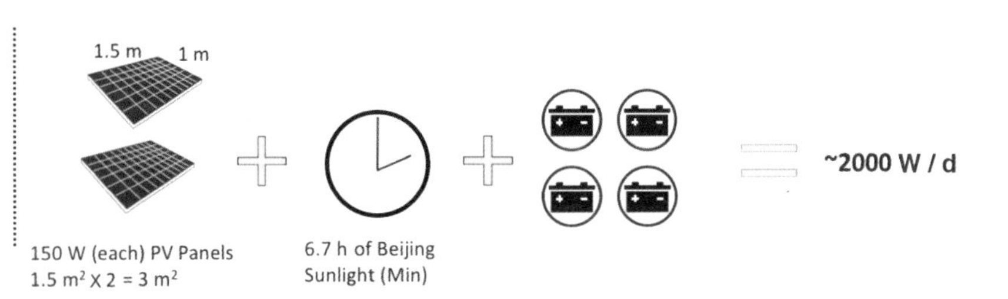

Wattage Chart

Appliance											
Starting Watts	50 - 150	100 - 200	150 - 500	800	200	25	200	250	850	600	1000
Running Watts	50 - 150	100 - 200	150 - 500	800	200	25	200	250	400	200	1000

设计方案

Phase 2: Remodel

Through an analysis of existing functions and contextual conditions, public washrooms will be converted into service hubs. Service hubs retain the initial sanitary service, with new needed amenities such as wash sinks, showers and private stalls. At least one additional service (laundry, battery charging etc), will be added to service hubs, to not

阶段二：改造

通过对于目前功能及分布状况的分析，公共卫生间将被改造成为基础设施中心。中心保留原有的公厕功能，并增加新的设施，如洗手池、淋浴设施、私密浴室等。至少一项新的服务设施（洗衣、充电等）会增加到基础设施中心，不仅服务于邻近的居民，而且服务整个白塔寺社区。设施中心潜在的位置、加

PROPOSAL

PHASE 2: SERVICE HUBS

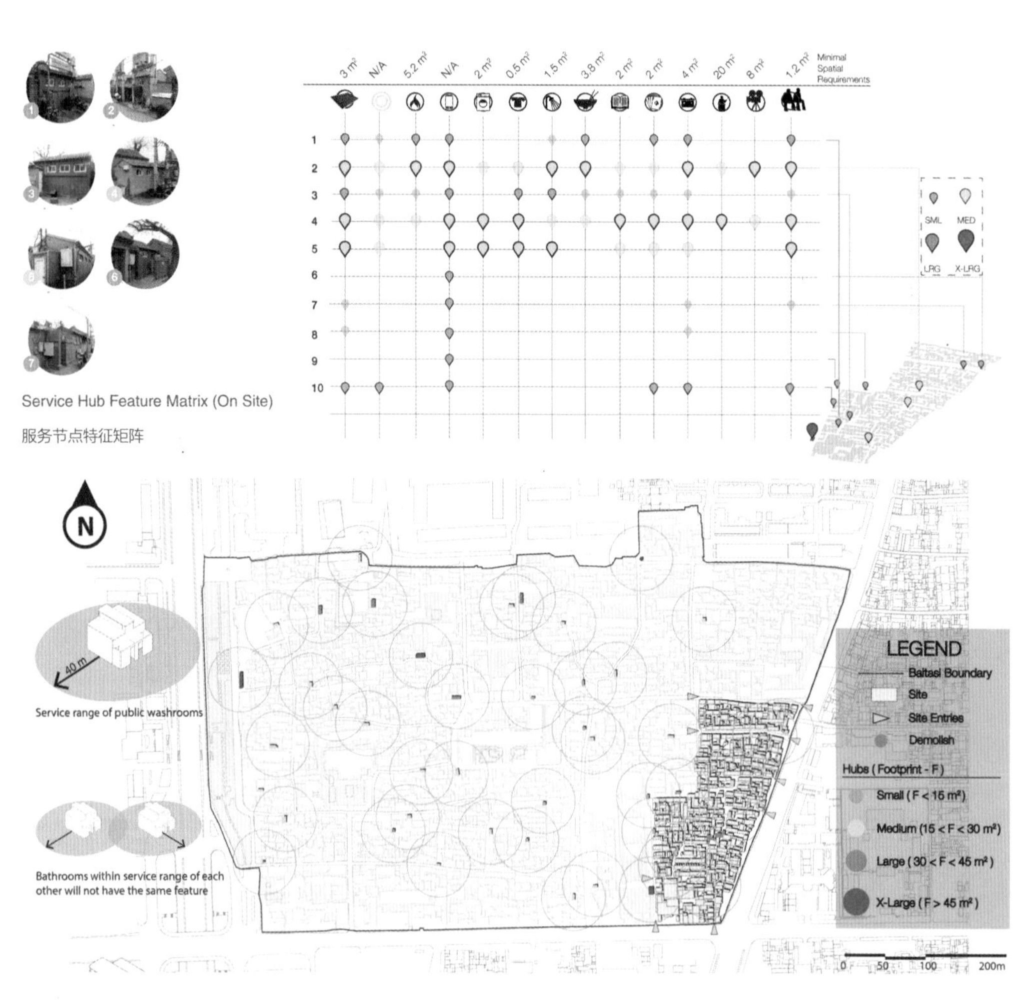

Service Hub Feature Matrix (On Site)

服务节点特征矩阵

only serve the adjacent households, but also the Baitasi community at large. The Hub's potential footprint, potential second storey and adjacent public spaces, all determine the type of features to be added to the service hub. The hubs are primarily positioned to provide adequate sanitary coverage to Baitasi via new construction, renovation or removal of existing washrooms.

设二层的可能、临近的公共空间都决定了附加在设施中心的属性。设施中心的选址考虑通过新建、修复、改造现有卫生间的方式为白塔寺地区提供充足的公共卫生设施。

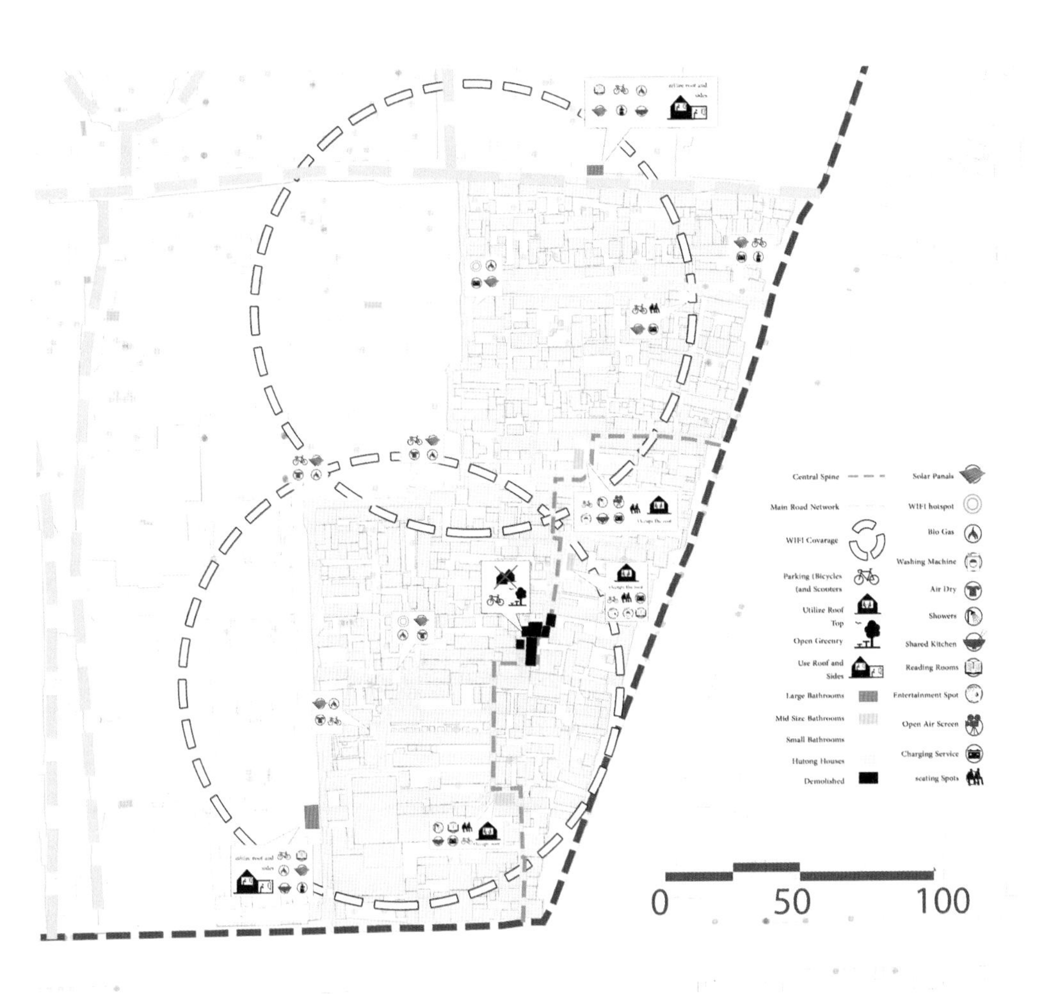

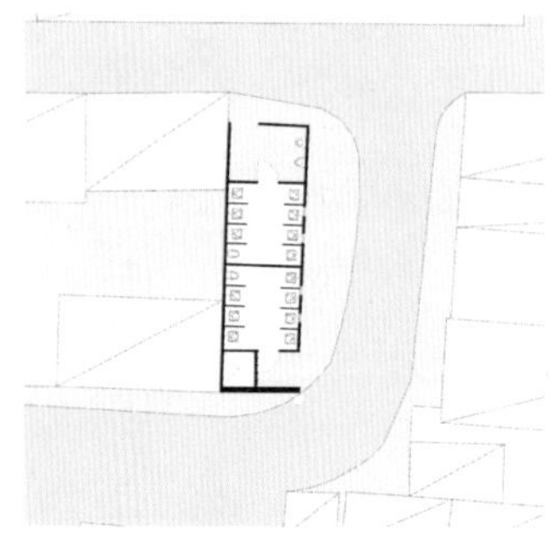

Large Hub Ground Floor Plan (Existing)

大型基础设施中心首层平面（现状）

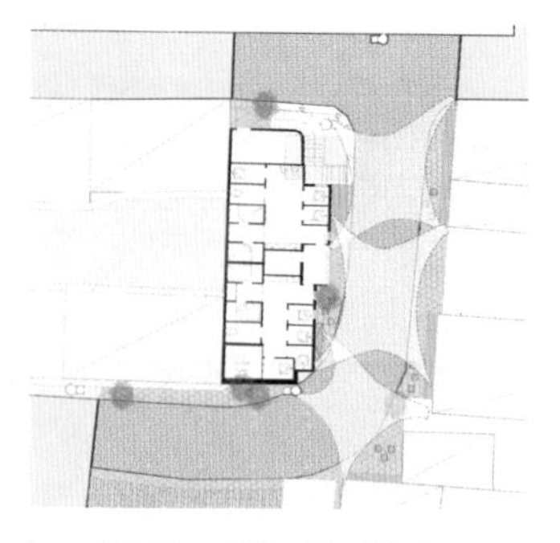

Large Hub Ground Floor Plan (After)

大型基础设施中心首层平面（改建）

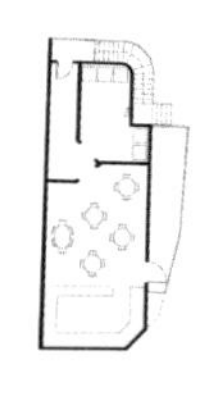

Large Hub First Floor Plan (After)

大型基础设施中心二层平面（改建）

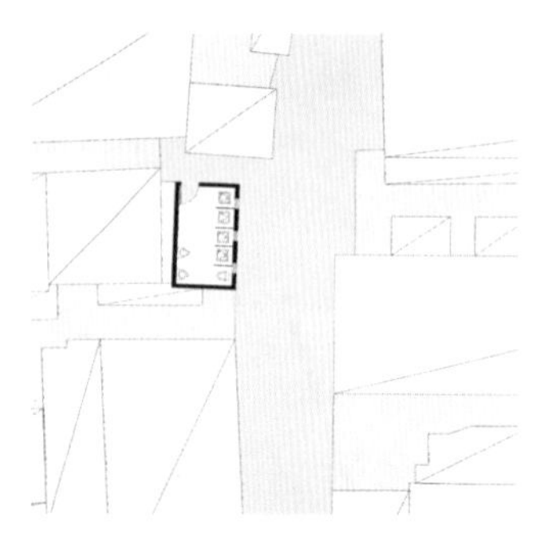

Small Hub Floor Plan (Existing)

小型基础设施中心首层平面（现状）

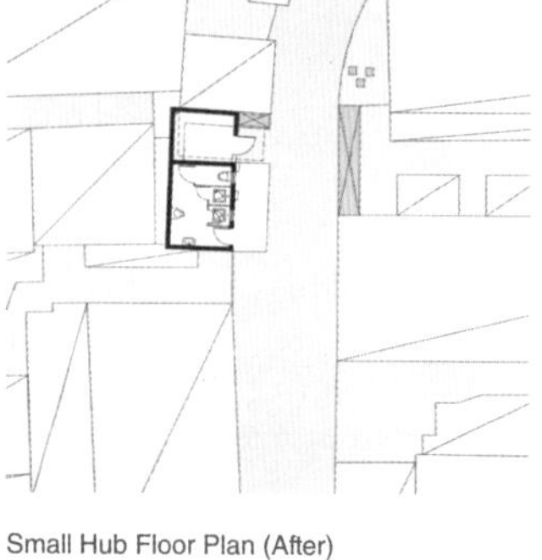

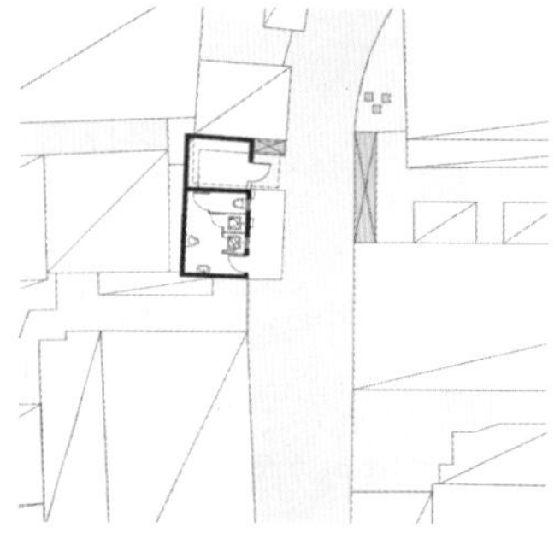

Small Hub Floor Plan (After)

小型基础设施中心首层平面（改建）

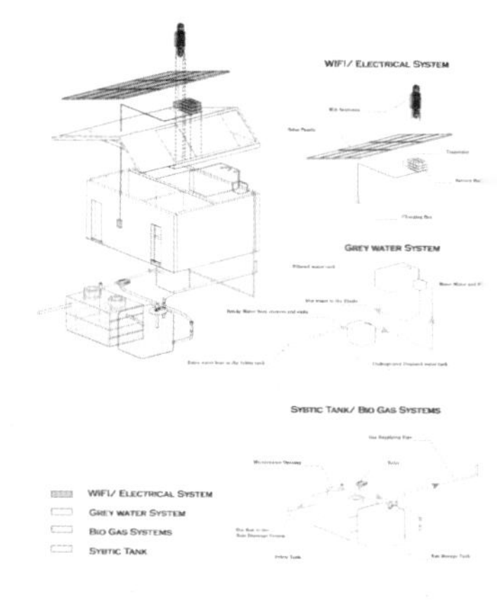

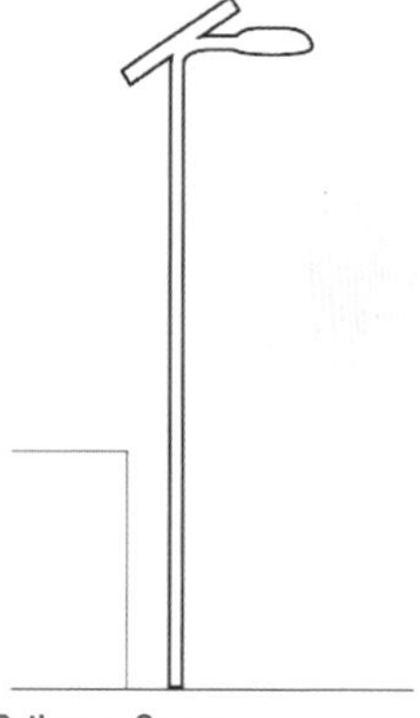

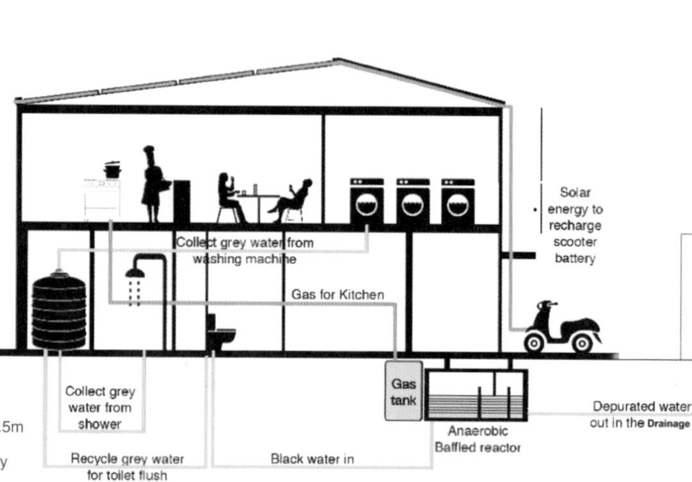

Bathroom Coverage
80 users per bathroom
Septic Tank Capacity
6.2m^3 Therefore the dimension 1.2m x 1.7m x 2.5m
Gas tank
5 m^3 covers the cooking need of 22 people / day

Phase 3: Integration

An application (APP) will be created to map all washrooms and their related features, which is facilitated by an existing digital database of public washrooms. The APP will also allow users to remotely book convenient services like showers within the service hubs.

阶段三：集成

在现有的公共卫生间电子数据库的基础上，开发一个包含所有卫生间位置信息和附加属性的应用程式。这个应用程式允许使用者远程预订基础设施中心中的淋浴等服务设施。

PROPOSAL

PHASE 3: BAITASI APP ECHO SYSTEM

:asi APP Demo

塔寺手机应用示意

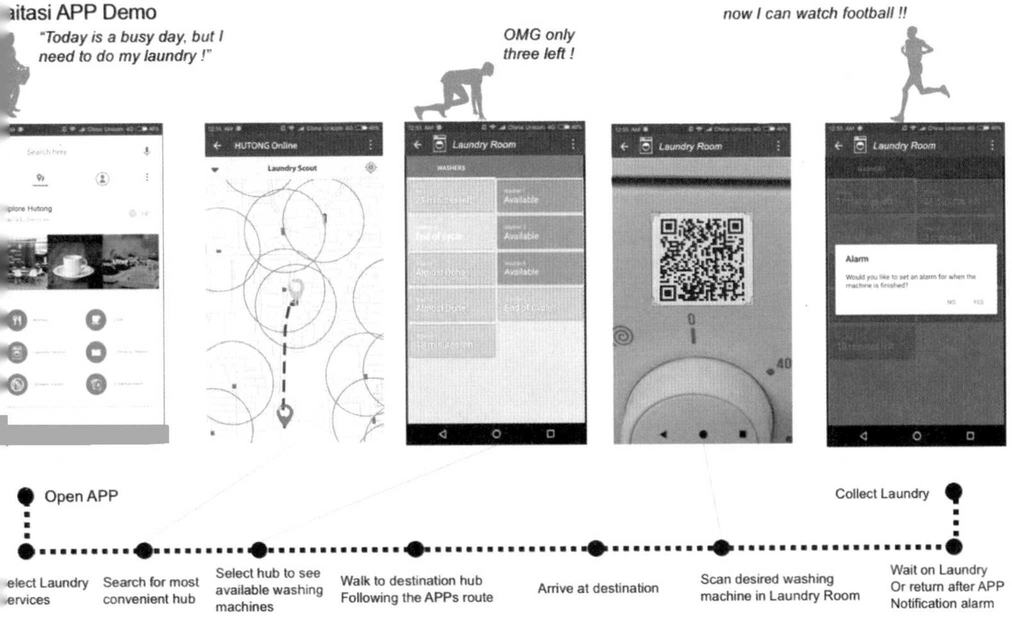

aitasi APP Demo

"Today is a busy day, but I need to do my laundry !"

OMG only three left !

now I can watch football !!

Open APP

Collect Laundry

elect Laundry ervices

Search for most convenient hub

Select hub to see available washing machines

Walk to destination hub Following the APPs route

Arrive at destination

Scan desired washing machine in Laundry Room

Wait on Laundry Or return after APP Notification alarm

SHARING AS *JUST*-SUSTAINABILITIES

共享与可持续发展

As a response to the privatisation and enclosure imposed by the neoliberalism, sharing is arguably becoming one of the defining characters of future cities. This studio hence takes on the sharing paradigm as a potential way to regenerate existing urban areas by proposing various design ideas as part of a collective system that is sustainable. The key challenge of this exercise refining the urban and architectural environment such that it adapts effectively to the emerging new lifestyles defined by sharing.

As such, the studio selected Joo Chiat, a multi-cultural traditional neighbourhood located close to the heart of Singapore, as the site to support design exploration. Its rich architectural heritage, fine-grained urban fabric and concentration of residents of various cultural backgrounds and socio-economic statuses provide a complex but fertile test bed for experimenting with new ideas of sharing. A systems approach is adopted

近年来，新自由主义加深了私有化和封闭的现象。因此，分享主义逐渐成为未来城市的特征之一，以作为对这个现象的某种回应。因此，这个设计studio通过共享范式，提出各种设计构想实现一个可持续集体系统，来重建现有的城市地区。这个设计方案的关键挑战是如何通过改善城市和建筑环境，使其适应新兴生活方式。

因此，设计studio选择专注于如切，一个位于新加坡的多元文化传统社区，将其作为探索设计起点。其丰富的建筑遗产、细致的城市布局和各种文化背景及社会阶层居民，是新的分享方案的良好试验地。系统式的调查引导设计过程，分为两个阶段进行。首先，是针对社区的某一环节开发的分享系统，接下来是将系统转化为城市设计和建筑解决方案。我们坚信，一个促进分享资源的实体环境是基于共享系统而设计的。

按照这样的概念，设计studio探讨并构想出四个共享系统和相应的设计方案，如：基础设施，公共设施与服务，经济生产与消费，文化与生活。接着，再将各个系统组合协调，以描绘新的分享形式如何在未来的城市中实行。最

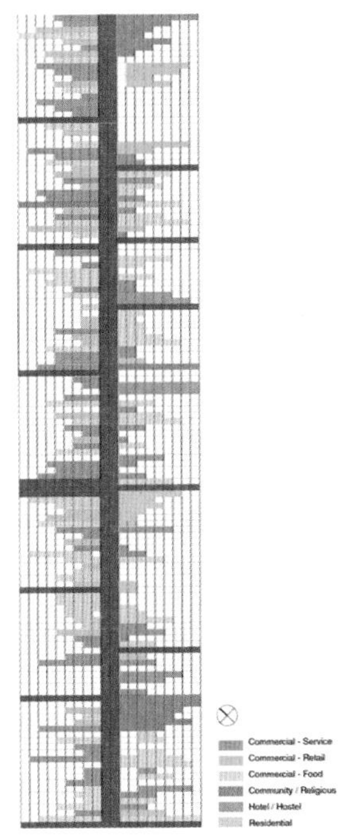

Top
Traffic flow analysis of Joo Chiat illustrates the congested traffic conditions along main axes; there is potential to distribute this to the network of smaller roads.

上图
根据对如切所做的交通流动分析结果显示， 主要公路的拥挤情况是可以被分散到网络中的其他支路上。

Bottom
An overview of the programme mapping shows that F&B (yellow) is a major economy in Joo Chiat.

下图
地图上的资料显示，餐饮业成了如切区重要的组成部分。

to guide the design investigation, which is primarily consisted of two phases. The first focuses in particular on developing a system of sharing, tackling a certain dimension of urban neighbourhood, and the second aims at translating the system into an urban design and architectural solution. This approach is also a proposition in itself that design for a physical environment to facilitate sharing practice is largely predicated on design of a sharing system.

Following this approach, the studio explored and developed four systems of sharing and corresponding design solutions, namely infrastructure, public facilities and services, economic production and consumption, and culture and everyday life, and how in combination the system can provide a concerted and coherent imagination of what kind of new forms of sharing may take in future cities. Finally, the studio proposes architectural and urban design in order to produce conducive spaces that enable and enhance the practice of sharing.

The four systems and respective design proposals are to be read whole as a sharing masterplan superimposed and weaved into the complex pre-existing conditions of Joo Chiat. This exploration could then provide insights to the future of sharing, as one may apply similar strategies to ease the paradigm into practice in cities, embracing their complexities no less.

后，设计studio提出不同的建筑和城市设
以创造能实现和增强共享生活的空间。

四个系统和各自的设计方案将作为共享方
总体规划，并将其注入如切现存复杂的环
中。这种探索与设计可为未来的分享蓝图
先见与建议，以便采用类似的策略实践
范式。

REFERENCES

George, Cherian. 2000. Singapore: The Air-conditioned Nation. E
on the Politics of Comfort and Control, 1990-2000. Singapore: Lan
Books.

Harvey, David. 2012. Rebel Cities: From the Right to the City to the
Revolution. Verso.

Lee, Hsien Loong. 2014. "Transcript of Prime Minister Lee Hsien Lo
speech at Smart Nation launch on 24 November." National Info
Awards. Singapore: Prime Minister's Office Singapore.

McLaren, Duncan, and Agyeman, Julian. 2015. Sharing Cities: A Ca
Truly Smart and Sustainable Cities. Cambridge: The MIT Press.

See, Bridgette. 2014. Co-Creating Singapore: Hear What Citizens Ha
Say. May 14. Accessed April 2017. https://www.challenge.gov.sg/print/
story/co-creating-singapore-citizens-have-their-say.

Shatkin, Gavin. 2013. "Reinterpreting the Meaning of the "Singapore M
State Capitalism and Urban Planning." International Journal of Urba
Regional Research 116-137.

Smart Nation Singapore. 2017. Enablers. February 28. Accessed April
https://www.smartnation.sg/about-smart-nation.

1900 - 1940
LATE SHOPHOUSE STYLE

- Also known as "Singapore Eclectic" or "Chinese Baroque"
- Lavish ornamentations with mix of ethnic styles
- Blend of classical elements with Chinese symbolism
- Three window arrangement
- Use of bright colours

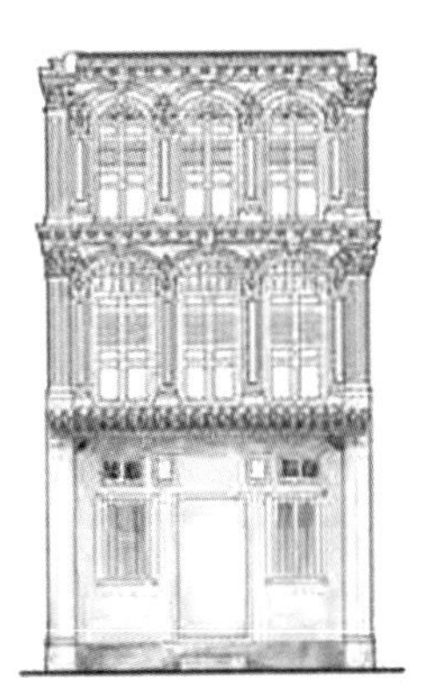

LATE 1930s
SECOND TRANSITIONAL SHOPHOUSE STYLE

- Simple design with some decorative elements
- Late-style motifs like colourful ceramic tiles are combined with art deco motifs like geometric balustrade designs

1930 - 1960
ART DECO SHOPHOUSE STYLE

- Inspired by classical motifs first developed in Europe and the United States in the 1920s
- Characterised by streamlined, geometrical design
- Most shophouses of this style feature a date plague

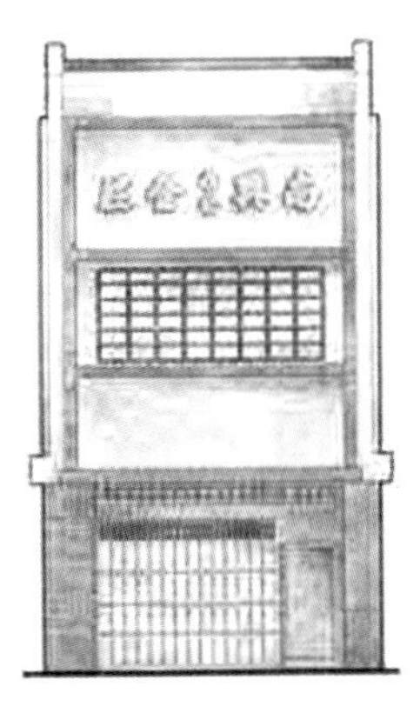

1950 - 1960
MODERN SHOPHOUSE STYLE

- Features thin concrete fins on their facades that doubles as air vents and simple decorations
- Shift towards more functional design and modern materials like concrete

The shophouse typology is dominant in Joo Chiat, and thus the studio seeks to preserve this in various proposals while establishing a new system in the masterplan.

如切区形形色色的店屋类型学是工作坊在设计许多新城市总蓝图的同时，也致力于保留的重要部分。

01

共享基础设施
SHARING INFRASTRUCTURE

PLAYGROUND

BY GRACE KOH
KAH SHIN &
ONG CHENG SIANG

游乐场

02

共享设施与服务
SHARING PUBLIC FACILITIES & SERVICES

THE LEARNING STREET
BY YUAN YIJIA

学习街道

PLAYSCAPE
BY HUNG YU-SHAN

游玩场景

03

共享经济生产与消费
SHARING ECONOMIC PRODUCTION & CONSUMPTION

REPAIRING VILLAGE
BY KENNY CHEN HAN TENG

合苑

共享文化与生活
SHARING CULTURE AND EVERYDAY LIFE

JOO CHIAT 2037
BY TAN JIA YU

如切 2037

PAYA LEBAR SQUARE
SINGAPORE POST OFFICE
SHARING INFRASTRUCTURE
CHANGI RO
TANJONG KATONG COMPLEX
GEYLANG ROAD
JOO CHIAT COMPLEX
JOO CHIAT ROAD
JOO CHIAT PLACE
KUAN IM TNG TEMPLE
CRANE ROAD
HAIG GIRLS' SCHOOL
HAIG ROAD
CEYLON LANE
DUNMAN ROAD
TANJONG KATONG SEC SCH
TANJONG KATONG GIRLS' SCH
TANJONG KATONG PRI SCH
ODEON KATONG SHOPPING COMPLEX
SHARING CULTURE AND EVERYDAY LIFE
SHARING PUBLIC FACILITIES AND SERVICES
SHARING PUBLIC FACILITIES AND SERVICES

JOO CHIAT MASTERPLAN

如切总体设计

Detailed site analysis are first conducted in groups based on the four identified dimensions of sharing to develop integrated urban design schemes addressing the issues.

With their proposed urban design schemes, individuals focus on the development of the above and production of architecture to develop an architecture proposal using selected building blocks and/or areas of the group work.

学生们首先分组按照四种不同的共享方式进行详细的场地分析，以制定解决这些问题的综合城市设计方案。

城市设计的方案中，学生们各自选定自己想研讨的设计范围或建筑以提交个人设计方案。

A NEW LEASE OF LIFE

By Grace Koh Kah Shin &
Ong Cheng Siang

新的租赁生活

Infrastructure and public facilities and services systems form the backbone of this studio's design exploration. The former seeks to integrate power, waste and transport systems into one and transforming them into resources contributed by the neighbourhood itself, which will serve public facilities and services. As such, this system of public contribution and distribution creates an all-accessible pool of resources that is equitable and just. It transforms the way the city produces and consume public infrastructure by creating a sustainable loop, constantly renewing resources and according waste a new lease of life. Furthermore, the involvement of the everyday people in infrastructural production potentially educates the masses and inculcates the culture of sharing in the neighbourhood.

基础设施和公共设施与服务系统的设计是该小组设计方案的核心。前者将电力、废物和交通系统整合为一体，将其转化为社区的公共资源。同时，新系统也为公共设施和服务提供基本资源。方案注重于集体贡献和分配制度，以创造一个公平公正的资源分配。通过可持续发展的循环体系，不断更新资源、废物利用，改变了城市生产和消费公共基础设施的模式。参与基础设施的生产也有着教育群众、传播邻里文化的效应。

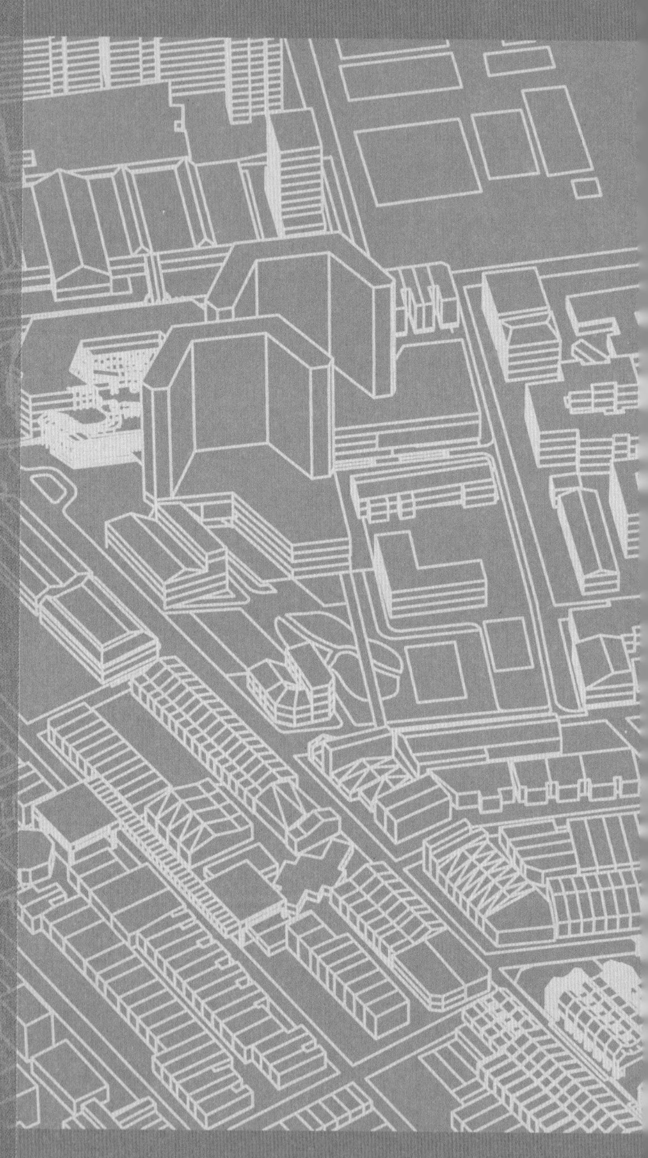

SHARING INFRASTRUCTURE

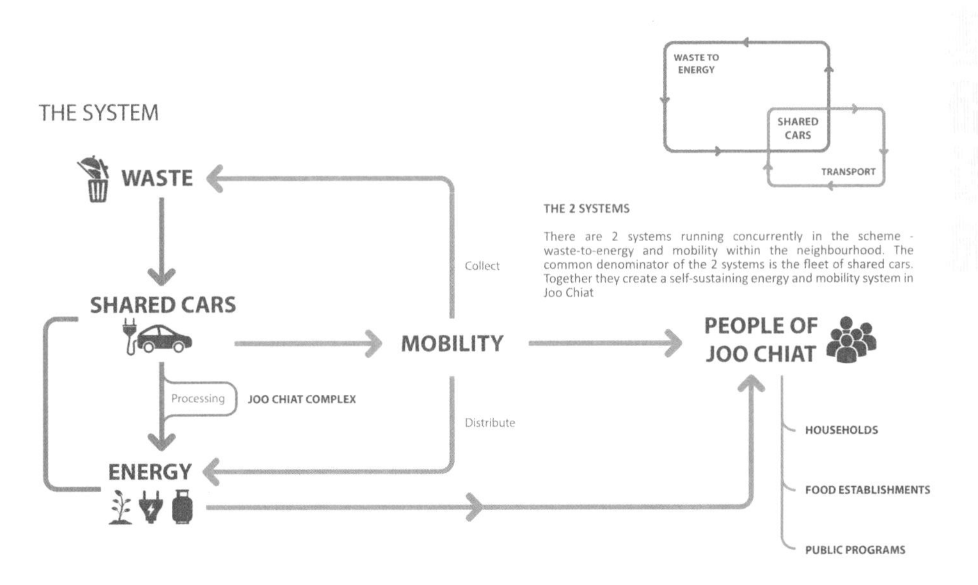

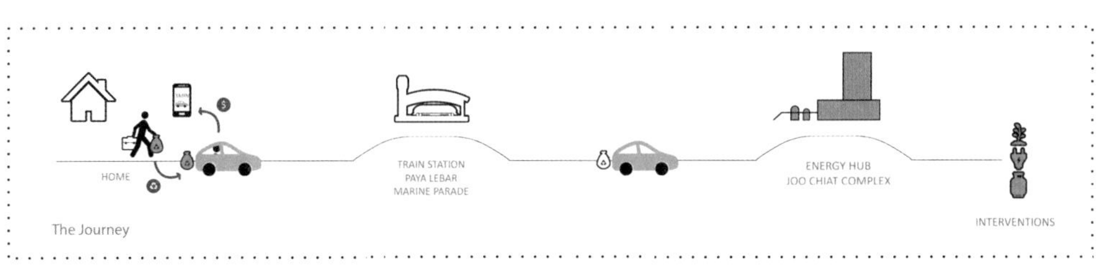

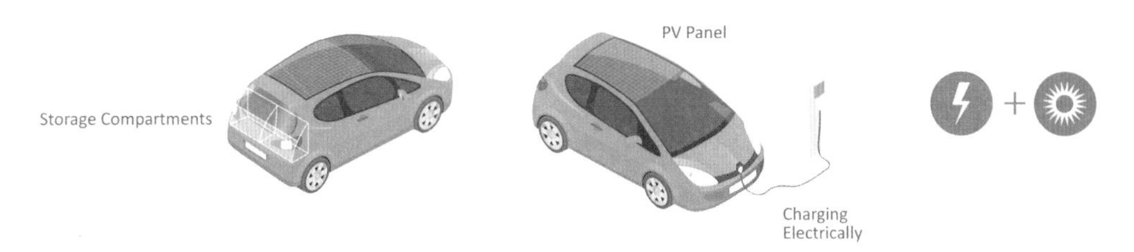

Top
The system: a 3 sector linear relationship from waste to cars to energy and fro, illustrating the flow of resources.

上图
系统：废物到汽车到能源的三类线性关系，说明了资源的流动。

Middle
The journey: the typical journey the shared car takes and its role in the system.

中图
旅程：共享汽车的旅程及其在系统中的作用。

Bottom
The car: how the car works as a carrier and transport vehicle; its energy requirements.

下图
汽车：汽车如何作为承运和交通工具，以及其能源需求。

This core of this system is converting food waste to energy for sharing; and this energy is primarily used to power a localized shared car system that in turn facilitates waste collection and energy distribution. There is hence a reciprocal relationship between transport and energy which both rely on each other.

At the centre of the system are the shared cars because it provides mobility for the system to run: collects waste, distributes recycled energy and also as a internal transport system for residents. in this scheme, this system materialises as a loop, a dedicated lane within the neighborhood for shared cars to run.

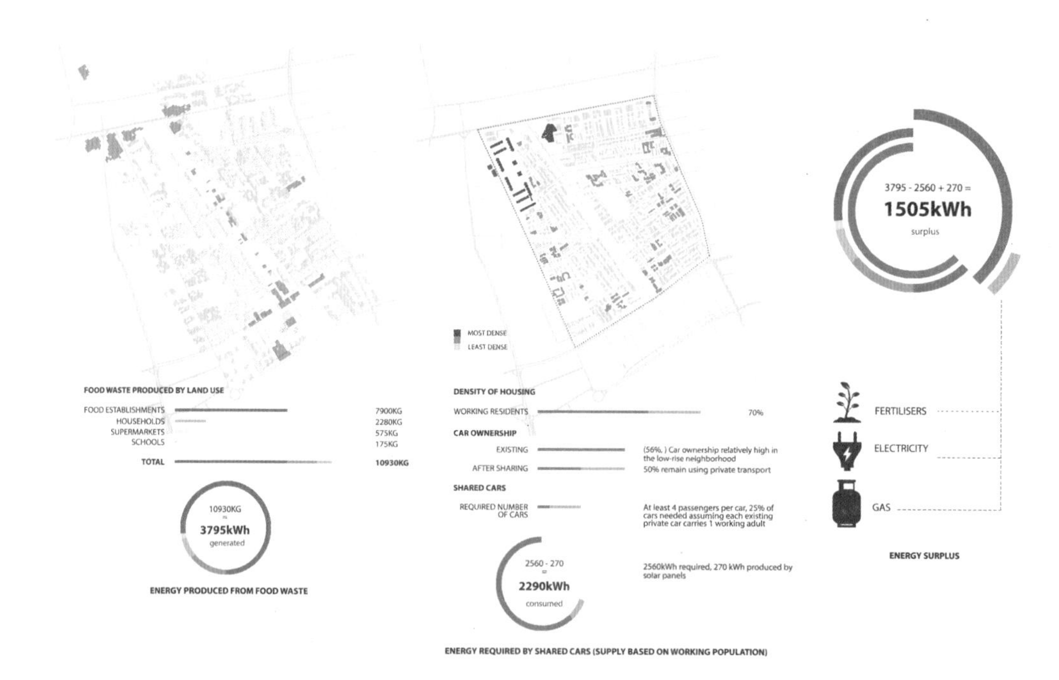

Viability of the system: surplus energy could be derived from food waste collected even after expending it on shared car system.

系统的可行性：在供应公用汽车燃料之余，仍有剩余的回收能源。

系统的核心概念是将厨余转化为共享能源，主要供公用汽车作为燃料，使得这些汽车能进行废物收集和能量分配。因此，运输和能源之间是相互支助的互惠关系。

系统的中心是公用汽车，它使系统拥有机动性而顺畅运行。其功能包括收集废物，分配回收能源，还可以作为居民的社区交通工具。在该方案中，社区被设计成为循环体系中的一条专用车道为共享汽车服务。

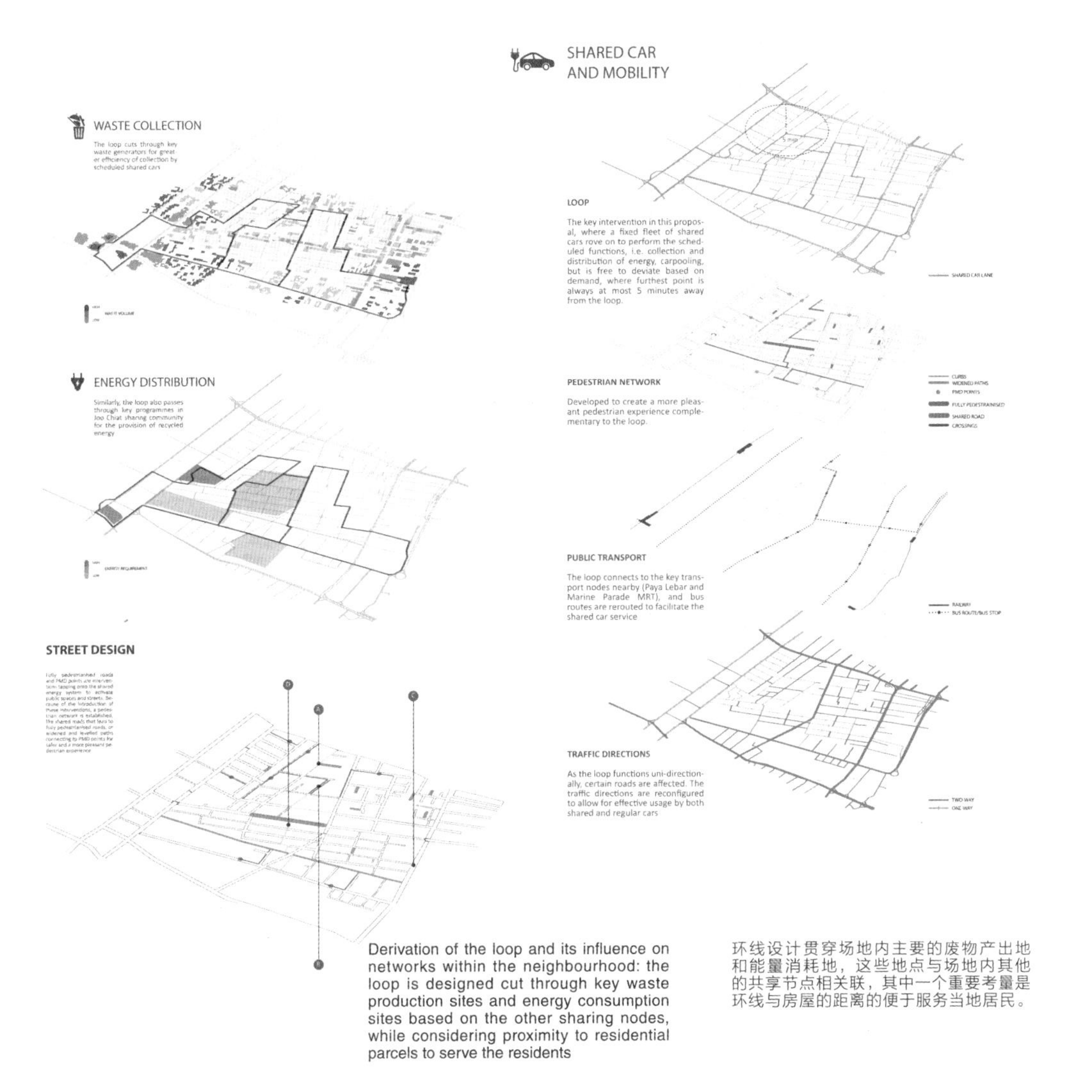

Derivation of the loop and its influence on networks within the neighbourhood: the loop is designed cut through key waste production sites and energy consumption sites based on the other sharing nodes, while considering proximity to residential parcels to serve the residents

环线设计贯穿场地内主要的废物产出地和能量消耗地，这些地点与场地内其他的共享节点相关联，其中一个重要考量是环线与房屋的距离的便于服务当地居民。

A. FULLY PEDESTRIANISED + SHARED STREET

New type of public space tapping onto opportunities on site, such as "courtyard conditions". The shared street creates a pleasant introductiotn into such spaces

B. LOOP + PMD SECTION

Typical loop section that value-adds the street upon introduction of interventions like PMD points

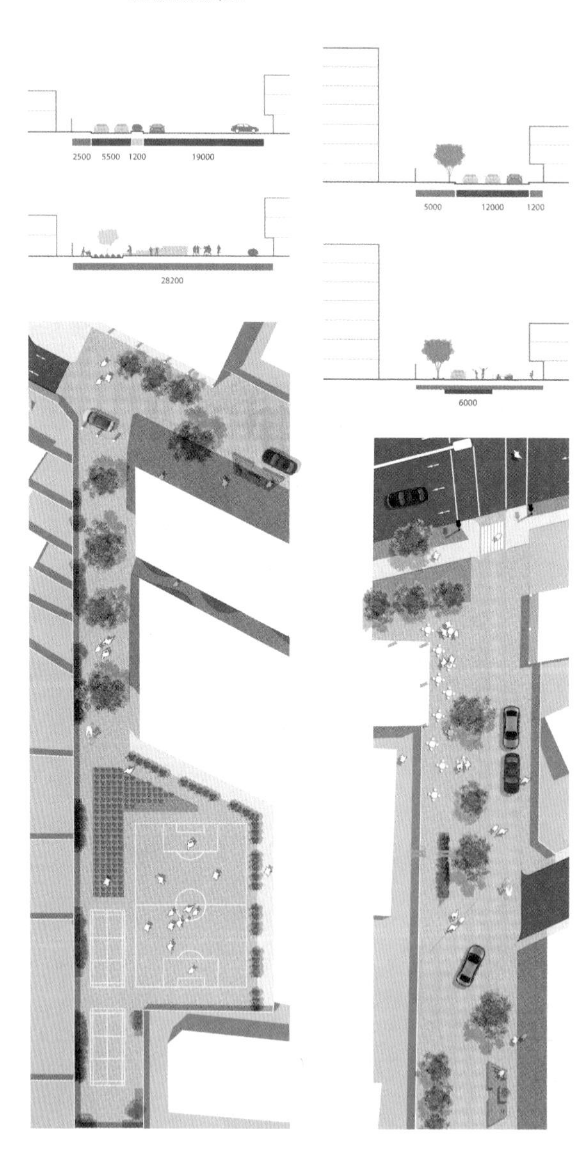

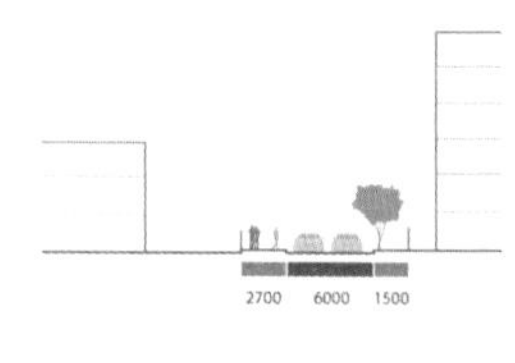

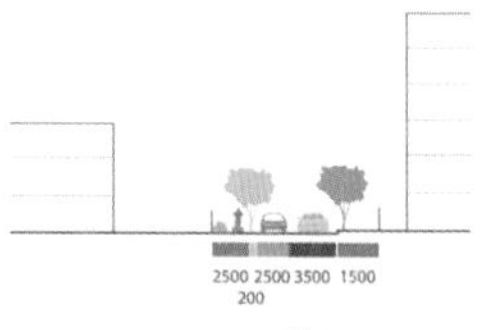

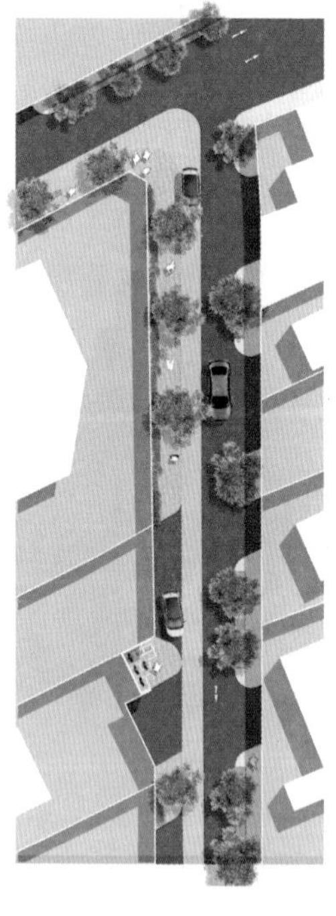

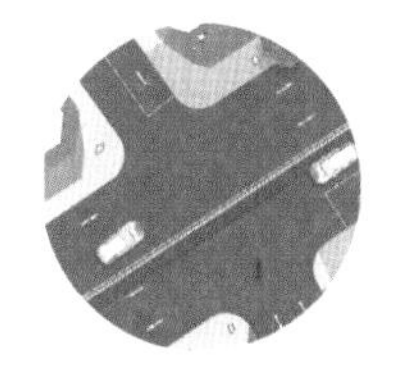

C. JUNCTION

Traffic signals are added upon junctions to facilitate shared car movement

D. CENTRALISED WASTE DISPOSAL

Central waste disposal point serving the F&Bs along Joo Chiat Road

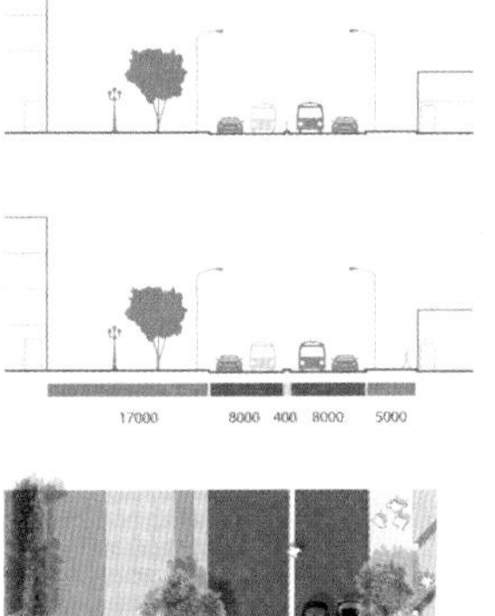

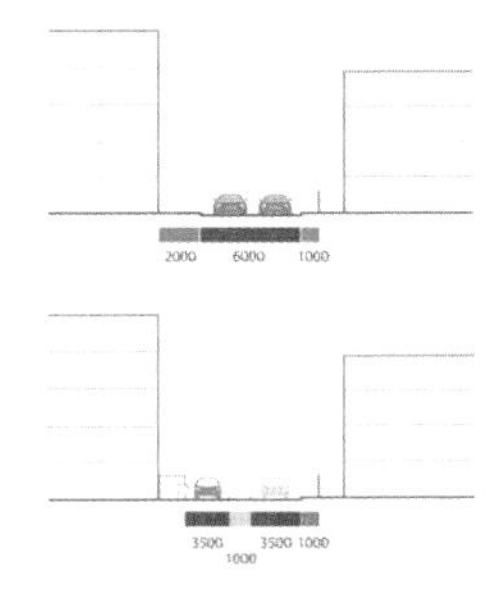

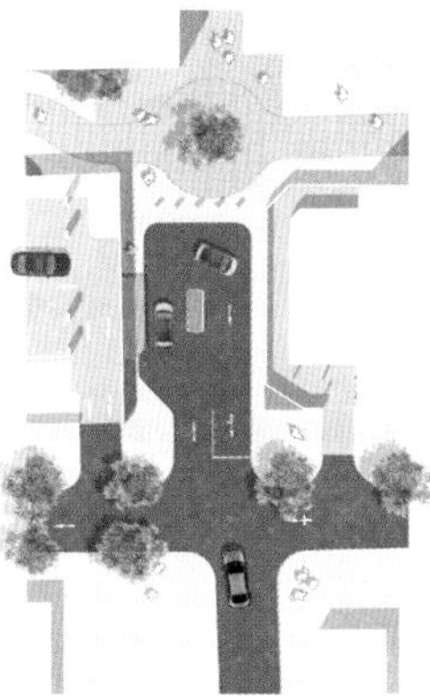

Surplus energy gets packaged into batteries distributed by shared cars to new interventions derived from opportunities sparked by the new infrastructure, like the loop. Street level interventions that are now made more attractive and viable as they are powered by shared energy: i.e. Community farming, outdoor class rooms.

剩余能量被注入电池，由公用汽车分配到新的公用场所。共享能源使包括社区耕作、户外课室等在内的场地能够运作，且更具吸引力和可行性。

游乐场

"Playground" retrofits new functions into the existing Joo Chiat Complex, and presents itself as the energy hub of the neighbourhood, symbolic of the new system in place. This proposal explores how to soften hard infrastructure, creating a refreshed gathering space for the residents while sensibly retaining its rich culture.

PLAYGROUND

THE DIGESTER LANGUAGE

The digester system gives opportunities to create usable spaces and programs around the respective parts. The **levels are varied** for practical purposes, however, it thereafter provides a more interesting experience for the visitor, engaging Joo Chiat Complex beyond the ground floor.

It repeats in **two scales** in this design - the larger system powers the neighbourhood, while the smaller system at the food centre powers the complex.

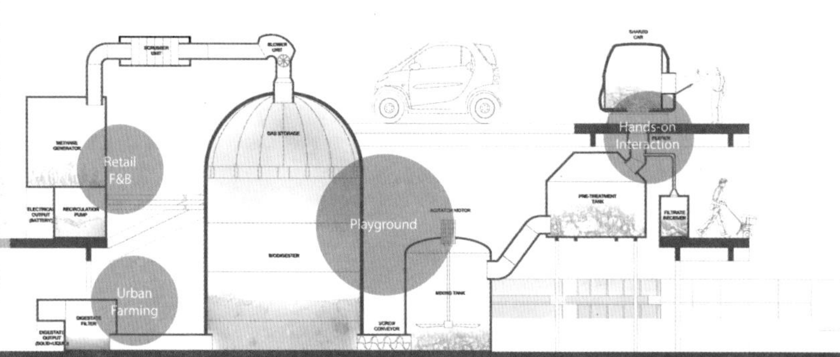

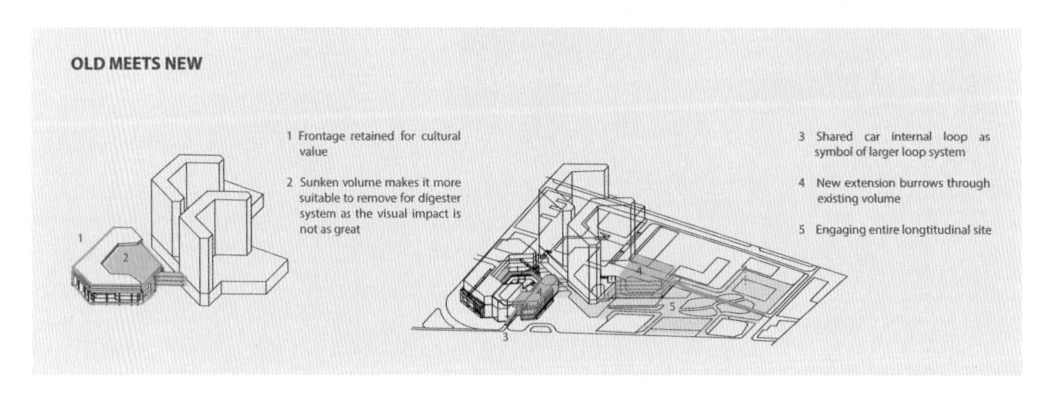

OLD MEETS NEW
1 Frontage retained for cultural value
2 Sunken volume makes it more suitable to remove for digester system as the visual impact is not as great
3 Shared car internal loop as symbol of larger loop system
4 New extension burrows through existing volume
5 Engaging entire longtitudinal site
1
2
3
4
5

Top
The digester system directs the main program of Joo Chiat Complex, providing opportunities for programmes to be created around the parts.

上图
蒸煮器系统指引着如切综合体的主要程序。

Bottom
The form and facade of Joo Chiat Complex is carefully manipulated to retain its cultural value amidst the insertion of new programmes.

下图
如切综合体的形式和立面通过精心的设计，在改造之余，保留其文化价值。

“游乐场”将新功能注入现有的如切综合体，并将其设为社区的能源中心，象征着场地内的新系统。这个设计方案探讨了如何将基础设施开放于大众，为居民创造一个新的聚集场所，同时保留着原有丰富的文化色彩。

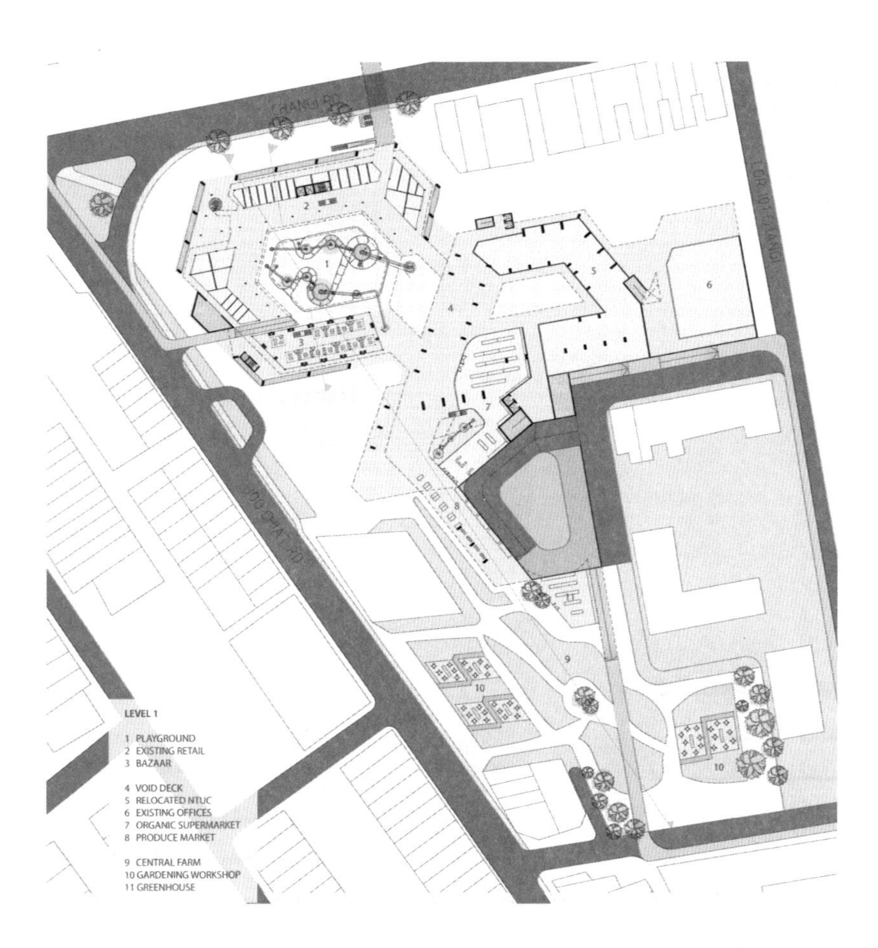

Level 1 Plan: an important part of the scheme as it activates the entire plot with various community programmes such as the playground and urban farming.

一层平面图：该方案的一个重要部分是如何通过游乐场、城市农业等各种社区活动运用整块土地。

Top Left
Perspective of Approach

左上图
建筑外观透视图

Top Right
Perspective of View from HDB

右上图
从组屋望外的透视图

Bottom
Longitudinal Section

下图
纵剖面

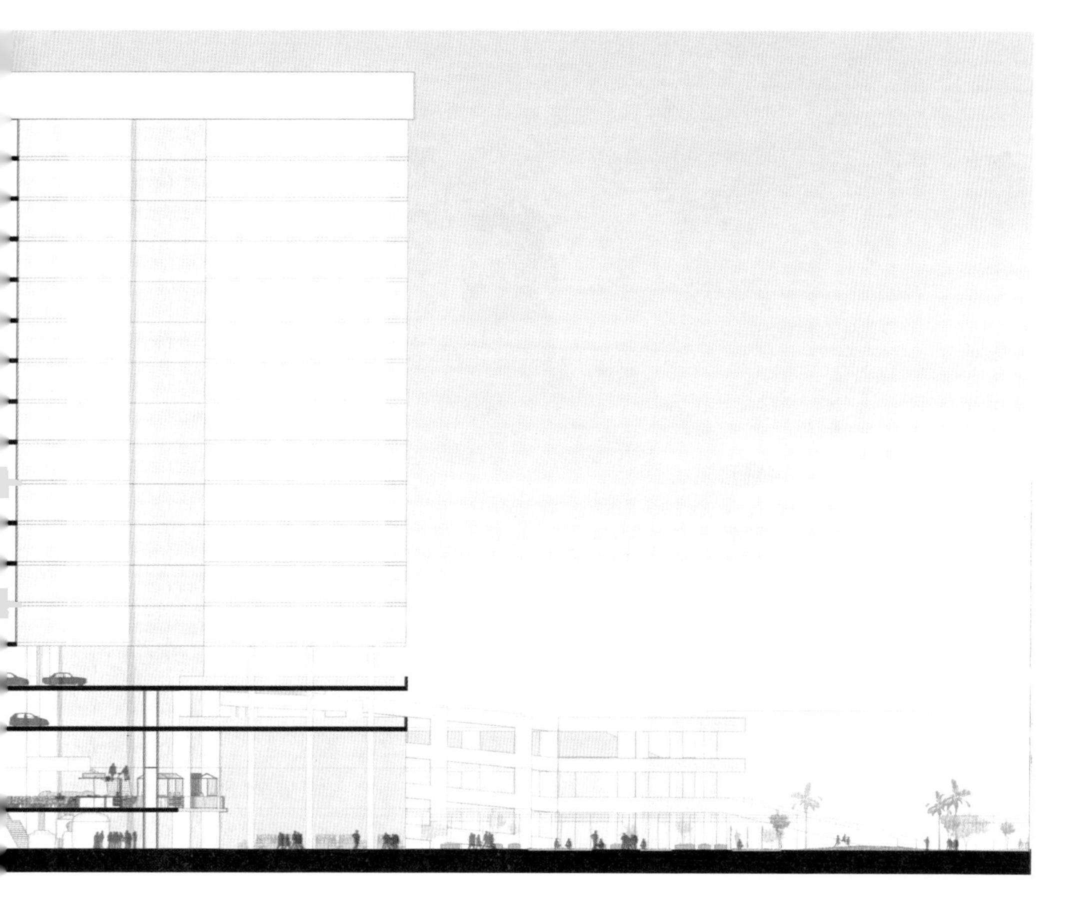

The result is to create a gathering space that is vibrant and playful. While functioning practically as an energy hub, "Playground" becomes a new public space, freshly injected into the neighbourhood. This is made possible through the negotiation between hard and soft infrastructure. This reconciliation created a new typology of space that introduced accessibility into a facility that is conventionally closed off.

Equally important is the consideration of the complex's cultural value. While moving forward with a new typology, the program that shaped the complex over the years remains and becomes a complementary element that enriches the new public program.

这设计项目最终希望创造一个充满活力又俏皮的聚会场所。除了作为社区的能源中心，“游乐场”也成了社区新的公共空间。基础设施和公共空间这两种空间的结合创造了一种新的空间类型，为常规被封闭的基础设施增加了可达性。如切综合体存在的文化价值同样重要。在设计新空间类型的同时，我们也考虑保留原有的活动并使其成为丰富新的公共活动中的要素之一。

NEIGHBOURHOOD LEARNING

By Hung Yu-Shan & Yuan Yijia

社区学习

Educating the neighbourhood of the daily workings of the system does not stop with observation and usage of infrastructure. A more cohesive program is put in place to encourage a neighbourhood-based learning, where the school is deinstitutionalised and expanded into the neighbourhood's premises. Teaching material is derived directly from public infrastructure and also residents themselves, creating a true sense of ownership within the neighbourhood. Education is no longer reserved for the select few, but freely accessible to all members of the neighbourhood.

在该共享系统的日常工作中，社区教育不仅仅停留在对基础设施的观察和使用上。一个更完善的教育方案需投放其中，学习不再限于学校范围内，而扩展到社区之中。教材也直接利用公共基础设施和居民本身。这做法不止让社区居民更有归属感，也让大众都可随时参与学习。

SHARING PUBLIC FACILITIES AND SERVICES

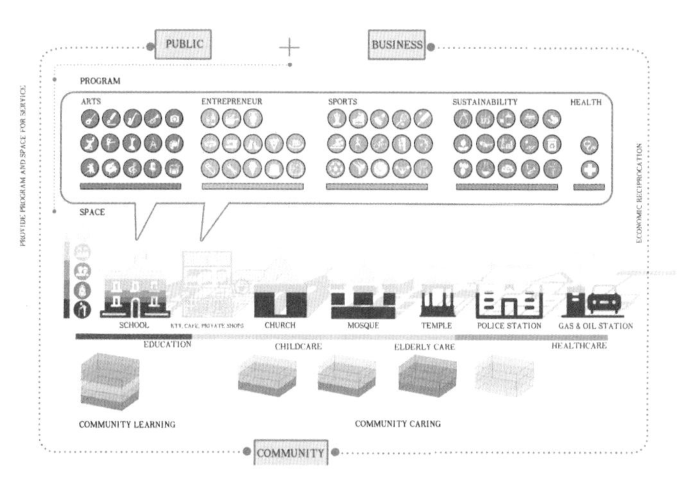

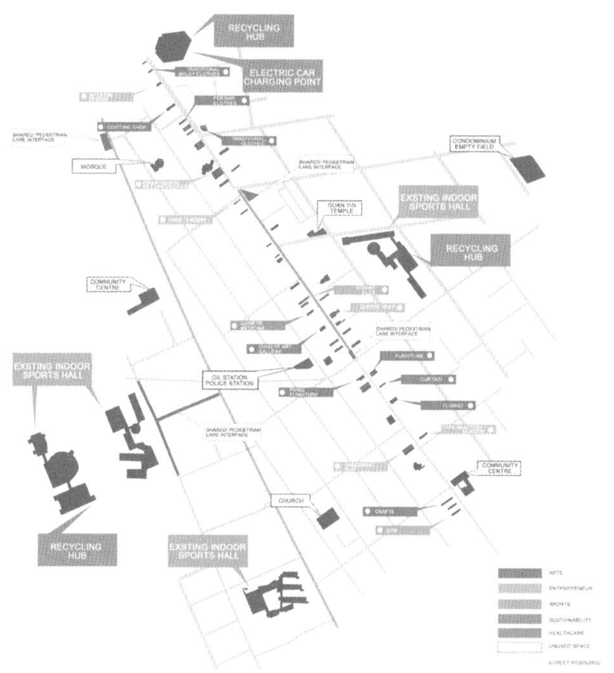

Top
Mapping showing suggestion of possible new programmes to existing infrastructure

上图
现有基础设施可获有的新程序

Bottom
Chart showing the overlap interests between public and business programmes

下图
图表显示公共活动和商业活动之间的利益重叠

Current infrastructure like schools, shop-houses and religious buildings usually carry only one-dimensional function. Hence there is potential to bring new activities to infrastructure beyond its intended function. This concept is being carried out by 3 main stakeholders: The community, public and businesses.

The hierarchy of our design intervention can be categoried into 3 levels: Primary node, secondary node and links. Primary nodes are the biggest design interventions which house the most number of programmes. Secondary nodes are smaller design interventions which houses few programmes to facilitate outdoor learning experience.

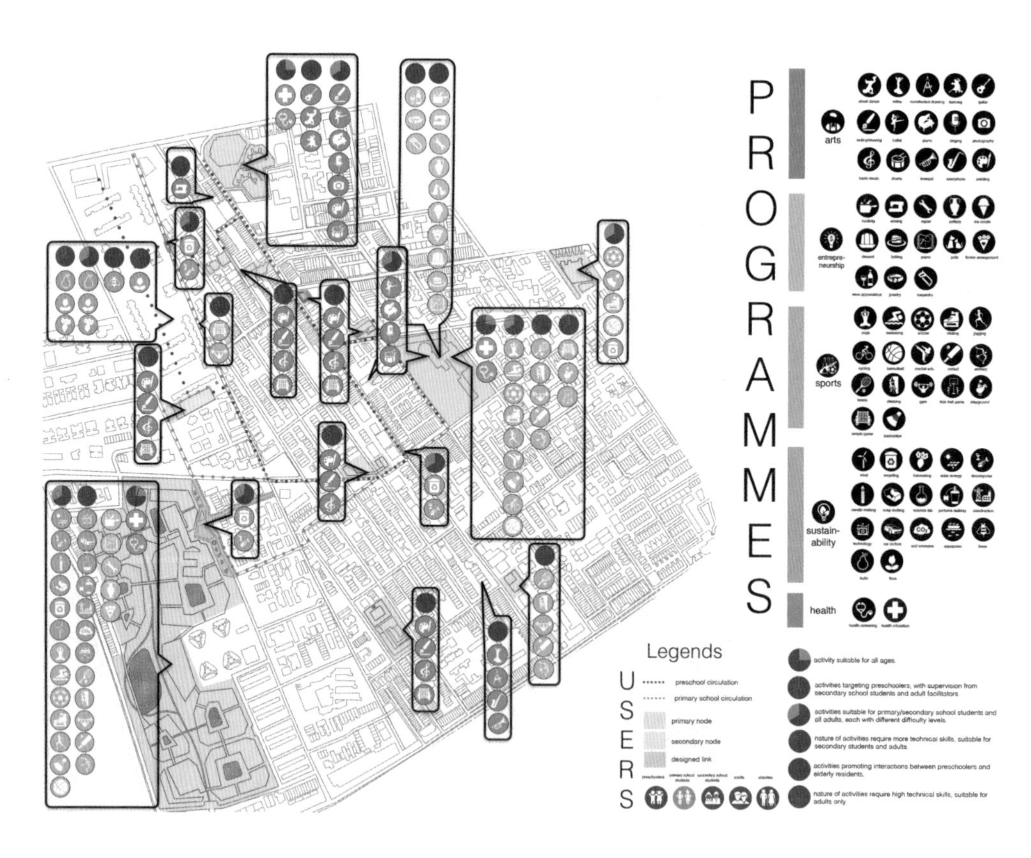

Detailed mapping of new programmes showing possible combinations and target audience at different nodes

不同节点上可有的程序组合和程序对象

如学校、商店和宗教建筑等类别的基础设施，通常只有单一功能。因此，为了更有效地运用这些建筑空间，大可注入新的活动与程序。这个概念涉及3个主要参与方：社区、公众和企业。

设计方案的结构可以分为3个层次：主要节点、辅助节点和连接。主要节点被着重设计为包含最多的活动程序。辅助节点相对设计干预较少以包含较少的活动程序，主要促进户外学习。

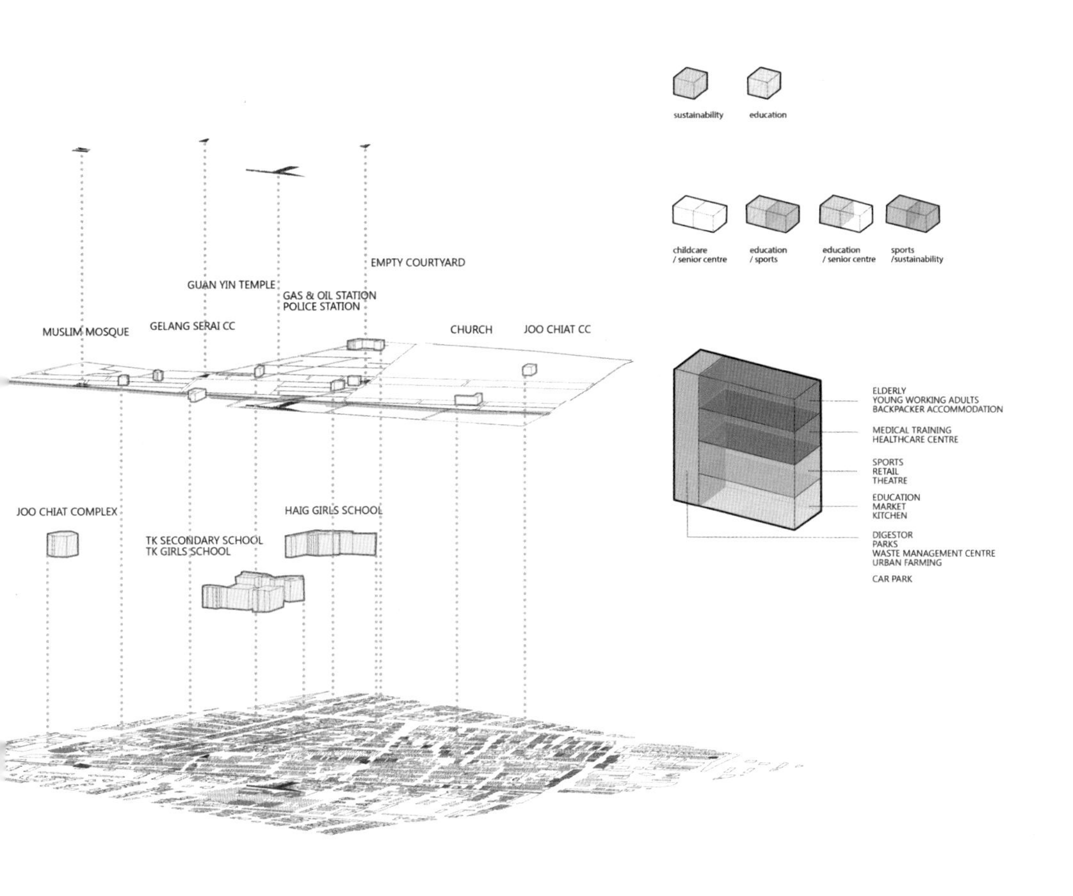

Left
Hierarchical assignment of selected nodes

左图
所选节点的分层分配

Right
Possible combinations of programmes

右图
可用的程序组合

知识街

The primary node sets the foundation of sharing of public services, where education, healthcare, elderly care and childcare all take place within one compound. The proximity of two public schools, Tanjong Katong Girls School and Tanjong Katong Secondary School, allows them to be combined to generate a large land plot for conceiving the primary node.

THE LEARNING STREET

Top
Perspectives showing learning street

上图
街头学习的环境与体验

主要节点为共享公共服务奠定了基础，教育、医疗保健、老年护理和儿童保育等都位于其间。两所靠近的公立学校，丹戎加东女子学校和丹戎加东中学可组合起来，在更大的空间范围内形成该方案的主要节点。

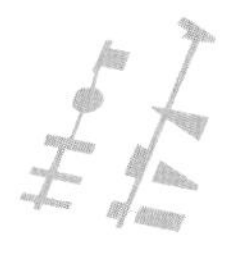

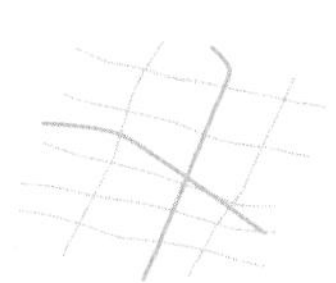

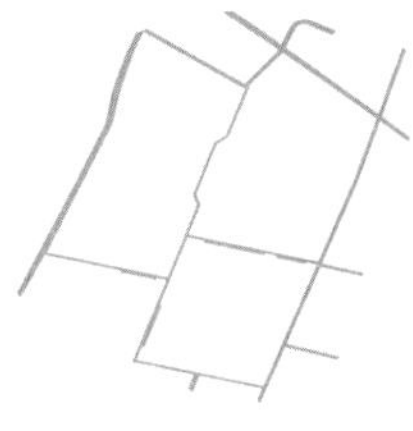

Bottom
New connections and networks within site

下图
地块内新的连接与网络系统

The proposal introduces the idea of "street learning". Existing school buildings are largely kept intact, re-adapted for healthcare and accommodation purposes. The design focuses on the NW-SE axis framed by the existing school building geometry.New vehicular and pedestrian routes are introduced for better connectivity, breaking down the large land parcel to bring convenience and contextualising the huge plot to its adjacent blocks.

New low-rise building blocks are used to define street edge, recreating the Joo Chiat street experience to reinvent the learning system. Apart from conventional compartmentalised learning, this design reimagines learning as more inclusive and spontaneous. Learning is now spread to the street and everyone can be a student and a teacher all at once.

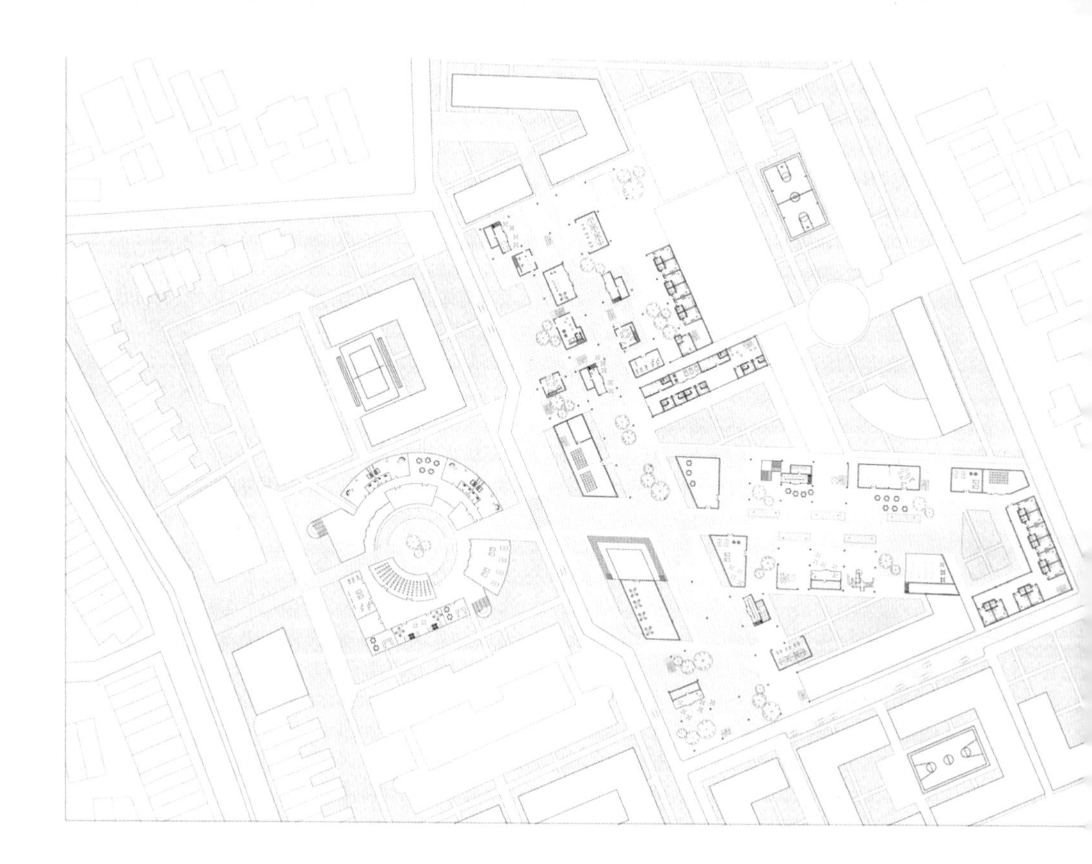

Site plan showing broken down street blocks

大块土地分隔后的面貌

该方案以“知识街”为主要创意。现有的校舍大多予以保持，重新植入医疗保健和住宿的功能。设计侧重于现有学校建筑所形成的西北—东南走向的轴线序列。在新的机动车和步行线路引入后，打破了原有大尺度地块的分割，为交通出行带来便利；同时加强了地块与相邻街区的联系。新的低层建筑被设计为界定街道界面，重新塑造如功街道的体验和学习系统。与传统的分阶学习模式不同，新的系统更重视包容性和自发性。学习扩展到街道，每个人都可以成为学生或者老师。

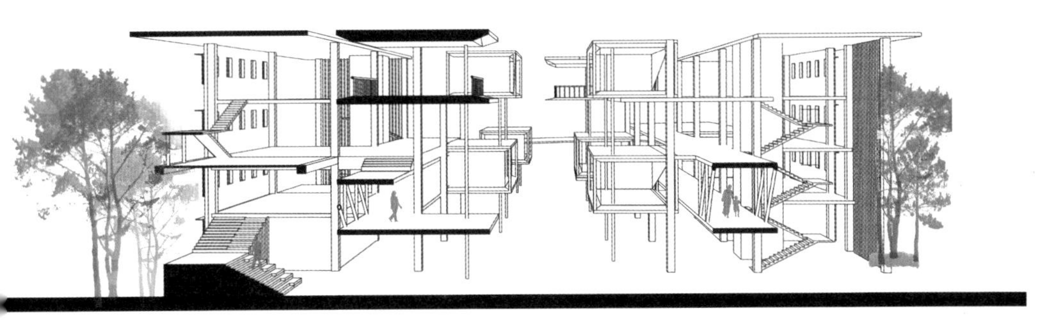

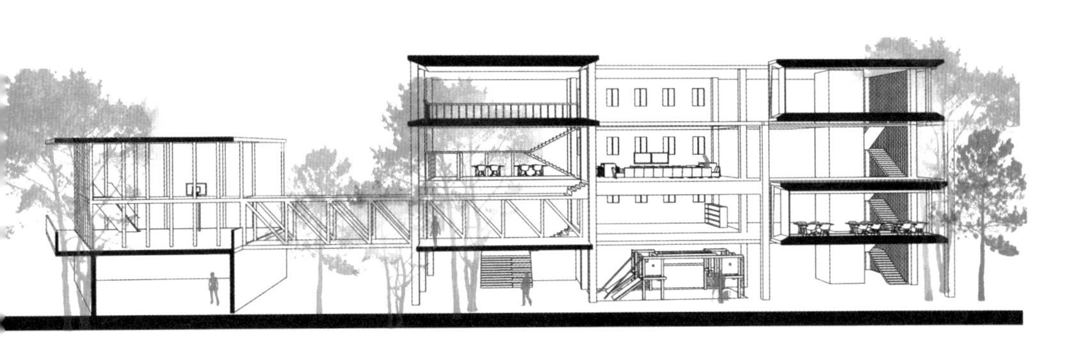

Sections showing building form and street learning experience

建筑形式和街头学习的环境。

游玩场景

The design of the secondary node takes on the concept of a playscape. where ramps and double volume, triple volume spaces make for differential and flexible spatial experiences. Open and well sheltered public spaces and public sitting areas and facilities encourage not only elderly and kindergarten children but the public and students to use the space as well.

PLAYSCAPE

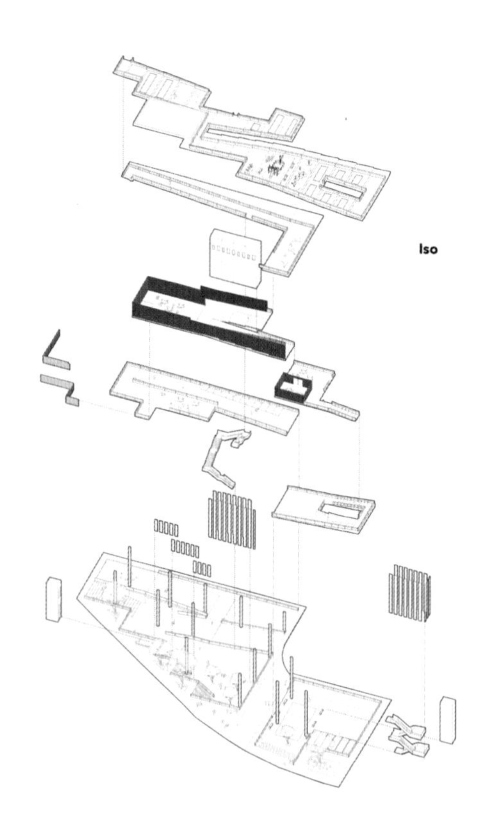

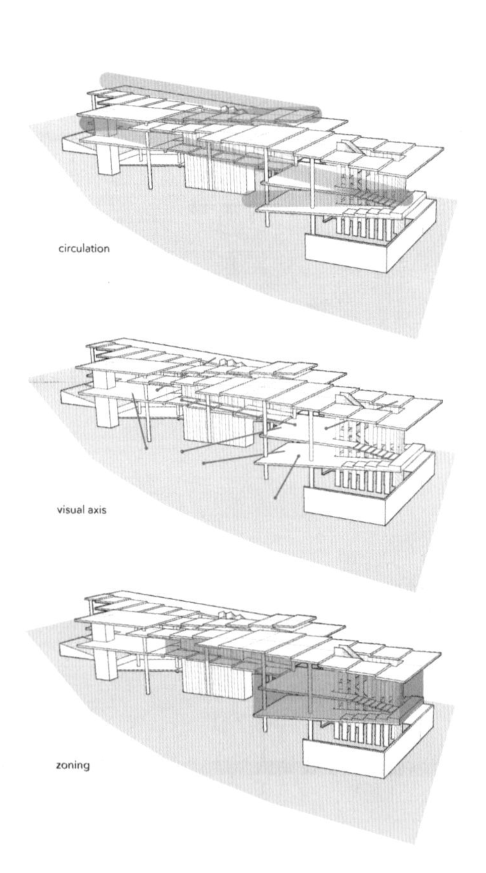

Left
Exploded isonometric drawing illustrating the openess of the structure

左图
空间结构的立体三维图

Right
Spatial analysis showing design's emphasis on visual exchanges and interaction amongst users

右图
表明设计者重视使用者视觉变化体验和互动体验的空间分析

辅助节点的设计采用了“游玩场景”的概念。坡道和其贯穿的二层、三层空间带来不同的谈话空间体验。开放的和被良好遮蔽的公共空间、公共座椅等设施供老人、儿童等人群使用。

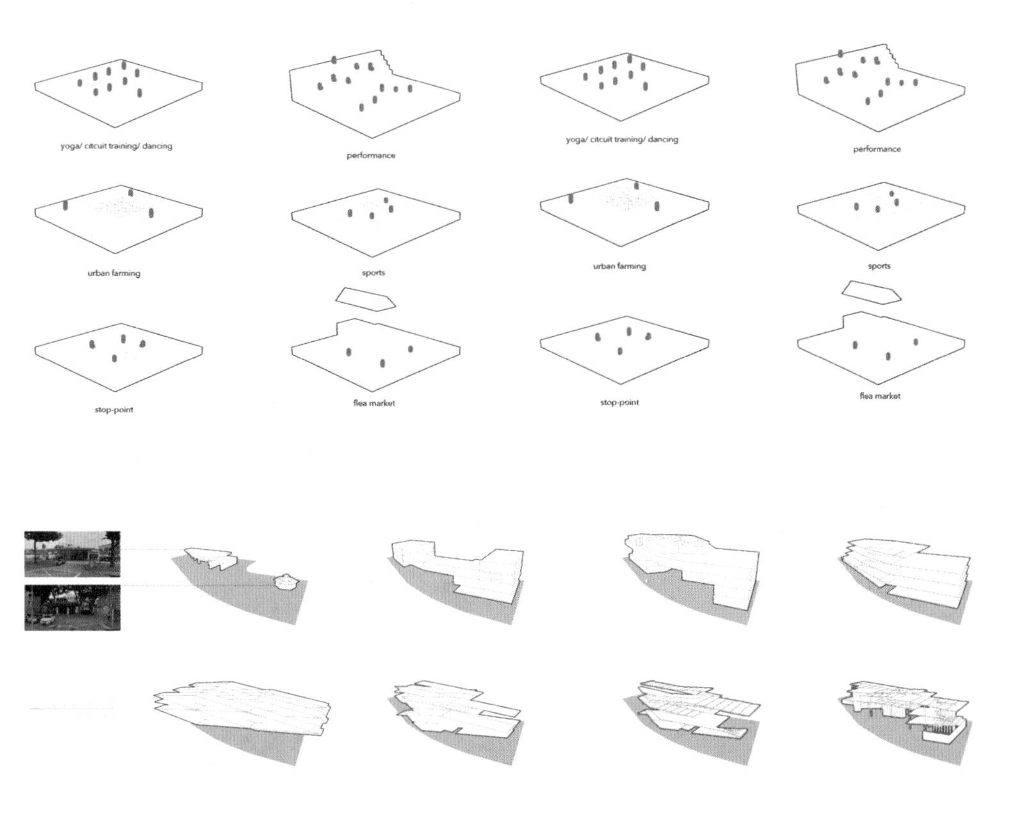

Top
Typologies of space within design and respective usage patterns

上图
空间设计和使用模式的类型

Bottom
Massing studies and refinement

下图
建筑群体分析和改进

The building can be loosely divided into 2 zones, the active zone and the quiet zone. The active zone encompasses elderly/ children daycare area with supporting programme such as commercial, urban farming and playground. The quiet zone encompasses the library and NPP, with supporting programmes such as seminar and learning spaces.

The ramp traverses the zones and hugely defines the design. It not only serves as a circulation, but also holding space for programmes, encouraging spontaneity. The ramp running throughout the building opens up the space, and it being continuous to the roof makes for its accessibility, maximising usable space for the public and intended users.

建筑可大体分为两个区域，即活跃区域和安静区域。活跃区域包含了老人、儿童的日间照料及商业、都市农业和游戏场地等支持项目活动；安静区域包含了图书馆等伴随着讨论和学习空间的相关支持性项目安排。

坡道穿过不同区域，在很大程度上界定了该设计的特点。它不仅仅是一个交通线路，更是为不同项目提供空间、鼓励自发性的系统。坡道穿越建筑，使空间开敞，与屋顶相连增加其可达性，最大化地为公众和目标用户提供可用空间。

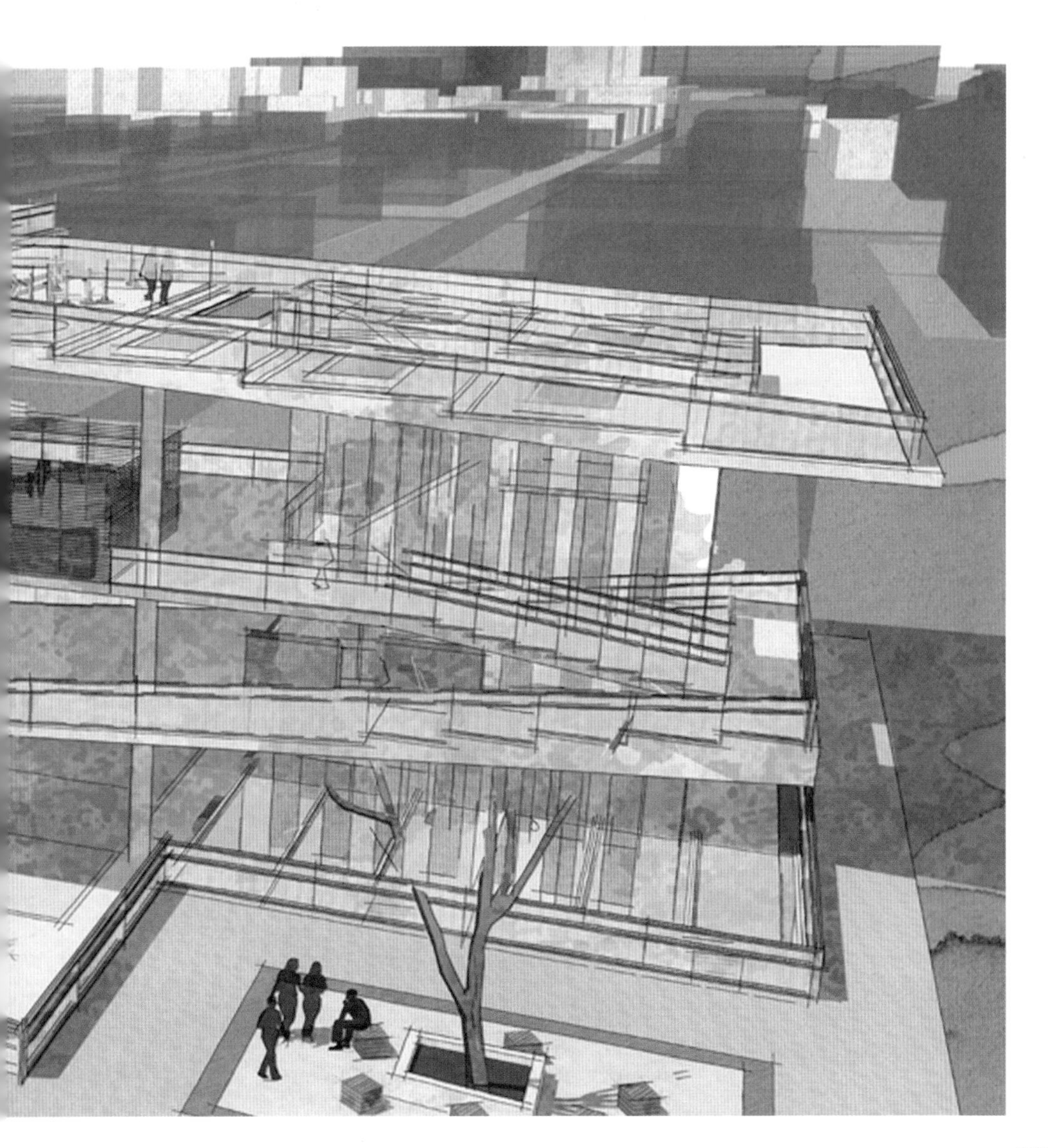

CO_4 JOO CHIAT
如切 CO_4

By Cherie Chan Wan Qing,
Kenny Chen Han Teng,
Yin Aiwei

In establishing a sharing system for the basic mechanisms the neighbourhood runs on, the studio is able to subsequently build on a system of economic production and consumption departing from capitalistic endeavors and one that furthers the sharing paradigm. Economic production in the new sharing neighbourhood is fuelled by a common pool of resources borne out of shared infrastructure and facilities, and therefore less concerned with profits than the intrinsic value of resultant goods and services. This set of ethos follows through with consumption, as consumers and users become less concerned with the possession of goods and services than consuming such, as seen from the slew of sharing programmes such as co-working and co-living.

当共享资源为邻里社区建立基本机制时，设计studio可以在经济生产和消费体系的基础上，以资本的角度来进一步扩大其共享范例。由于新共享资源社区的经济收益是源自于共享基础设施和设备等资源，较原有的商品和服务而言，较少的关注其利润，这种新思潮将随着消费量的增加而扩散开来。从一系列共同工作与共同生活的活动模式可以看出，相对使用权而言，人们已没那么在意这些资源和服务的所有权了。

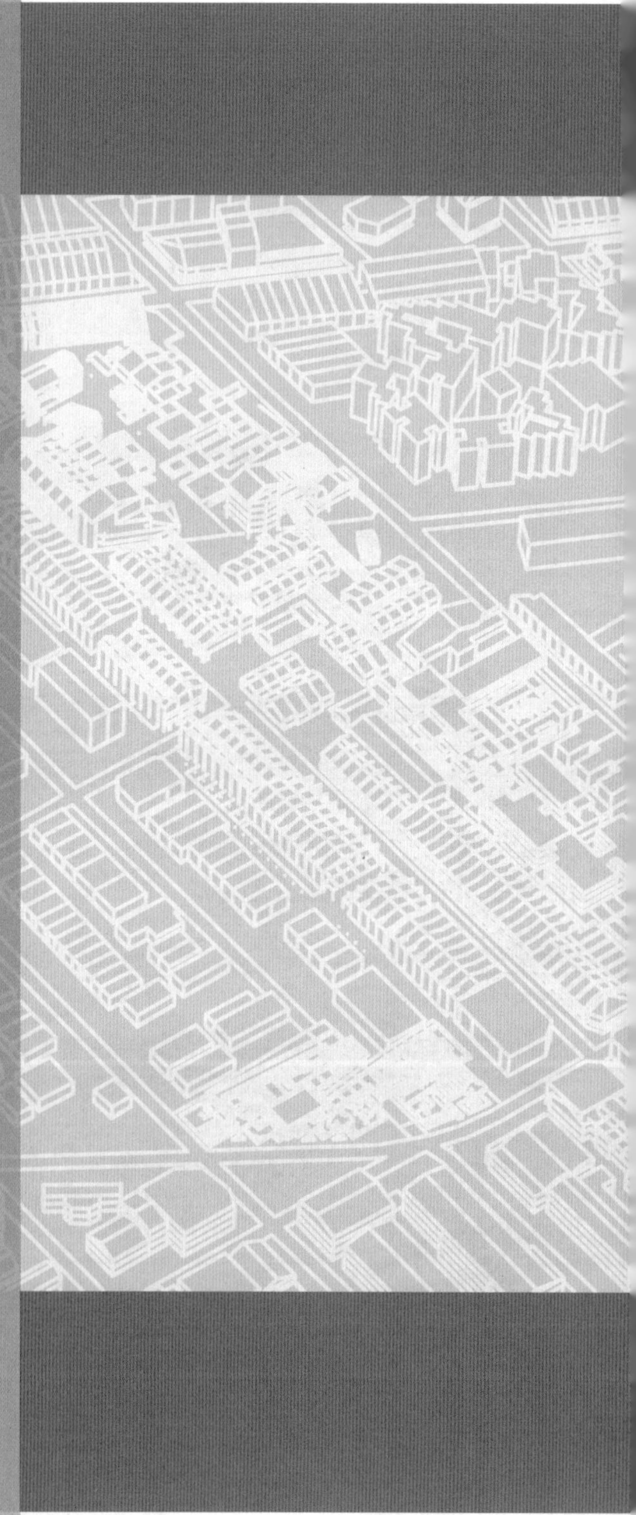

SHARING ECONOMIC PRODUCTION AND CONSUMPTION

TIER 03 >

PEDESTRIANIZED STREET
CO-PRODUCTION
COLLABORATIVE CONSUMPTION

C04 SQUARE

TIER 02 >

SERVING THE IMMEDIATE
PRECINT AREA WITH THE
REQUIRED SECONDARY PROGRAMS

DISTRIBUTED C04 HUBS

TIER 01 >

FACILITATED BY CLUSTERS (NORTH, SOUTH, EAST, WEST)
SERVING THE WIDER JOO CHIAT NEIGHBOURHOOD
WITHIN WALKING DISTANCE FOR ALL

CONNECTED ECOSYSTEM NETWORK

Multi-tiered approach towards achieving sharing economic production and consumption.

透过多层次途径去实现共享经济生产与消费模式

In this dimension of sharing, a CO_4 Joo Chiat Vision Map is proposed. CO_4 here refers to Co-Production, Co-Consumption, Community and Collaboration. It is hoped that the CO_4 system can harness the potentials of the overarching sharing network in the neighbourhood through its multi-tiered approach, integrating aspects of hard and soft infrastructure in its workings.

Four central hubs dubbed the CO_4 Hubs - CO_4 North Hub, CO_4 South Hub, CO_4 East Hub, CO_4 West Hub, are strategically located across the neighbourhood through re-programming of the selected existing buildings within walking distance from households. Programs are characterized according to the needs of the community within the vicinity.

A stretch of Joo Chiat Road is pedestrianised to become the central CO_4 Square, as a form of closing the loop between the distributed

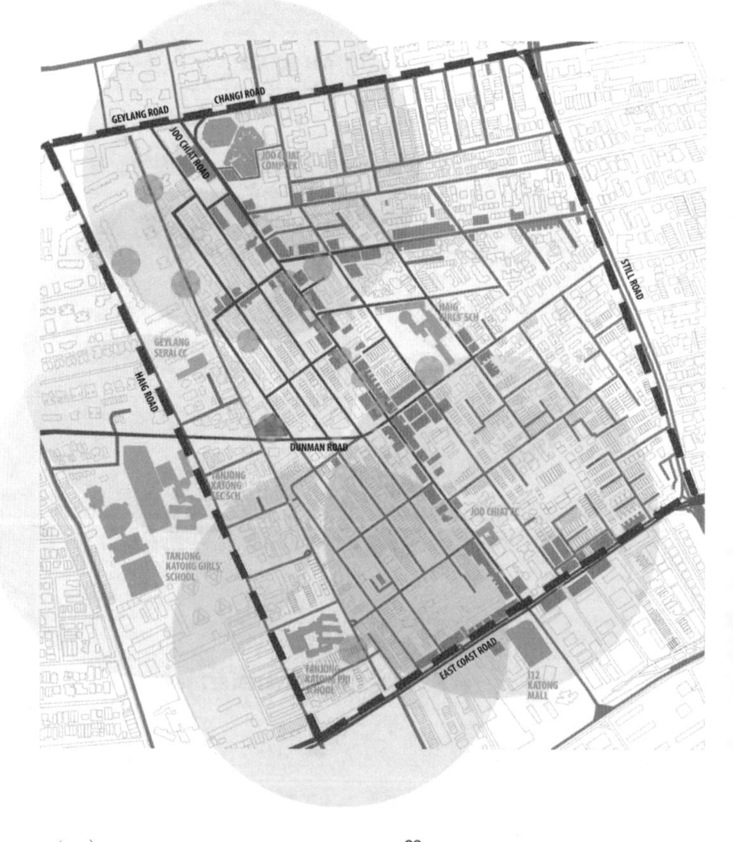

01.
A connected ecosystem network forms the fundamental backbone with four CO_4 Hubs distributed across the neighbourhood within walking distances. Contribution from households and small businesses ensures participation facilitated by the shared transportation.

（一）
CO_4的四个区域中心有效紧密地结合成一个系统，彼此间只隔着步行的距离。共享交通资源促进了小型企业和居家人士的积极参与。

02.
Distributed hubs with customized services provide immediate facilities and services to the precinct areas, integrating shared infrastructure with facilities.

在上述共享范畴中，我们推出了“CO_4如切”的计划。在这里，CO_4指的是共同生产、共同消费、社区和协作。我们希望“CO_4系统”可以利用邻里社区共享网络的潜力，透过多层次的途径，将各种硬基础设施和软基础设施整合在它的运作模式中。

“CO_4中心”是由四个中心枢纽组成：CO_4北部枢纽、南部枢纽、东部枢纽和西部枢纽，它们策略性地分布在从家里便能步行到达的距离范围内，让现有的建筑物重新组织于该社区附近。同时，方案的特点和所提供的服务在于根据附近社区的需要来设计。

如切路的一段将延伸成行人专用区，以作为CO_4的中央广场，并将所有的区域中心联成一环。所有保留的住宅、店屋和酒店建筑，将会分布在CO_4广场范围内。

在这里，城市和建筑环境将被规划成三种不同空间，划分后的空间用以适应

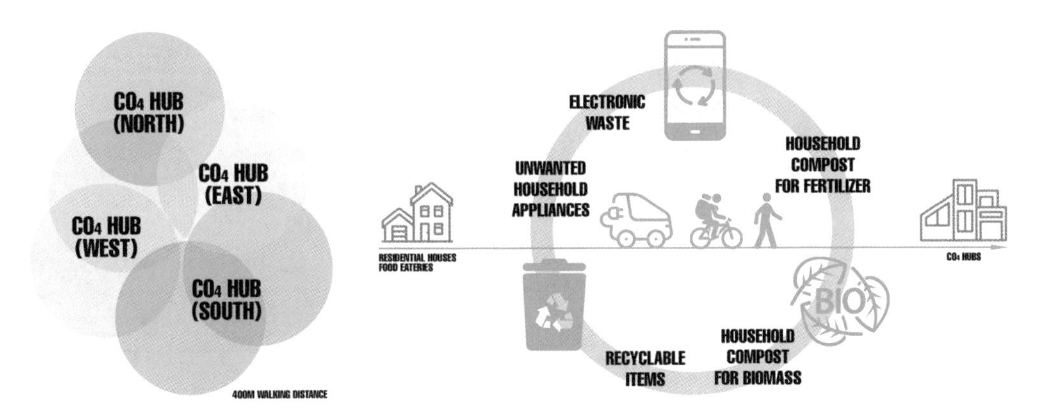

（二）
具有特定服务的区域中心分布网结合了共享资源的基础设施，能为人们提供快捷方便的设备与服务。

03.
Within the CO_4 Square is a myriad of conserved heritage shophouses, residential buildings and hotels, providing the necessary collaborative and participatory activities to occur.

（三）
由于多数保留的建筑遗产，包括店屋、住宅和酒店都处在CO_4广场范围内，为促进人们的参与及合作提供有利条件。

regional hubs. Within this CO_4 Square are the conserved heritage shophouses, residential buildings and hotels.

Here, the urban and architectural environment is manipulated and refined through dividing the area into three different plots to adapt for the different emerging new lifestyles defined by sharing, namely in the dimensions of sharing production through urban farming, sharing consumption through new residential living, and finally how community collaboration in a repairing village through physical and virtual marketplace for work could potentially provide the linkage to create a holistic approach in addressing the four proposed COs.

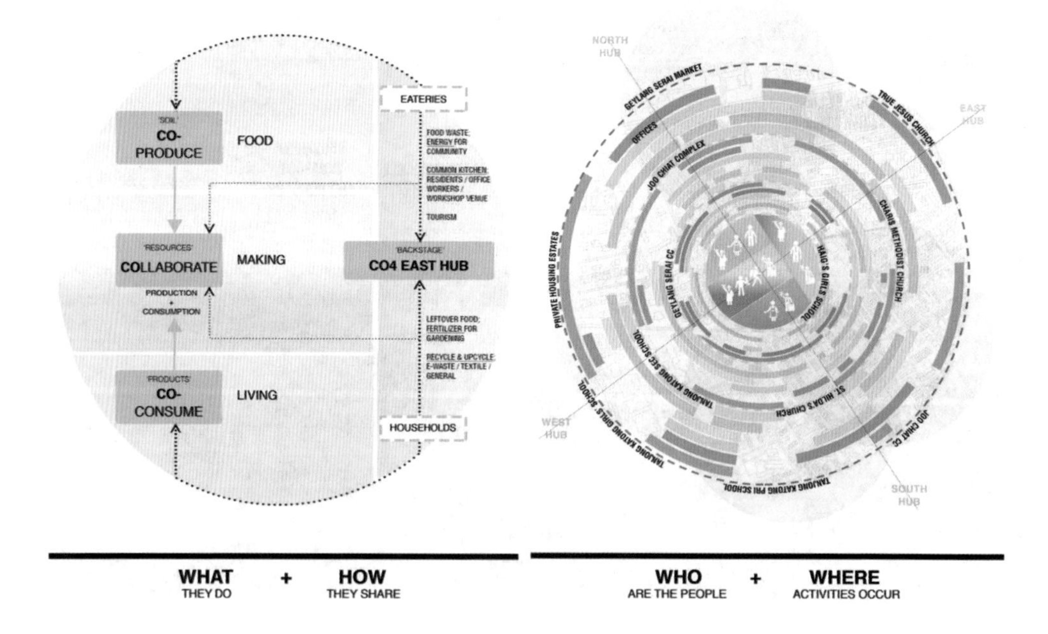

Within the dimension of sharing economic production and consumption, it is further narrowed down into co-production, co-consumption and collaboration.

从共享经济生产与消费的基础延伸下去的，将会是合作开发、共同生产与消费。

The possible sharing activities within the respective narrowed classifications are proposed, including the targetted communities.

共享资源活动将在其他目标社区发展开来。

新兴的共享生活，主要是通过都市农业的共享生产与通过新居住方式的共享消费，以及最终社区如何通过物质和虚拟的市场，在修缮的村庄中营造强调COs的整体途径。

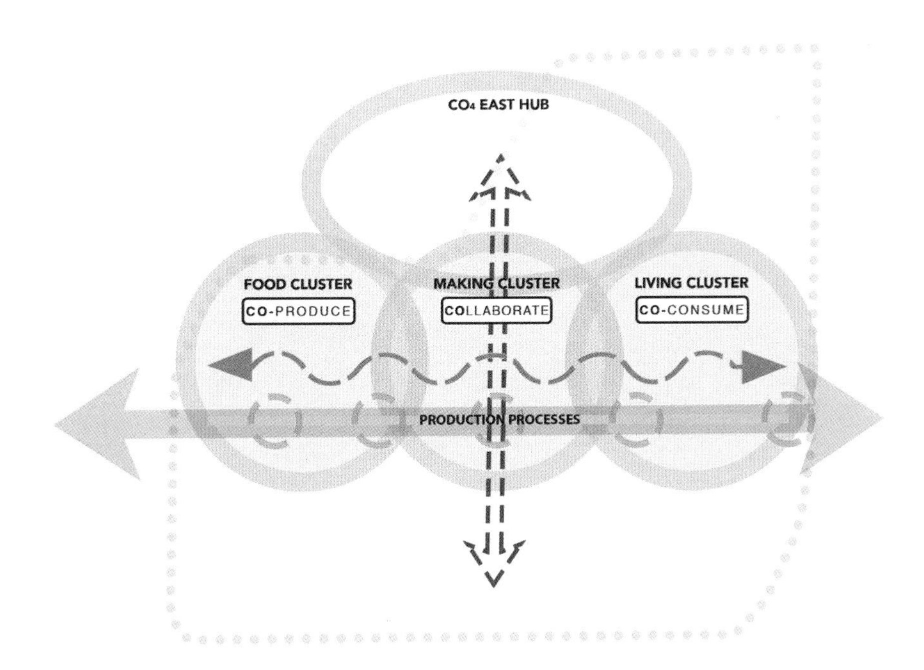

Spatially, the respective three classifications of co-production, co-consumption and collaboration are inter-related and backed by the CO_4 East Hub.

在空间上，共享生产、共享消费和合作这三类要素相互联系、支撑，分布于CO_4东节点上。

REPAIRING VILLAGE

Repairing Village seeks to provide a physical and virtual market place for work within the Joo Chiat neighbourhood.

The rich architectural heritage, fine-grained urban fabric and concentration of residents of various cultural backgrounds provided the basis for sharing practice within the dimension of sharing economic production and consumption.

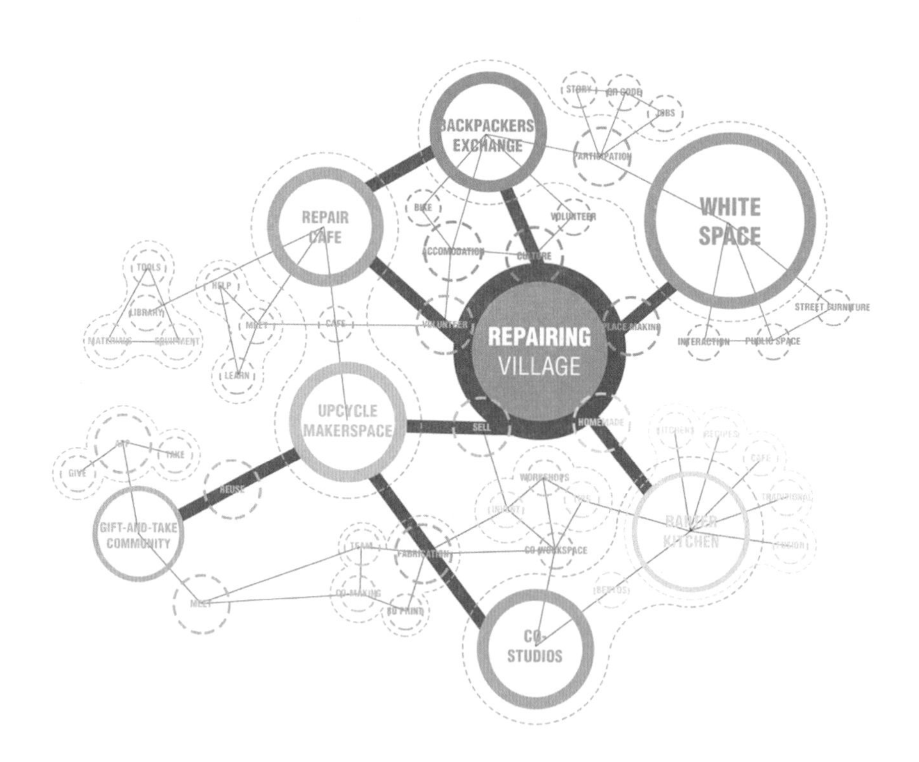

合苑旨在为如切社区附近提供实体和虚拟的交易场所。

丰富的建筑遗产、细致的城市格局以及各种不同文化背景的居民聚集，为共享资源提供了经济生产与消费的实践基础。

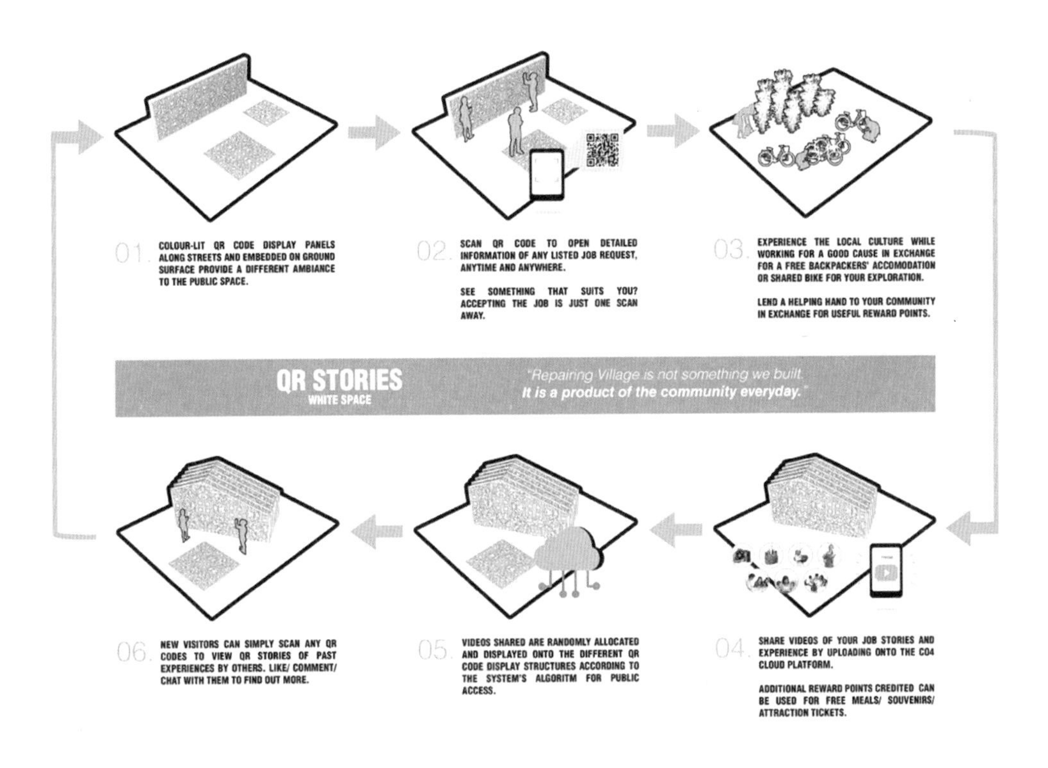

Knowledge and facilities sharing is encouraged through different work processes related to repair and upcycling in a promotion towards a circular economy redesigned to accommodate new lifestyles and enhance exchange within the neighbourhood.

It is hoped that the mentality of a throw-away culture can be improved through inspirations on the diverse opportunities available with repairing and upcycling, and how in combination the system can provide a concerted imagination of a collaborative lifestyle.

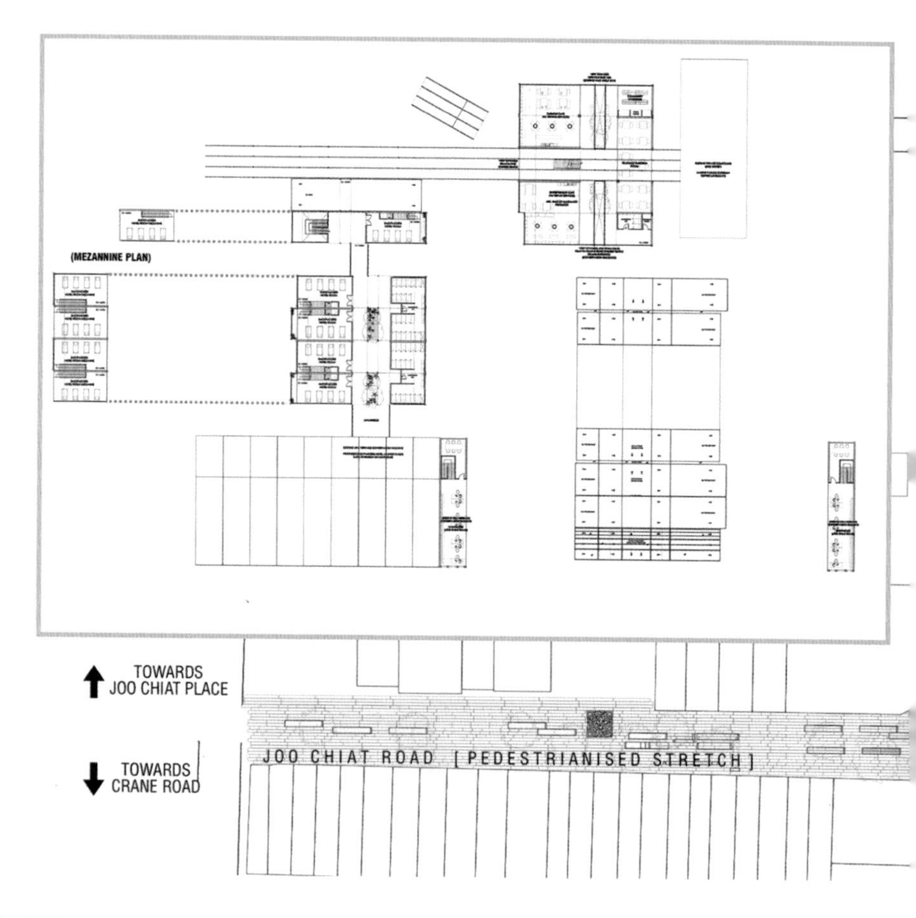

Second Storey Plan 1:200

二楼平面图 1:200

修复与再造的不同工作流程鼓励了人们共享知识和设施，从而促进循环经济适应新生活的方式，并加强邻里间的交流与合作。

此设计方案希望通过修复和循环使用的各种机遇所带来的启发，来改善人们原有的用后丢弃的传统，以及展现它如何在整体系统上综合协作，提供整体式的生活方式。

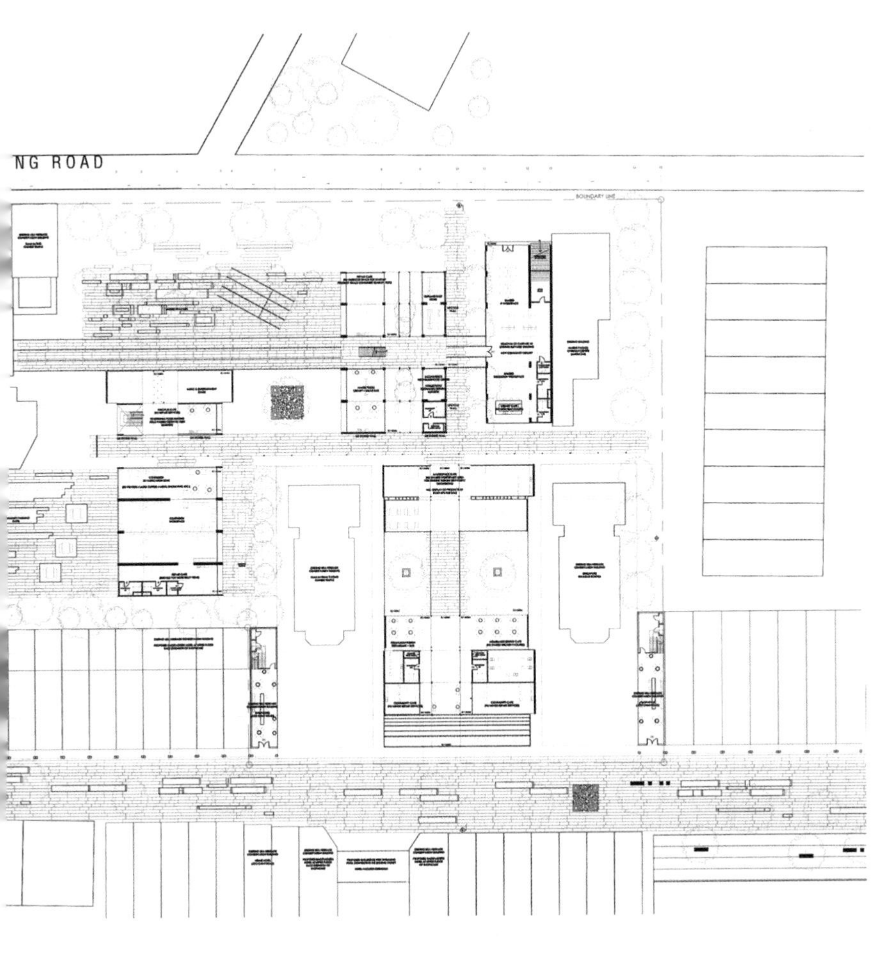

Site Plan / First Storey Plan 1:200

底层平面图 1:200

Top
Design Impression

上图
设计示意图

Bottom
Section A-A

下图
剖面图 A-A

AUGMENTED REALITY

By Tan Jia Yu & Ye Baogen Josiah

增强现实

On an individual level, the everyday living routine of a resident will evolve with the emergence of the sharing paradigm, and hence domestic environments should adapt to such. The studio therefore sought to envisage new modes of private living that complements the sharing neighbourhood. This is inexplicably linked to the culture existing in Joo Chiat as well. Where the shophouse used to be the basic unit of domestication, its roots are anchored deep in the collective consciousness of the residents. With the means of augmented reality, the studio attempts at preserving heritage and culture even as modes of living evolve and change with the emergence of the sharing culture.

在个人层面上，随着共享范例的出现，居民的日常生活作习将会随着急剧改变的环境而调整。因此，设计studio也需谋求如何为当今新的私人生活方式做出相应的补充，以便于配合共享社区。这也对如切现有的文化带来冲击，例如商屋以前是居民生活的基本单元，这一集体意识早已根深蒂固。运用增强现实的方法，设计Studio尝试保护地区遗产和文化传统，并将其融入随着共享文化兴起的生活变化模式之中。

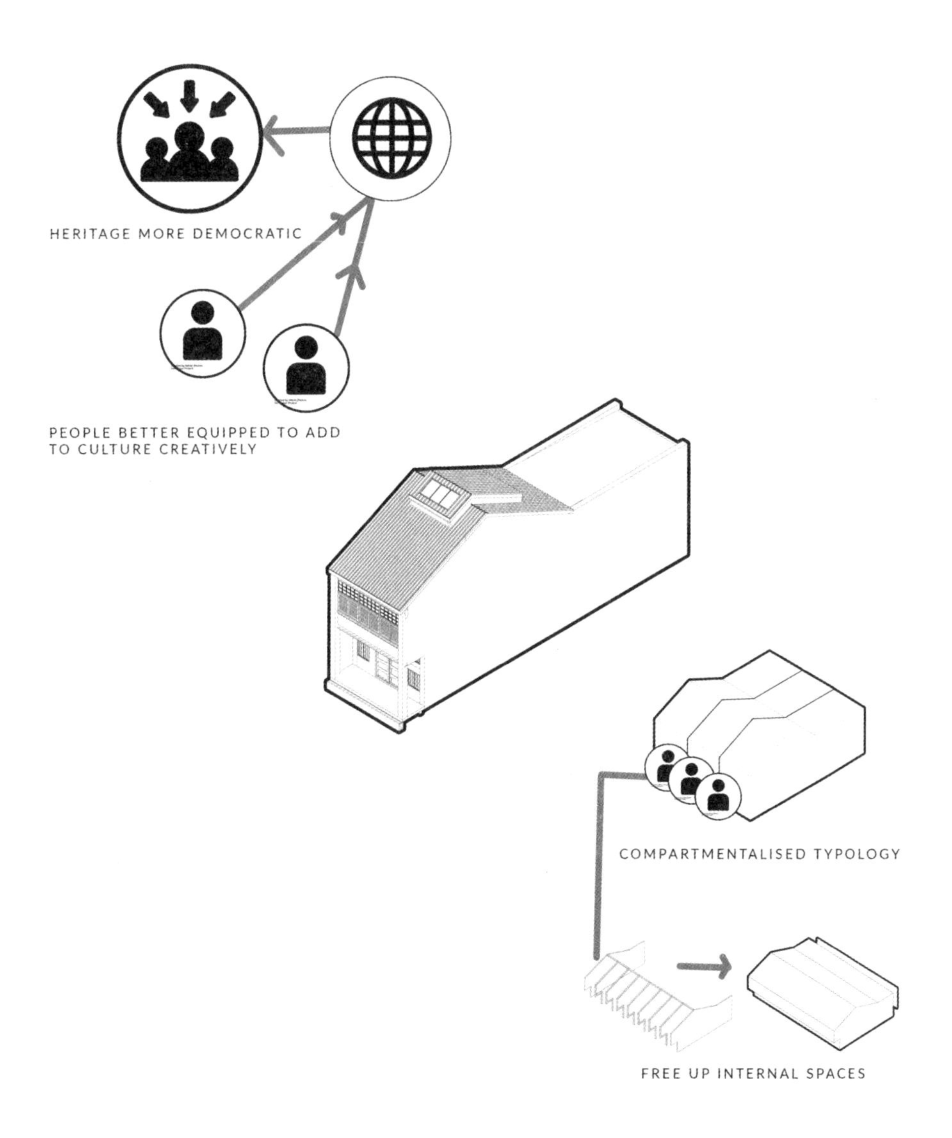

Why augmented reality?

增强现实的理由

In this dimension of sharing, The future of how a culture and heritage of a particular location can be shared with others through the modification of the shophouse typology found abundantly in Joo Chiat and commonly around Singapore is proposed.

The project seeks to cease the immortalisation of the conserved shophouse by proposing the next step to its evolution, enabled by the advent of augmented reality.

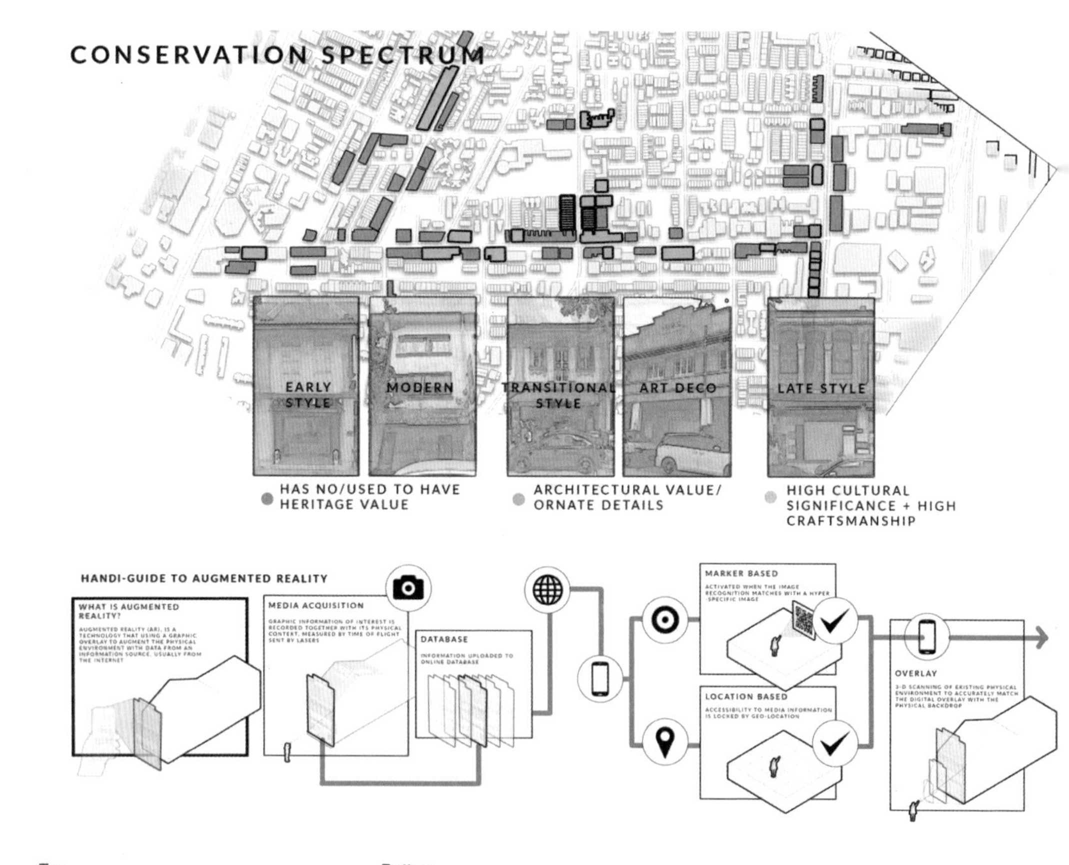

Top
Proposed conservation spectrum

上图
建筑保留意图

Bottom
Proposed Handi-Guide to augmented reality

下图
增强现实设计意图

在分享的层面上，我们提出了通过修复如切不同类型的商店所累积的丰富经验，将来也能与他人分享如何保留新加坡其他特定地点的文化遗产。

该项目旨通过增强现实技术支撑，提出新的演化方式，终止传统商屋一成不变的保护方式，为其注入新的内容。

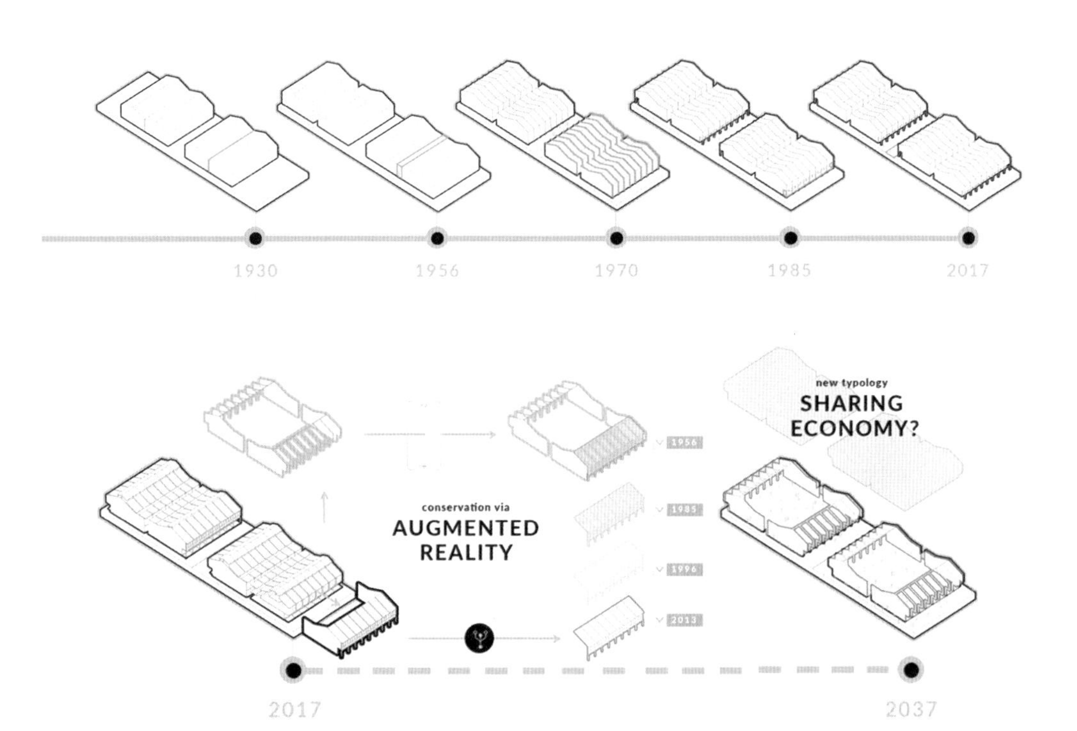

Timeline of shophouses and Conservation via augmented reality

店屋的时间线和保留与增强现实的对比

如切2037

JOO CHIAT 2037

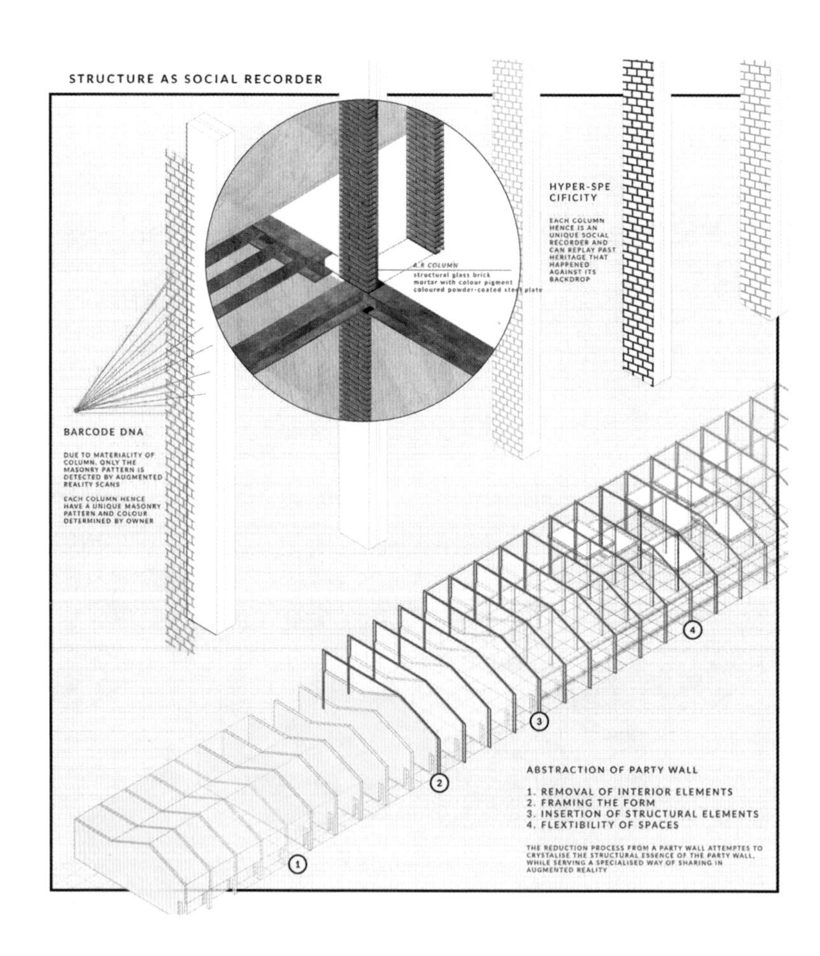

Structure as social recorder

作为社会记录的结构

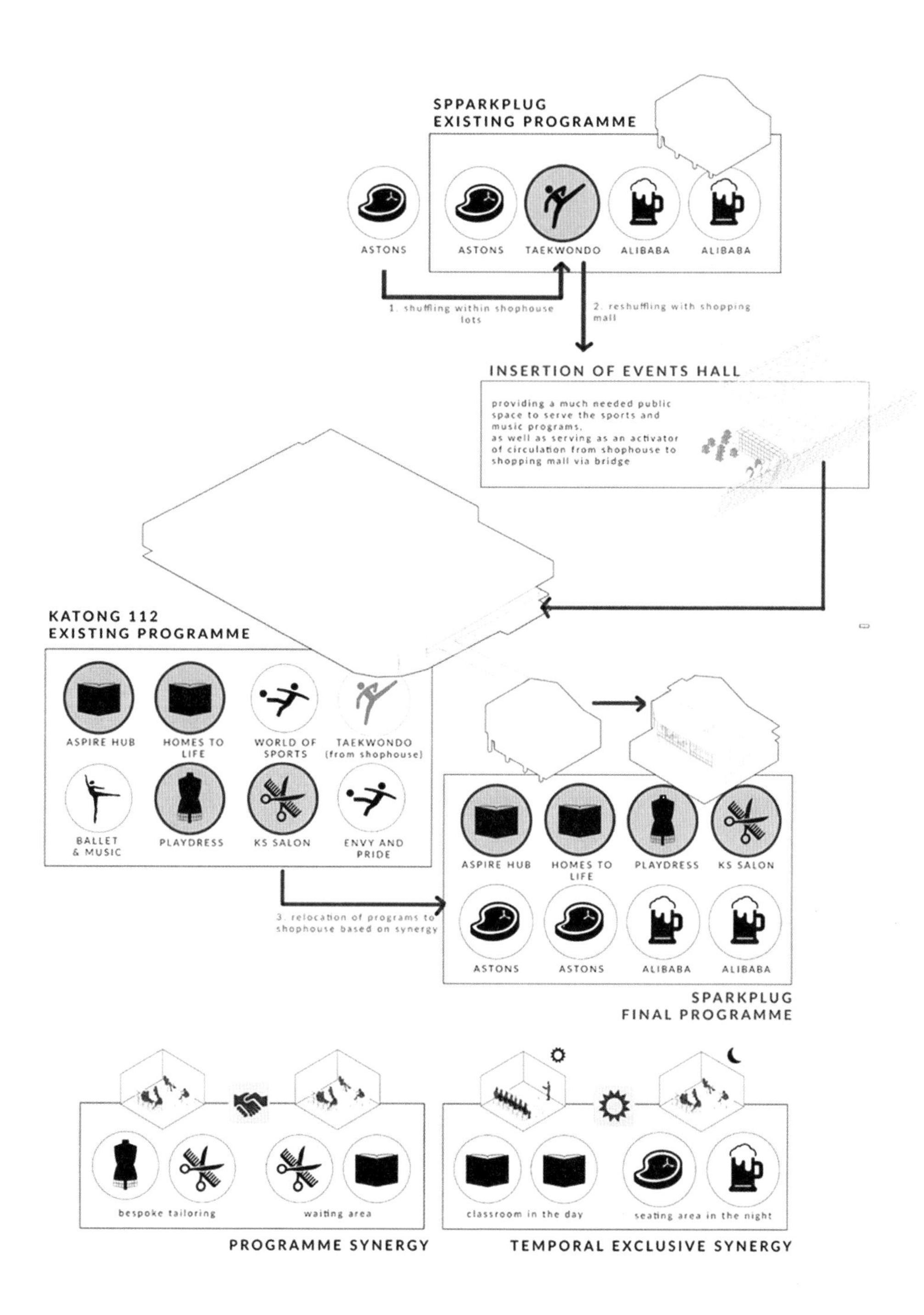
SPPARKPLUG
EXISTING PROGRAMME
ASTONS
ASTONS
TAEKWONDO
ALIBABA
ALIBABA
1. shuffling within shophouse lots
2. reshuffling with shopping mall
INSERTION OF EVENTS HALL
providing a much needed public space to serve the sports and music programs,
as well as serving as an activator of circulation from shophouse to shopping mall via bridge
KATONG 112
EXISTING PROGRAMME
ASPIRE HUB
HOMES TO LIFE
WORLD OF SPORTS
TAEKWONDO (from shophouse)
BALLET & MUSIC
PLAYDRESS
KS SALON
ENVY AND PRIDE
3. relocation of programs to shophouse based on synergy
ASPIRE HUB
HOMES TO LIFE
PLAYDRESS
KS SALON
ASTONS
ASTONS
ALIBABA
ALIBABA
SPARKPLUG
FINAL PROGRAMME
bespoke tailoring
waiting area
classroom in the day
seating area in the night
PROGRAMME SYNERGY
TEMPORAL EXCLUSIVE SYNERGY

The project hopes to solve two issues: the gated and heavily institutionalised resources of the existing heritage, and the spatial limitations of the compartmentalised shophouses.

By reducing the shophouses to its abstract elements, spatial flexibility is enabled and heritage is hence delegated to the digital realm and democratically shared onto the database.

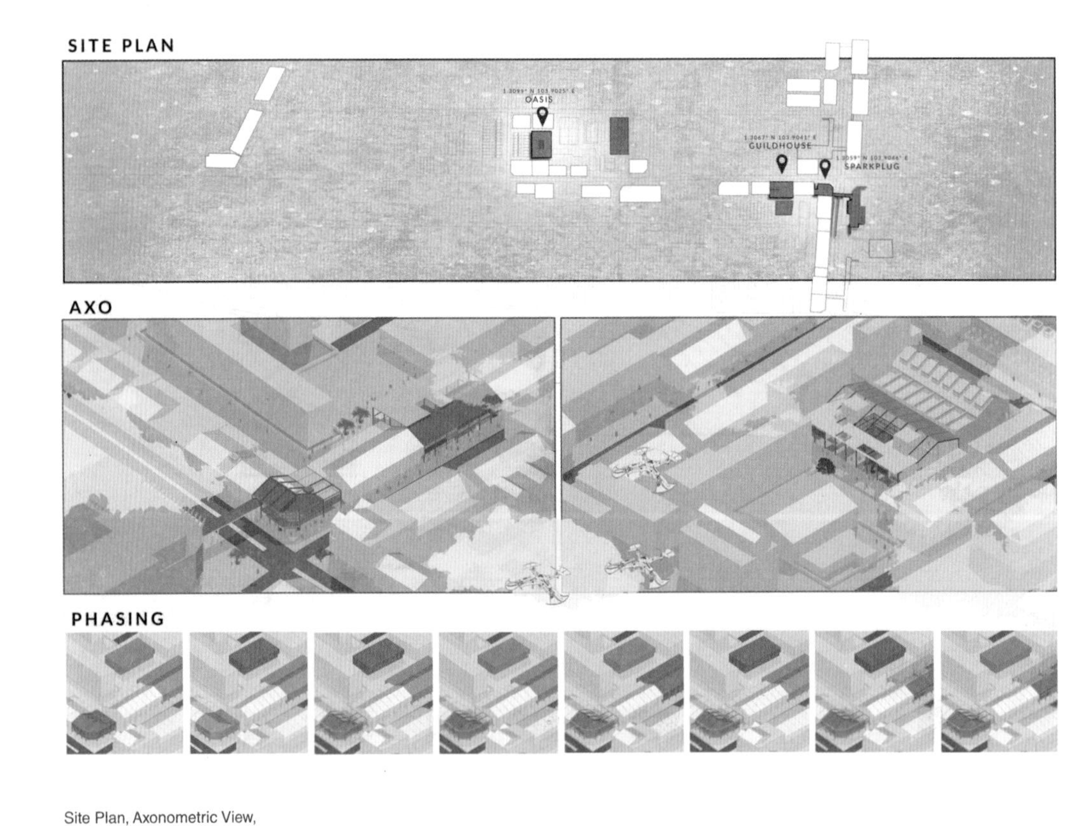

Site Plan, Axonometric View,
Design Phasing

平面图和轴测图（设计阶段）

如切2037希望能解决两个问题：现有遗产的隔离和大量制度化资源的消耗，以及分区化商店的空间限制。

通过将对商屋的保护缩小至最关键的建筑要素，可使空间更加灵活，遗产也能录入数字系统而让更多的人自由分享。

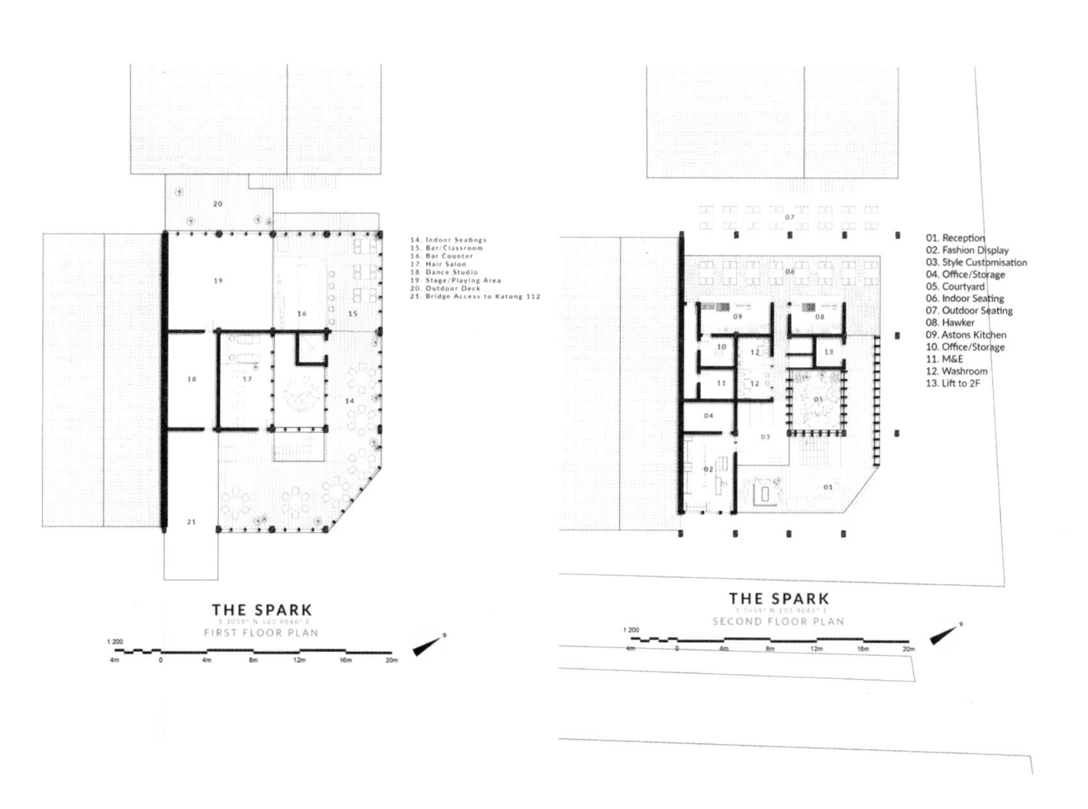

The Spark: Floor Plans 1:200

The Spark 平面图 1:200

At the same time, consumers and residents can actively contribute, relief the ever-compounding scenes of everyday life of Joo Chiat, recorded through the architectural barcode of the abstract structure.

Hence, the new typology fufills not just a physical and spatial purpose but an important role in introducing a new sharing culture (of sharing culture) by acting as a social recorder.

Design Impression

设计示意图

与此同时，消费者和居民可以通过抽象结构的建筑条形码，为记录如切日常生活中日益复杂的场景作出贡献。

因此，新的类型学不仅仅体现在对物质和空间的新定义上，它同时也借助人与人之间的交往为推广新的共享文化发挥重要的角色。

URBAN COMMONS IN JURONG EAST

裕廊东之城市共同体

Known for its efficiency and growth-orientated planning policies, the urban development of Singapore has a history of being dominated by state-led capitalism (Shatkin 2013). While the government sees itself as a custodian of people's welfare, the people entrusts major decision-making processes to the state (George 2000). In this social contract established between the government and the people which premises on the negotiation between comfort and control, the prevalence of low levels of civic participation is of little surprise where the socio-political and urban development of Singapore is concerned.

However, as the civic society matures, both people and government have displayed an increasing awareness on the role and importance of active citizenship (See 2014). Recognising that strengthening communities play a crucial role in propelling the nation towards a socially more inclusive and sustainable city, the empowerment

长期以来，新加坡的城市发展以国有资本主义为主导，并以追求效率和力促经济增长的政策而著称。政府视自己为人民利益的托管人，而民众也充分信任政府在重大问题上的决策。在这样一种社会契约下，新加坡公民在社会及城市发展方面的参与度较低也就不足为奇了。

随着公民社会的成长，近年来，公民参与的重要性得到了政府与人民的关注。基于对公民参与在国家包容可持续发展过程中的重要性的共识，公众也意识到增强个体参与行动是应对当今及未来城市挑战的关键所在。这也预示着一个社会政治意识形态的范式变革，也即公众以及组织机构以合作共创城市将被推向前沿。因此，在社会文化层面，作为社会再生产以及社区关系的基础，共享与合作应该也必将成为未来民众日常生活的关键构成部分。

被大卫·哈维所定义的城市共同体大致上被定义为集体活动和意愿所引发的社群所致力于的共同生产、享用及管理的资源池。为了了解城市共同体的意义，我们必须理解“参与”和“协同制造城市共同体”之间的关联。“参

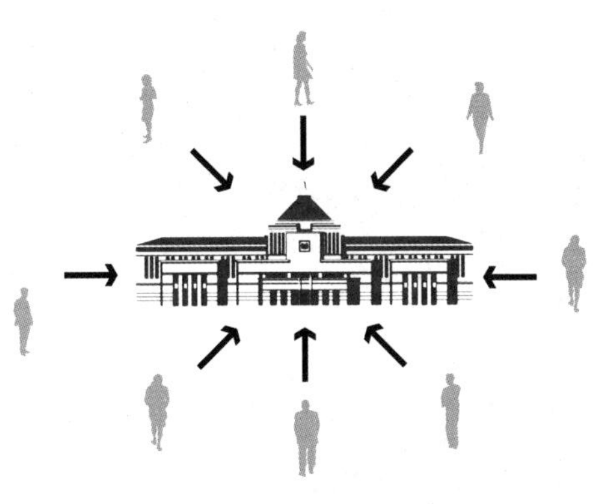

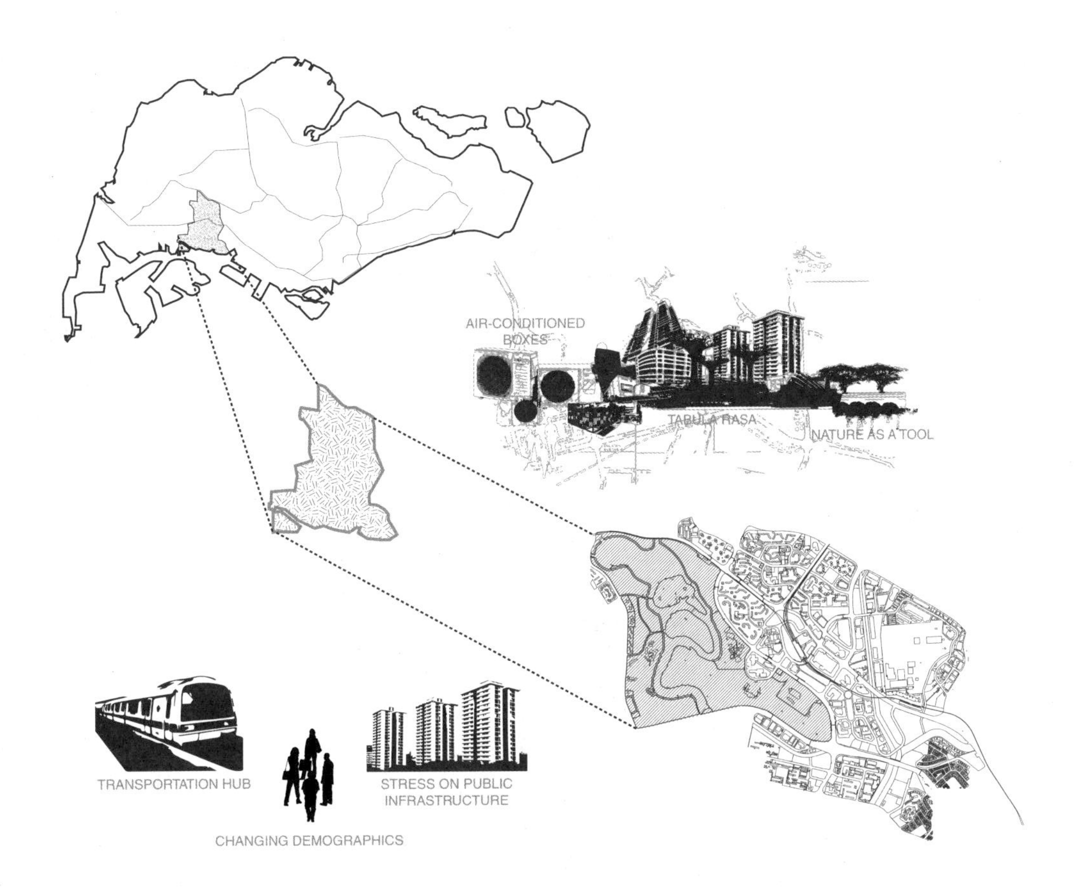

Top
The urban development of Singapore has a history of being dominated by state-led capitalism.

上图
新加坡的城市发展以国有资本主义为主导。

Bottom
Slated to become the next Central Business District, Jurong East is currently undergoing rapid changes.

下图
由于裕廊东将被发展成新加坡第二个中央商务区，它目前正在经历各种快速的改变。

of individuals to involve and act will be the key to addressing existing and future challenges of the city (Lee 2014). These challenges are related to the management of urban resources, as conflicting needs and evolving desires of different communities arise in view of the rising and increasingly diversified population. Therefore, this signals a paradigm shift in the nation's socio-political ideology where the collaboration of citizens, organizations and agencies in co-creating the city is pushed to the forefront. Socio-cultural practices of sharing and collaboration should and would become key aspects of the daily lives of people, as they underpin social reproduction and social relations between communities (McLaren and Agyeman 2015).

Collective activities and interests motivate the communal effort to produce, modify and sustain pools of common resource, and are defined by Harvey (2012) as the urban commons. To understand the significance of urban commons, it is first crucial to understand the relationships between participation and "commoning". While participation is to take part in or to be involved in an activity (Cambridge University Press, 2017), "commoning" is specifically the activity of either creating or establishing a commons (Chan, 2017). "Commoning" necessarily presume some sharing, and sharing entails the participation of individuals. However, not all sharing practices lead to a commons. The significance of participation with respect to "commoning" lies in the context of participatory design, where it goes beyond the involvement of a shared activity. While citizen control is at the highest rung of the ladder of citizen participation according to Arnstein (1969), power is also shared in this instance. Therefore, through sharing and exercising decision-making power, "commoning" entails a shared agreement on how to establish and manage the commons.

Within the "Sharing Paradigm" lies a spectrum of sharing practices ranging from the

与”基本含有个人参加其事或加入某种活动的意思，人们“协同制造城市共同体”的过程则需具体的活动。虽然人们必须通过“参与”具体的活动才能“协同制造城市共同体”，可是并不是所有具有参与性的活动都拥有“协同制造城市共同体”的意义。值得注意的是，“参与”在公民参与之中拥有更具体的含义。根据阿恩斯坦(1969)的见解，公民“参与”意味着人们在决策中享有更平等的决策掌权。因此，“协同制造城市共同体”不仅表现社群所达成的共识，从而建立城市共同体的过程，也是让人们拥有平等决策权的途径。

在所谓的“共享范式”之中，存在着一系列不同的共享方式。例如，有些是拥有商业动机的，有些则是以社会文化变革为动因；有些是有形的，比如物质和产品，有些则是无形的，比如服务和经验。无论哪一种形式，城市共同体都意味着在公共领域对于创造共同资源的社会长远投资。共享文化发展的趋势也意味着城市共同体将以不同的形式存在于新加坡未来的社会里，并且对于激发民众的积极参与及创造城市的未来都变得越来越重要。

由于裕廊东将被发展成新加坡第二个中央商务区，它目前正在经历各种快速的改变。在新加坡智慧国2025的计划下，裕廊东将从一个工业居住区转型成给当地及跨国企业提供资源协助的区域研发中心，并伴随着土地利用的改变和空间容量的增加，而以智能科技为主的城市发展计划也将更加注重公民的参与。因此，裕廊东可视为新加坡未来城市发展的试验平台。选择裕廊东作为案例能让我们更深入地了解城市共同体的潜力与弊端。

我们的设计小组所研究的专题主要是探索各个分享机构在促进社区合作、共同行动以及社会资本增长等方面的可能性。这些对于城市共同体的培养都极为重要。所有的设计方案以裕廊东为基础，针对不同社群所需，以城市共同体的概念作为设计的主题。每个设计方案都有独特的见解和结论。

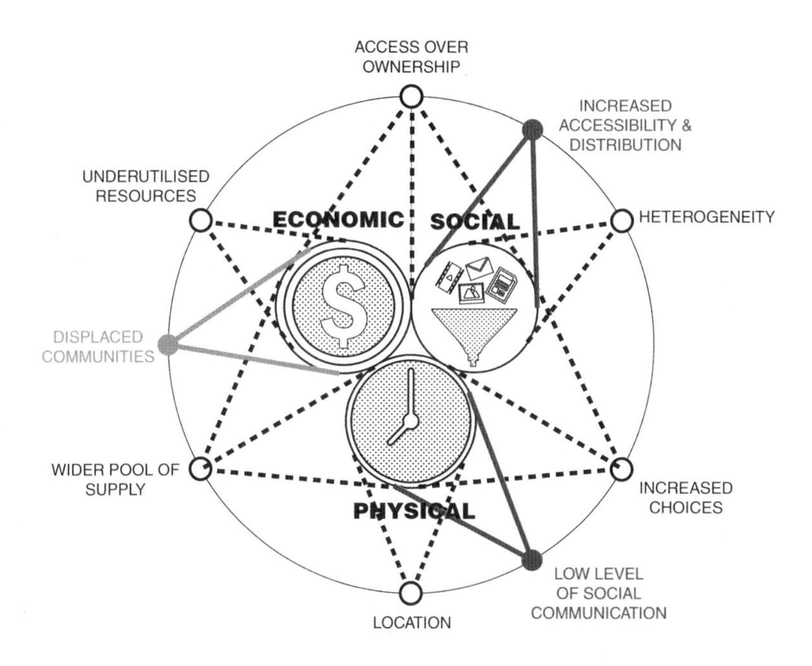

Top
As the civic society matures, both people and government have displayed an increasing awareness on the role and importance of active citizenship

上图
随着公民社会的成长，近年来，公民参与的重要性得到了政府与人民的关注。

Bottom
Jurong East presents timely opportunities for investigations into future potentials and challenges with respect to urban commons.

下图
选择裕廊东作为案例能让我们更深入地了解城市共同体的潜力与挑战。

commercially mediated to the socio-culturally evolved; from the sharing of the tangible, such as materials and products, to the intangible, such as services and experiences (McLaren and Agyeman 2015). Be it preconditioned by commercially-mediated or socio-cultural forms of sharing, the urban commons signify social investments into the co-production of resources in the public realm. In the ensuing urban development of Singapore where co-creation is concerned, the notion of urban commons becomes increasingly relevant and crucial in sensitizing citizens to partake in and effectuate the growth of the entire city.

Slated to become the next Central Business District, Jurong East is currently undergoing rapid changes. Its transformation from an industrial-residential estate to a Research and Development (R&D) hub for local and global businesses is manifested in the intensification of and changes in land uses. More significantly, it is a test bed for technology-led development in which the involvement of the civic society is heavily emphasized (Smart Nation Singapore 2017). As new infrastructure emerges and planning methodology evolves to favour communal participation, sharing and "commoning" may maximise the potentials for urban resources in Jurong East to be accessed and managed equitably. While the district provides a glimpse into the successive urban development of Singapore, its selection as a site presents timely opportunities for investigations into future potentials and challenges with respect to urban commons.

In this studio, projects by pairs and individuals illustrate the exploration of design possibilities for sharing institutions which facilitate collective efforts, actions and thus social capital to be enhanced. These processes are pivotal to the cultivation of the urban commons. With the vision of establishing urban commons in Jurong East, each project targets specific community groups and addresses their conflicting and complementary needs. These are achieved through identifying overlapping needs and possible contributions of the respective communities, and subsequently proposing sharing innovations amongst them. Various medium in the urban sphere, as elaborated in the projects, thus cater for the social production and consumption of tangible and intangible resources in Jurong East.

REFERENCES

Arnstein, Sherry. R. 1969. "A Ladder of Citizen Participation." Journal of the American Institute of Planners 35, no. 4: p 217.

Cambridge University Press. (2017). Participate. Retrieved 2017, from Cambridge English Dictionary: http://dictionary.cambridge.org/dictionary/english/participate

Chan, J. (2017, April). Distinctions Between Closely-related Concepts. Beijing.

George, Cherian. 2000. Singapore: The Air-conditioned Nation. Essays on the Politics of Comfort and Control, 1990-2000. Singapore: Landmark Books.

Harvey, David. 2012. Rebel Cities: From the Right to the City to the Urban Revolution. Verso.

Lee, Hsien Loong. 2014. "Transcript of Prime Minister Lee Hsien Loong's speech at Smart Nation launch on 24 November." National Infocomm Awards. Singapore: Prime Minister's Office Singapore.

McLaren, Duncan, and Agyeman, Julian. 2015. Sharing Cities: A Case for Truly Smart and Sustainable Cities. Cambridge: The MIT Press.

See, Bridgette. 2014. Co-Creating Singapore: Hear What Citizens Have To Say. May 14. Accessed April 2017. https://www.challenge.gov.sg/print/cover-story/co-creating-singapore-citizens-have-their-say.

Shatkin, Gavin. 2013. "Reinterpreting the Meaning of the "Singapore Model": State Capitalism and Urban Planning." International Journal of Urban and Regional Research 116-137.

Smart Nation Singapore. 2017. Enablers. February 28. Accessed April 2017. https://www.smartnation.sg/about-smart-nation.

01

奠定城市共同体的基础
SETTING THE FOUNDATION FOR URBAN COMMONS

THE MOBILE COMMUNITY CENTRE
BY LIN LEI
移动社区中心

CC2.0
BY KRISTA YEONG
社区交流中心

02

协同生产城市共同体
STIMULATING URBAN COMMONS

BIO-CYCLE
BY TAN CHIEW HUI
有机循环

STACKED KAMPUNGS
BY LIEW YUQI
立体村落

03

预测新兴城市共同体形式
ANTICIPATING EMERGING FORMS OF URBAN COMMONS

ARTSTREAM
BY GLORIA NEO
艺游

THE MOBILE COMMUNITY CENTER

By Lin Lei

移动社区中心

CC 2.0

By Krista Yeong

社区交流中心

The pair project aims to set the foundation of communication and thus connections amongst different demographic groups, so as to cultivate the social conditions to facilitate urban commons. The project is guided by the underlying principle that communication and subsequently participation, are pivotal to initiate communal bonding and active citizenship.

While "The Mobility Community Center" is the main community hub for residential artists to connect to the neighbouring communities, "CC 2.0" is one of the satellite hubs dispersed in different parts of the city to assimilate foreign backpackers with the local residents.

这两项设计方案以促进人与人之间的沟通方式为前提，从而建立起不同人群之间的关系。该设计方案希望通过制造沟通和谐的社区环境，奠定城市共同体的基础。该方案注重的是人与人之间的沟通和公民参与，并坚信这样能促进社群之间的关系，从而提高社群意识。

“移动社区”是方便居住于此的艺术家与周边社区互动的社区中心，而“交流中心CC2.0”则是分散在新加坡不同地区的卫星节点，目的是帮助外国背包客融入当地居民的生活。

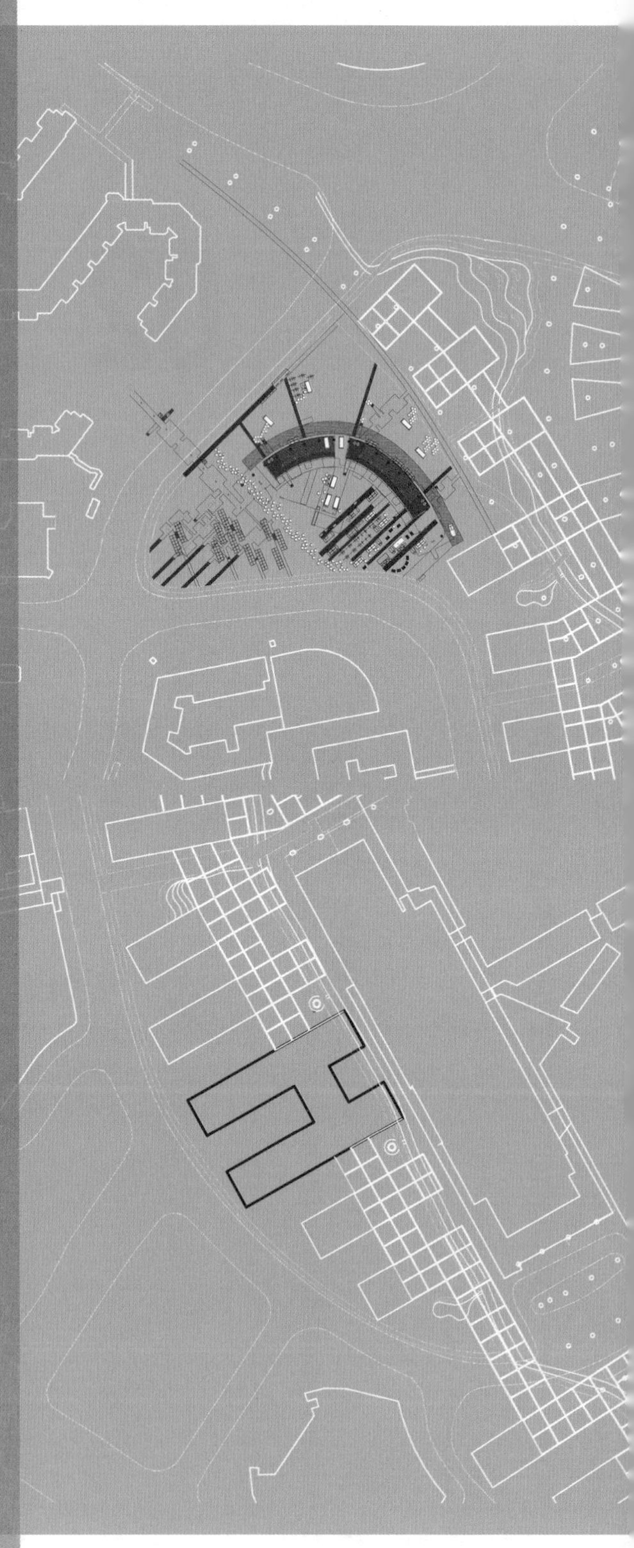

SETTING THE FOUNDATION

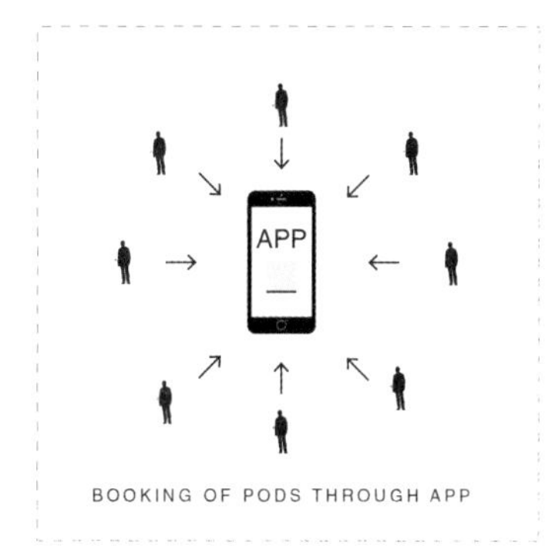

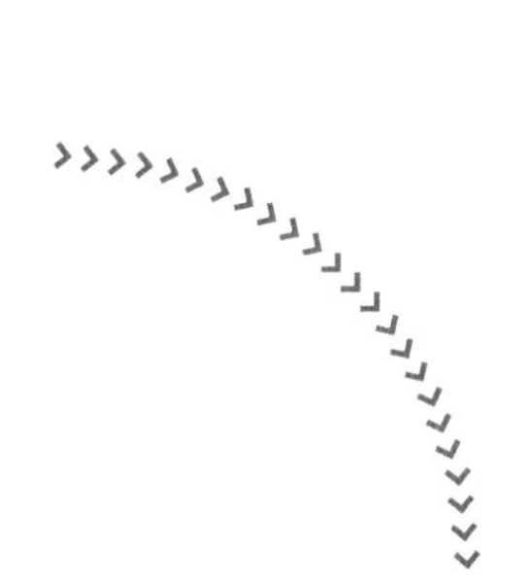

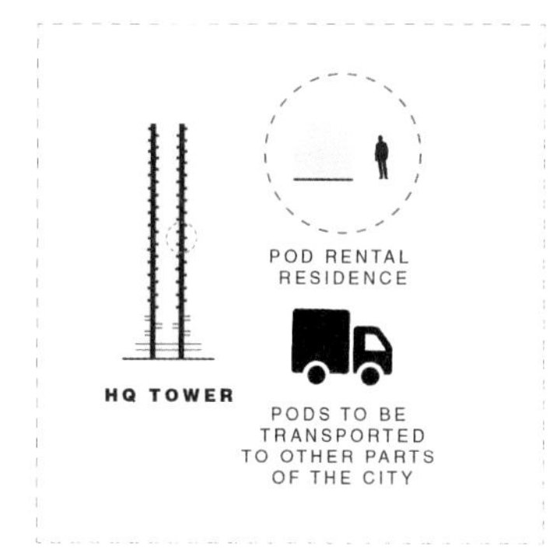

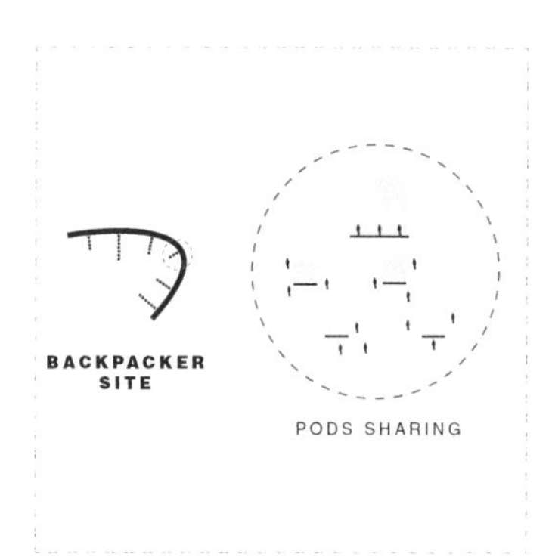

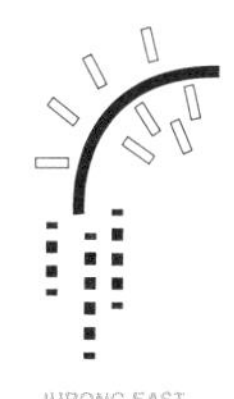

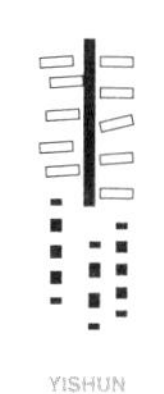

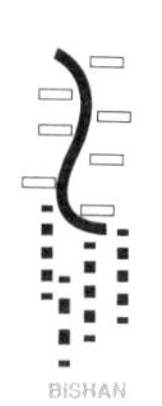

x 20
OTHER LOCATIONS IN SINGAPAORE

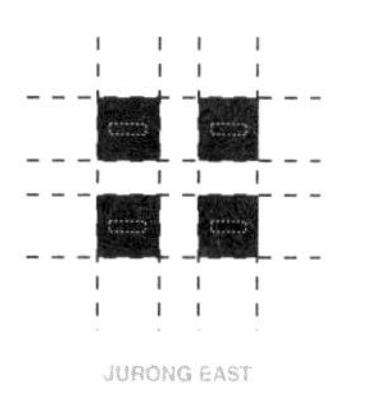

Top
Mobile pods from the HQ tower in each district may be booked through an app by people for different purposes. They are transported to other parts of the district, including those that are shared with other satellite communication hubs.

上图
每个地区总部大楼的移动单元可以通过应用程序预订。之后，它们将会被运送到区域里的住宅区，或其他区域里的卫星通信社区中心。

Bottom
Satellite communication community hubs are site adaptable and will potentially operate in other districts, across different parts of Singapore. Mobility pods are transported from and returned to the HQ Tower in each district.

下图
卫星通信社区中心将在新加坡各个地区，以不同的形式提供给社群各种服务。在每个区域里，移动单元将在所需要时往返于卫星通信社区中心与总部大楼之间。

While the mobile pods challenge the limits of geographical boundaries by reaching out to individuals, the stationary structures provide platforms for people to congregate, participate, and contribute; achieving a spatial equilibrium of the ephemeral and the permanent.

By examining the evolving means of communication and social interactions, these projects revolutionize community hubs into effective facilitators of intangible resources. They are managed and sustained by collective activities, which set the social foundation of inclusivity, participation and collaboration for the urban commons.

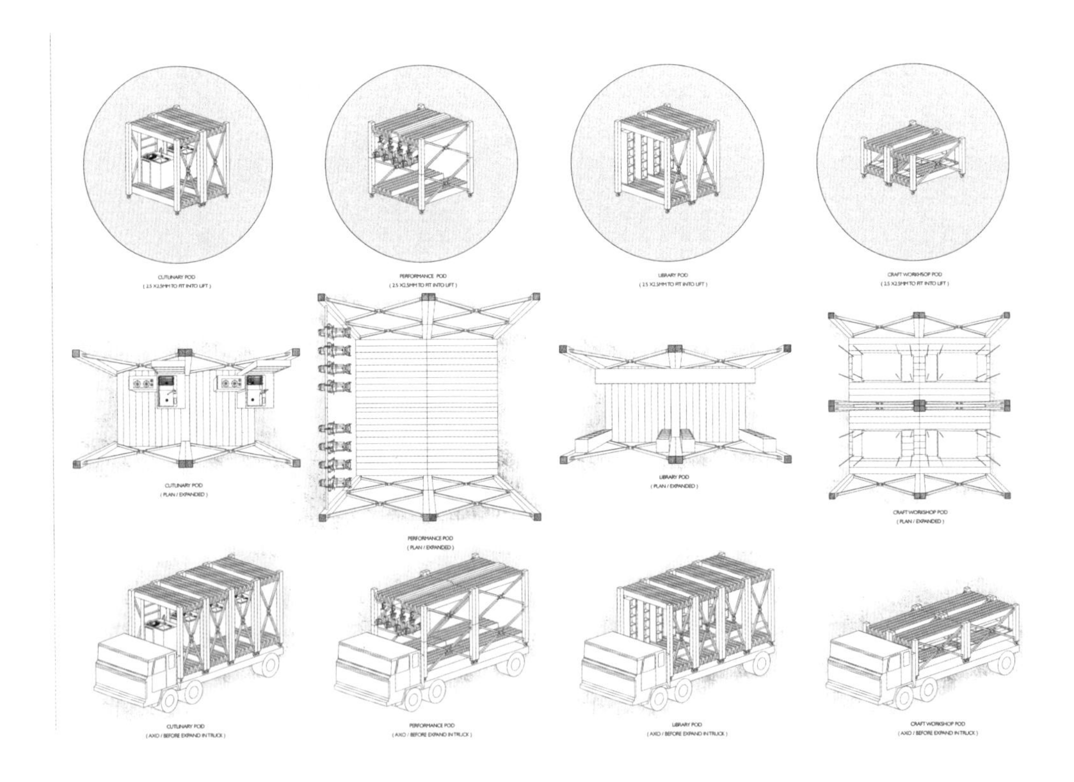

Mobility pods detachable from the headquarter tower, "The Mobile Community Centre", may be volumetrically reconfigured and collapsed into trucks.

来自地区总部的移动单元容易地变形、缩小以安装进卡车里。

通过可方便重整的移动单元，社区中心与居民之间不再受地理距离的限制，随时随地提供给人们聚集、参与和贡献的平台，从而获得了空间的平衡。

随着人们互动范式的改变，这两项设计方案提出了能有效分布非实体资源的新型社区中心。社区中心所提供的集体活动，也能提高人们之间的包容性和参与感，鼓励他们达成共识，从而奠定城市共同体的基础。

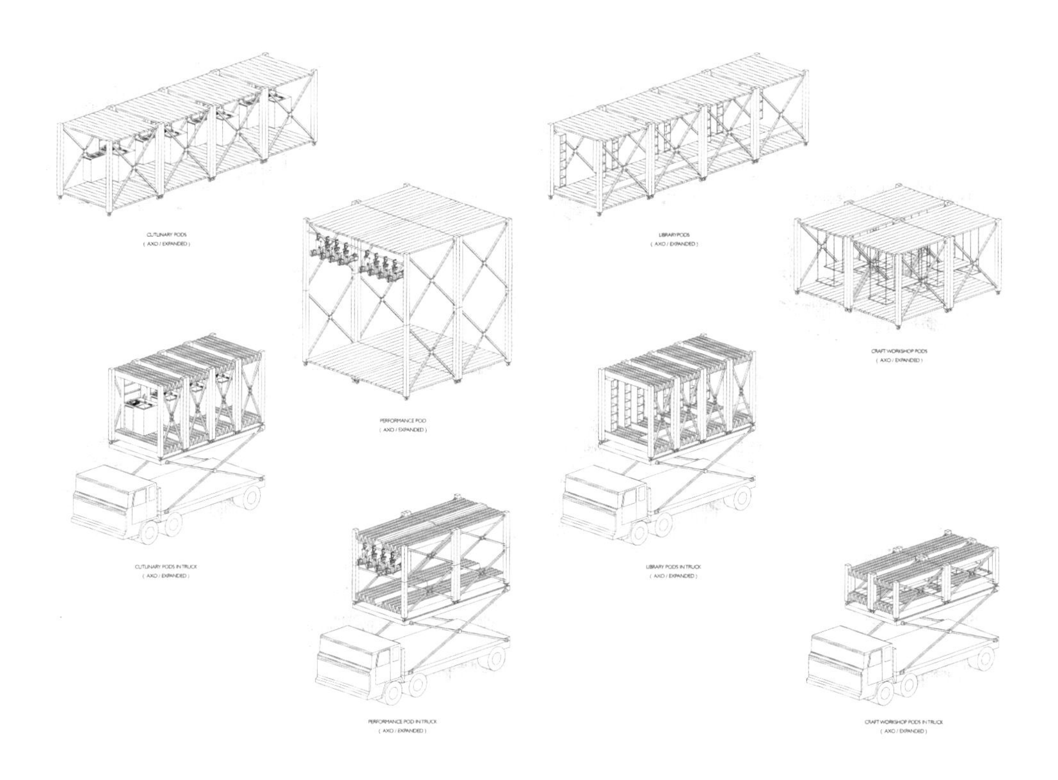

Subsequently, as these trucks transport the pods to other parts of the city, they may be expanded and reprogrammed for cultural experts to conduct acitivites and workshops for the residents.

卡车把移动单元运送到城市的其他地方之后，经展开与重组，文化专家可作为居民举办活动的场所和工作坊。

THE MOBILE COMMUNITY CENTRE

移动社区中心

"The Mobile Community Centre" is sited next to the MRT station, an underutilised site due to existing noise levels from the trains. The project is designed such that pedestrians from the two adjacent shopping malls are able to traverse through beneath the station and to the street level of the proposed mobile community tower.

Top
Perspective view showing how mobile pods are attached to permanent infrastructure

上图
展示移动单元停泊处的透视图

Bottom
Perspective view to permanent station for pods at the street level

下图
底层街道的透视图

“移动社区”位于捷运站旁边，这些捷运站由于经常性的噪声，周边至今未有任何建设。该设计方案考虑到来自两个相邻购物中心的行人，让他们能够方便地穿过捷运站，轻易地到达塔楼的底层街道。

Top
Perspective vlew to the street level.

上图
底层街道的透视图。

Bottom
Perspective view from the residential bridge level.

下图
住宅桥梁层面的透视图。

The tower predominantly houses residential units for the "creative nomads", defined as people with specific skillsets willing to share with the community. The mobile pods are designed according to four types of expertise and their specific practises - stage pods for performers, culinary pods for chefs, library pods for storytellers and craft pods for designers.

"移动社区"为那些拥有特别才能并愿意与社区分享的专家提供住宅。作为这些专家的住所，住宅塔楼让他们有机会通过移动单元与居民接触。移动单元是根据四种专业而设计的——表演者的舞台单元、厨师的烹饪单元、故事员的图书馆单元和设计师的工艺单元。

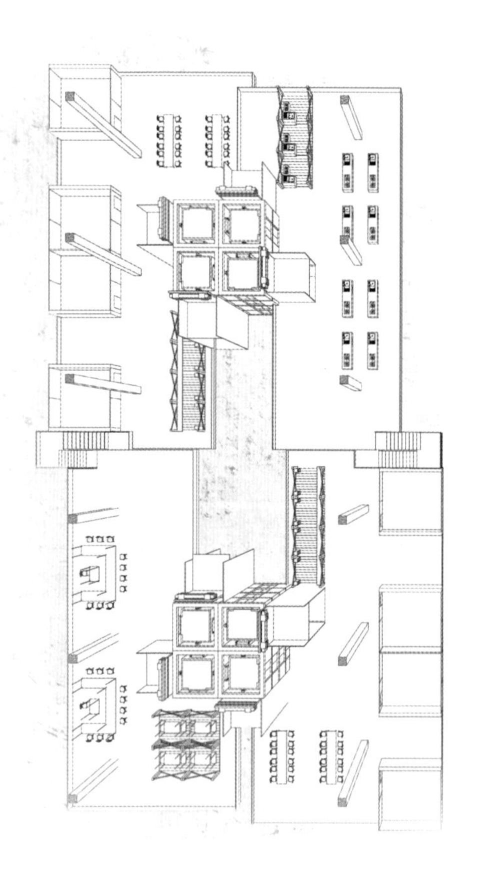

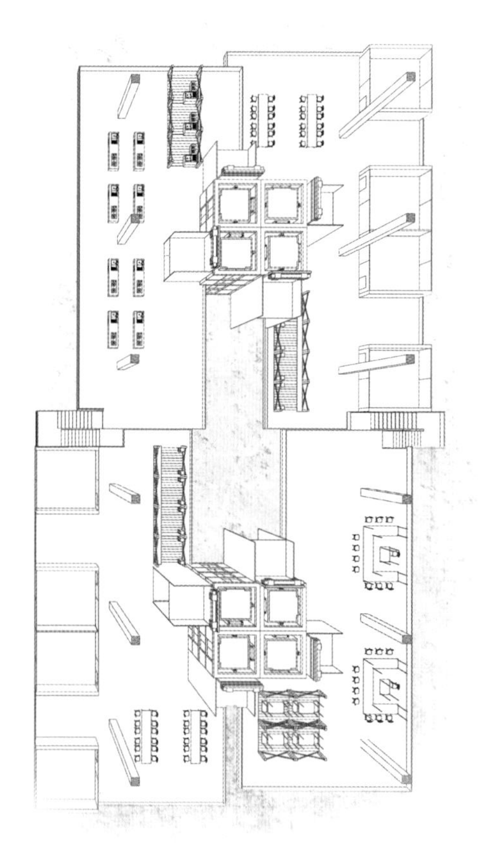

PODIUM FLOOR PLAN TYPE A

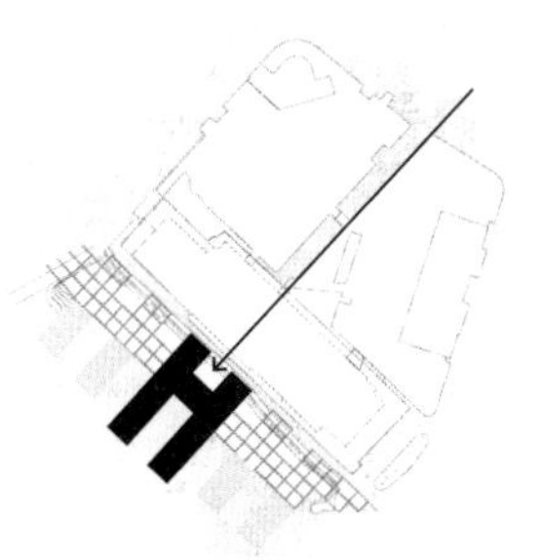

THOROUGHFARE
VISTA FROM THE SOFTSCAPE BETWEEN JEM & WESTGATE

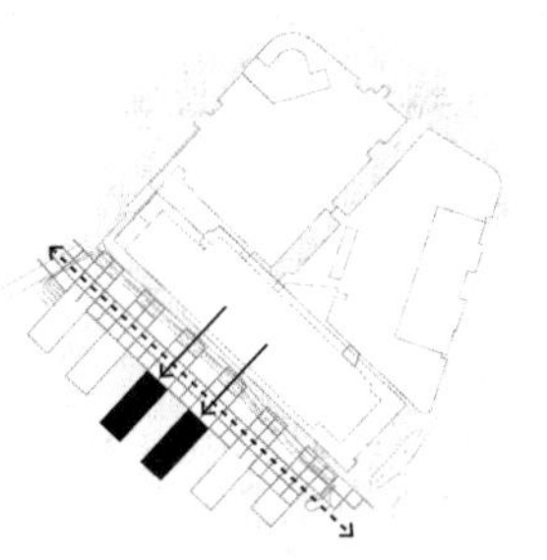

LINKAGE
LINKAGE FROM JE MRT STATION TO THE TOWER

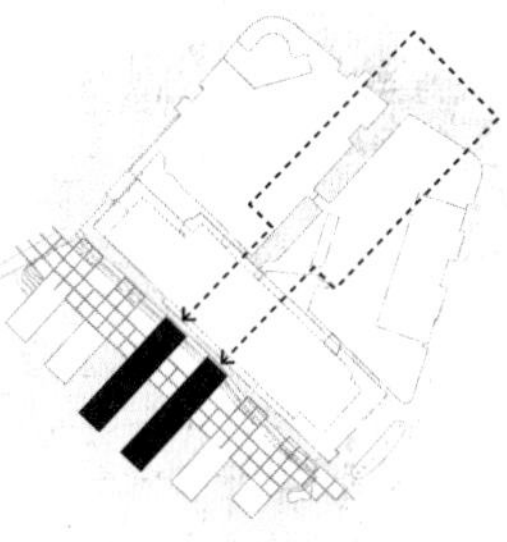

JOURNEY
CONNECT THE J-WALK TO THE TOWER

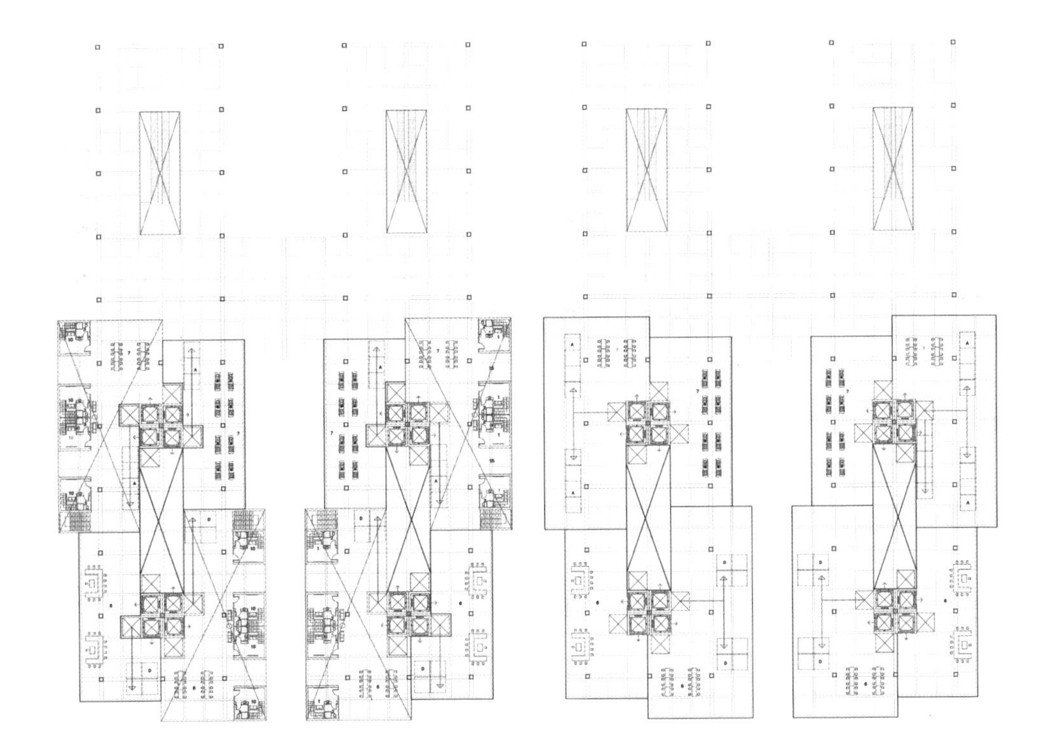

Podium Block Type Plan A
TEMPORARY SPACES: (A) Culinary Workshop Pods (B) Performance Pod (C) Library Pods (D) Craft Workshop Pods
PERMANENT SPACES: (6) Craft Artist Workshop (7) Chef Culinary Classroom (8) Outdoor Painting Area (9) Outdoor Dining (10) Nomad Office for Rent

讲台式区域平面图 A
临时空间：（A）烹饪车间单元；（B）表演单元；（C）图书馆单元；（D）工艺车间单元
永久空间：（6）工艺艺术家工作室；（7）厨师烹饪课室；（8）户外绘画区；（9）户外用餐区；（10）“游牧民房”出租

Additionally, the residential tower for the creatives also acts as the headquarters for the mobile pods, affording parking spaces as they return to the tower.

On the other hand, the podium block and street level spaces are meant to facilitate monthly events and festivals such as flea market and skills sharing events.

住宅塔也作为移动单元的总部，为它们提供停泊位置。

讲台式区域和底层空间则在每月都会举办各种活动和节日，如跳蚤市场和技能分享活动。

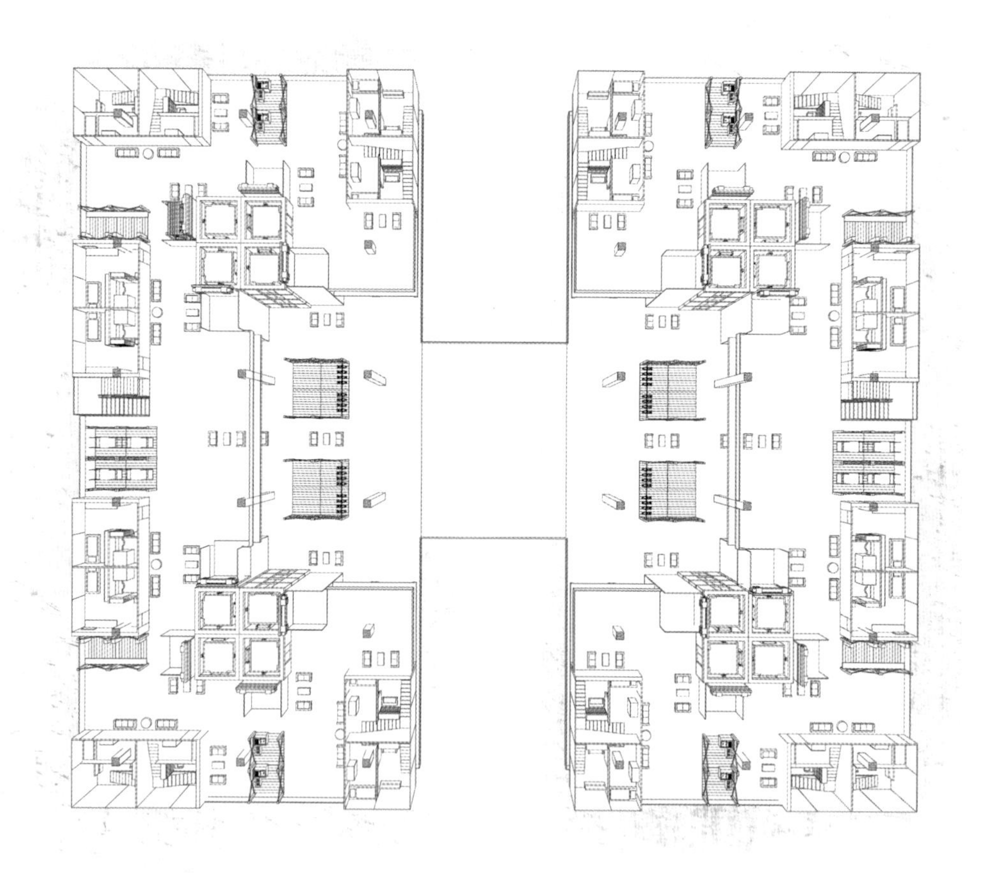

RESIDENTIAL FLOOR PLAN TYPE A

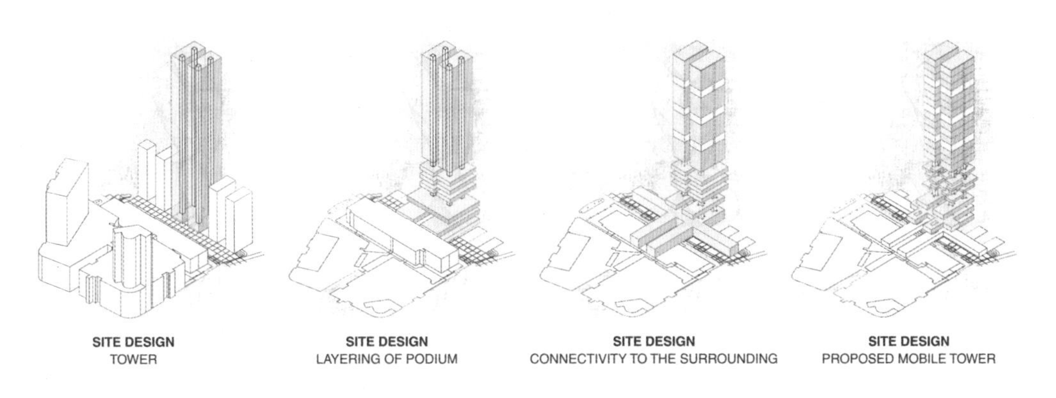

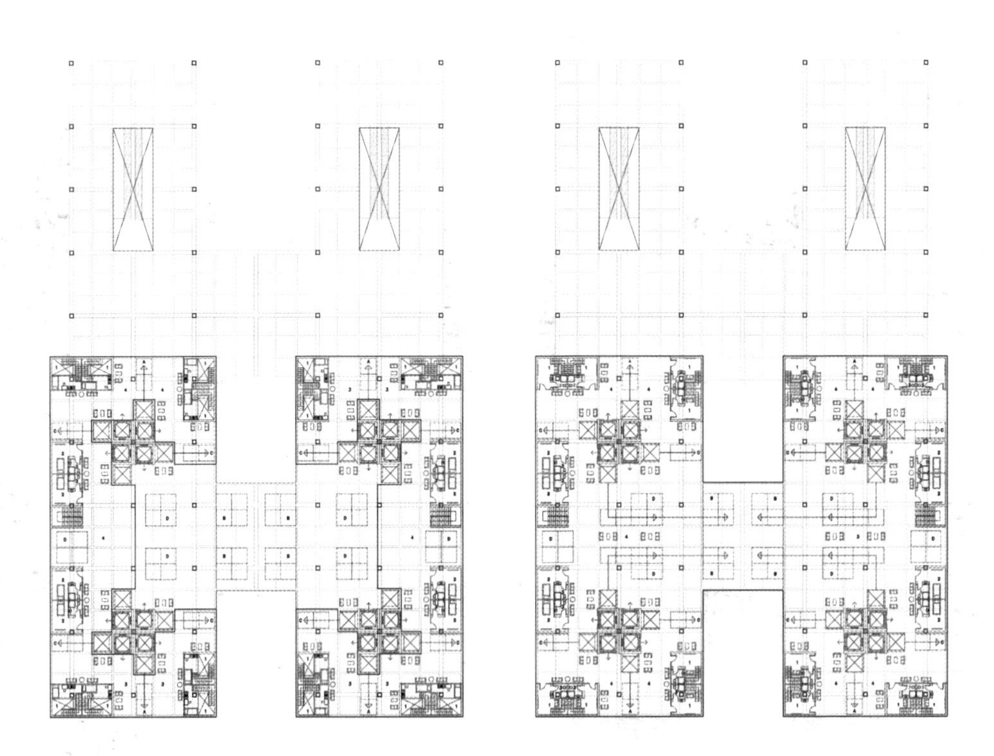

Residential Block Type Plan A
TEMPORARY SPACES: (A) Culinary Workshop Pods (B) Performance Pod (C) Library Pods (D) Craft Workshop Pods
PERMANENT SPACES: (1) Loft Studio Apartment (2) Studio Apartment (3) Washroom (4) Living Room (5) Shared Spaces

住宅塔楼平面图 A
临时空间：（A）烹饪车间单元；（B）表演单元；（C）图书馆单元；（D）工艺车间单元
永久空间：（1）阁楼一室公寓；（2）一室公寓；（3）卫生间；（4）客厅；（5）共享空间

Instead of designating separate floor levels for parking to store the mobile pods, they are embedded to the residential units, located in close proximity and adjacent to them. Therefore, such a design decision adds value to the leftover spaces between each of the residential units.

移动单元的停泊区与住宅单位整合在一起，而不是设计单独的楼层，一来方便居民，二来充分利用住宅单位之间的空间。这样的设计为住宅单元之间未利用的空间增加价值。

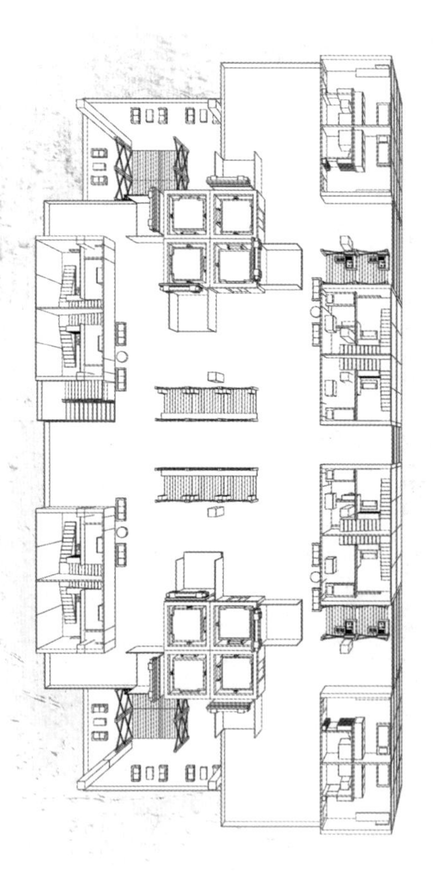

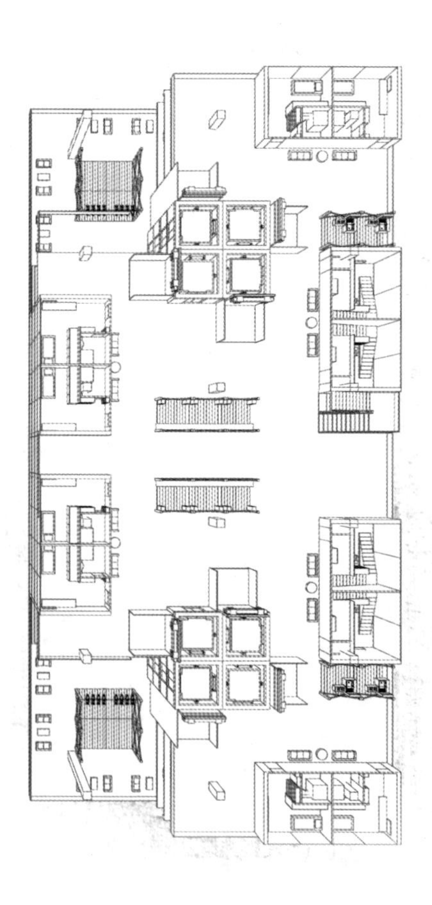

RESIDENTIAL FLOOR PLAN TYPE B

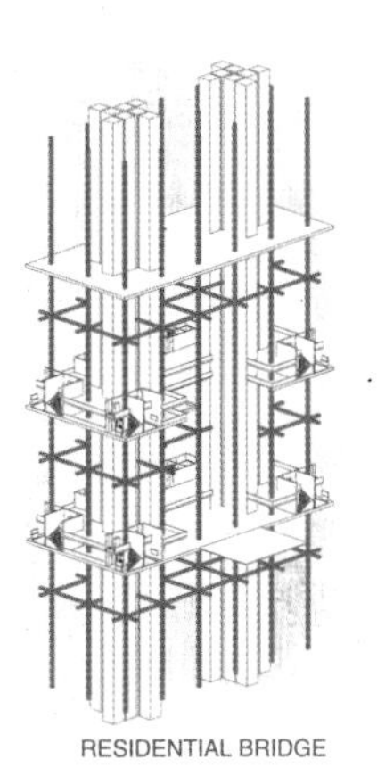
RESIDENTIAL BRIDGE

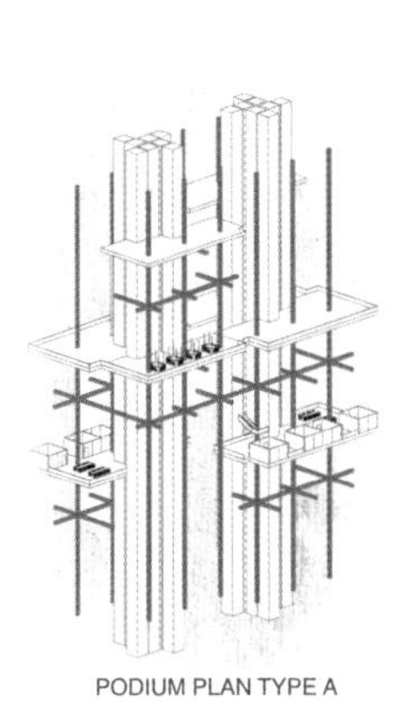
PODIUM PLAN TYPE A

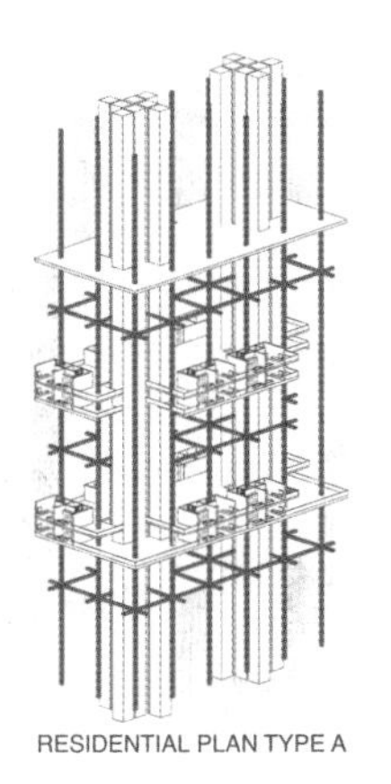
RESIDENTIAL PLAN TYPE A

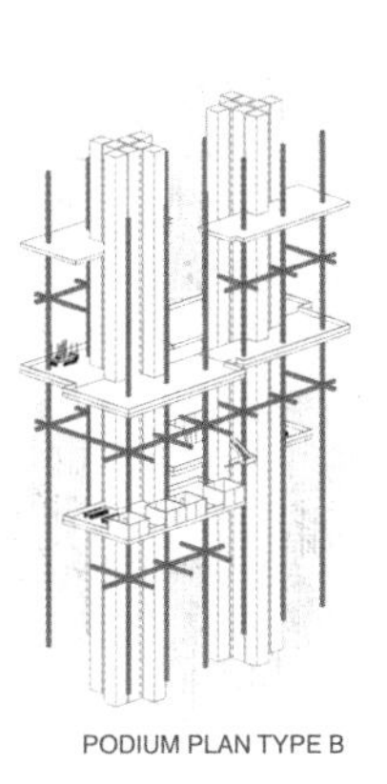
PODIUM PLAN TYPE B

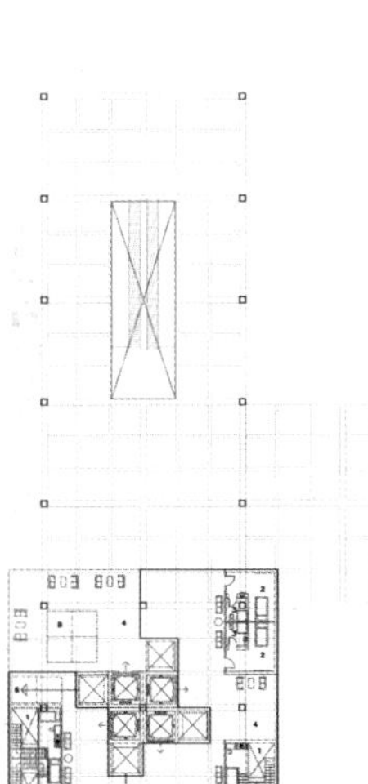
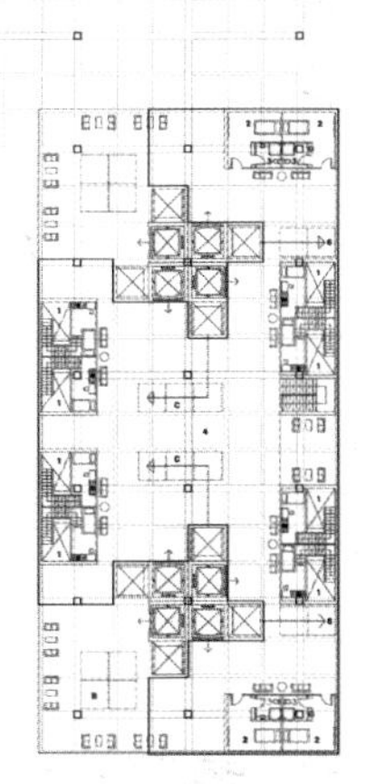
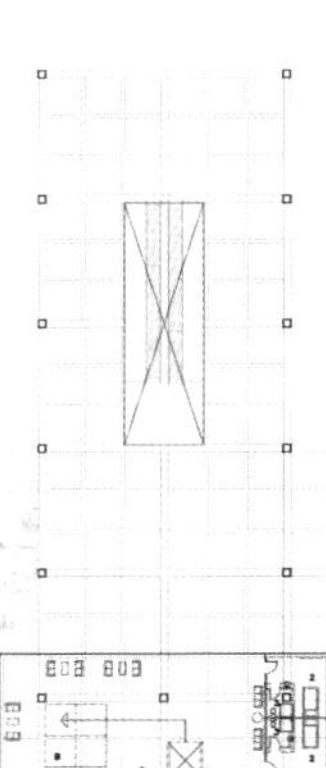
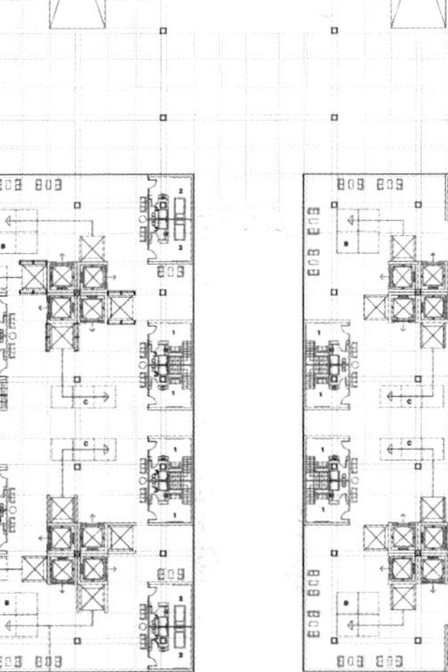

Residential Block Type Plan B
TEMPORARY SPACES: (A) Culinary Workshop Pods (B) Performance Pod (C) Library Pods (D) Craft Workshop Pods
PERMANENT SPACES: (1) Loft Studio Apartment (2) Studio Apartment (3) Washroom (4) Living Room (5) Shared Spaces

住宅塔楼平面图 B
临时空间：（A）烹饪车间单元；（B）表演单元；（C）图书馆单元；（D）工艺车间单元
永久空间：（1）阁楼一室公寓；（2）一室公寓；（3）卫生间；（4）客厅；（5）共享空间

In addition, there is an event level shared by every four residential floors. They are meant for creatives to station their mobile pods on that level, for occasions when they wish to share their skills with other cultural experts of different practices and skills. All in all, the towers serve mainly as integrated residential units and mobile pods parking, while the podium block and street level spaces are shared spaces for community activities.

Each of them are custom designed to suit their respective purposes. For instance, the podium block is designed to be connected to the existing elevated walkway "J-walk" while the street level spaces are open and accessible by the public.

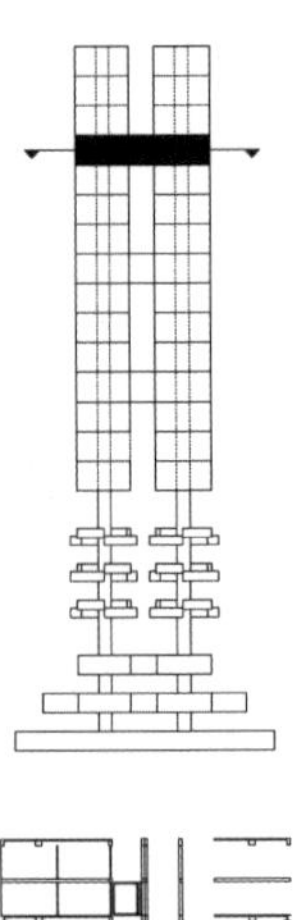

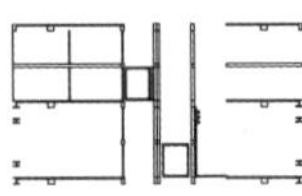

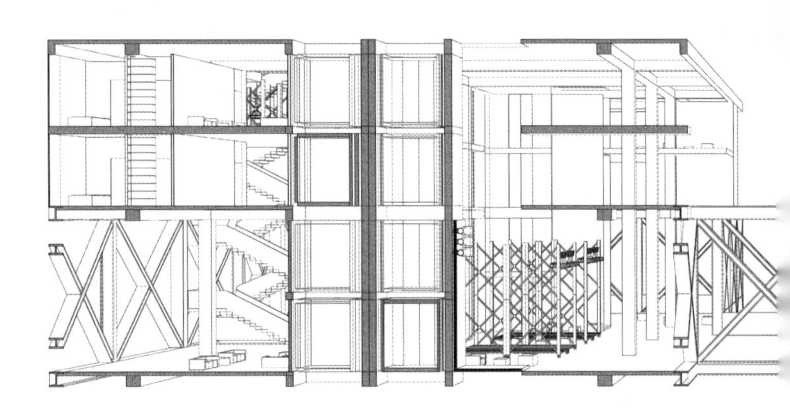

SECTIONAL PERSPECTIVE I
剖透视 I

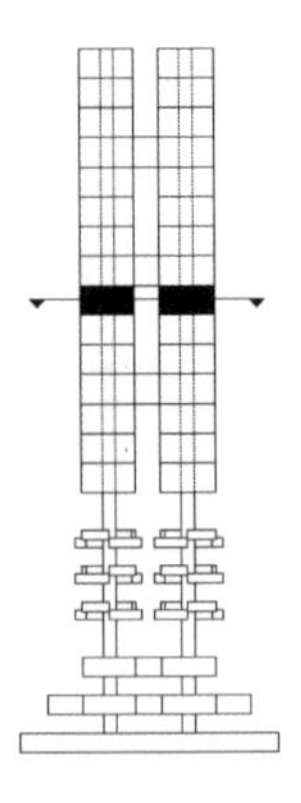

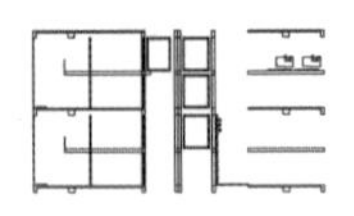

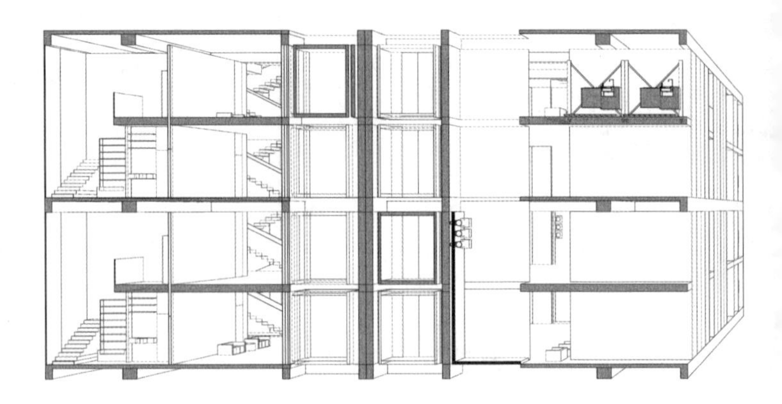

SECTIONAL PERSPECTIVE II
剖透视 II

每四层住宅楼有一个活动层，好让居住在塔楼的文化专家能与其他的专家分享心得、互相学习、交换技能。

整体上，建筑塔主要作为综合住宅单位和移动单元的停泊区，而讲台式区域和底层空间则设有社区活动的共享空间。

每个空间的结构都是以其用处而定制设计。例如，台式区域的设计与高架走道“裕廊步行连桥”有直接的连接，而底层空间是公共的，以方便公众自由进入。

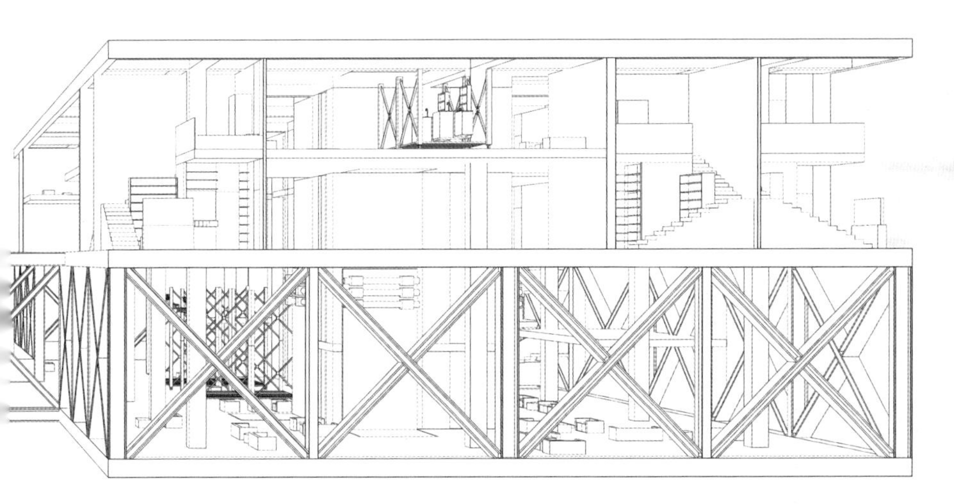

SECTIONAL ELEVATION I
剖立面 I

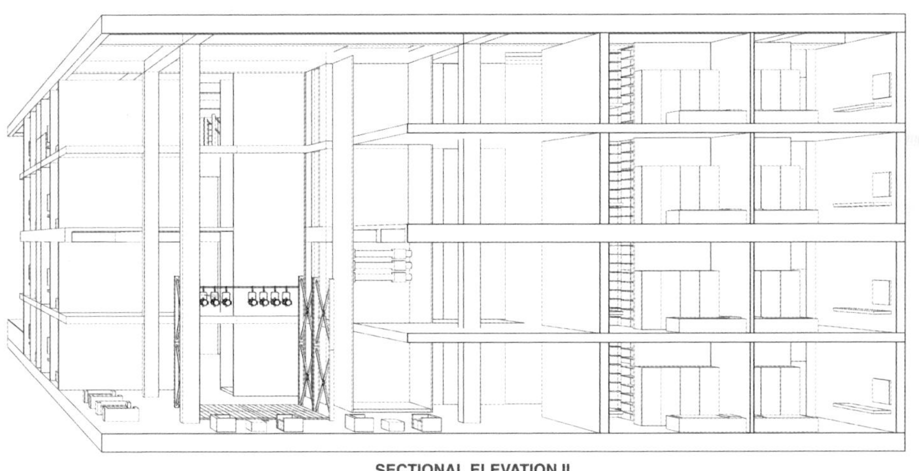

SECTIONAL ELEVATION II
剖立面 II

CC2.0

社区交流中心

"CC 2.0" functions as a community-communication hub that encourages back-packers and surrounding residences to engage in information and skills exchanges. Various facilities are provided along the extended "J-walk," in hopes of spurring bartering and other bottom-up initiatives.

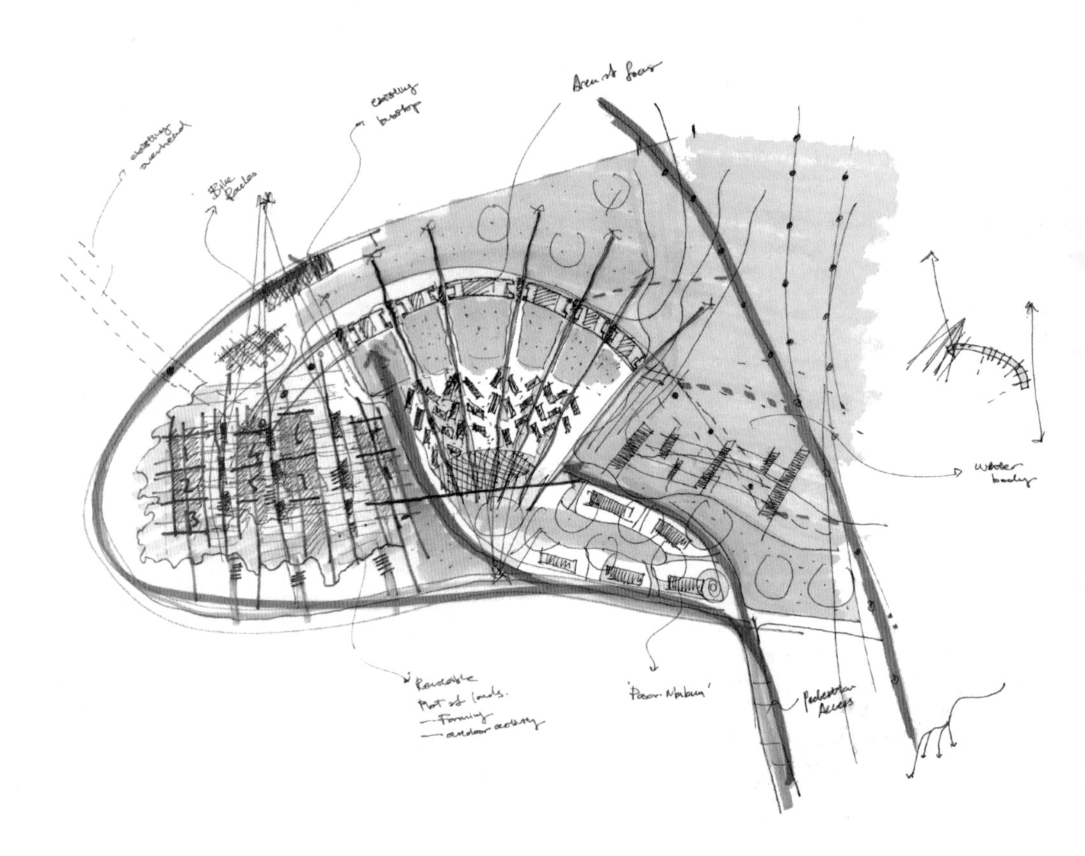

Conceptual sketch showing design decisions made in terms of overall form and spatial zoning.

整体形式与分区决策的概念草图。

The backpackers' inn does not just target travellers but is conceived as a socially minded communication hub which sensitively caters for the local residents and surrounding context.

社区交流中心主要是促进背包客和居民之间的沟通和技能交流。加上沿着现有高架走道“裕廊步行连桥”所提供的多样化设施，该设计方案也希望民众能与他们积极地互动。

背包客酒店的主要目的不仅是给予旅客住宿服务，也借此机会增加社区的凝聚力。

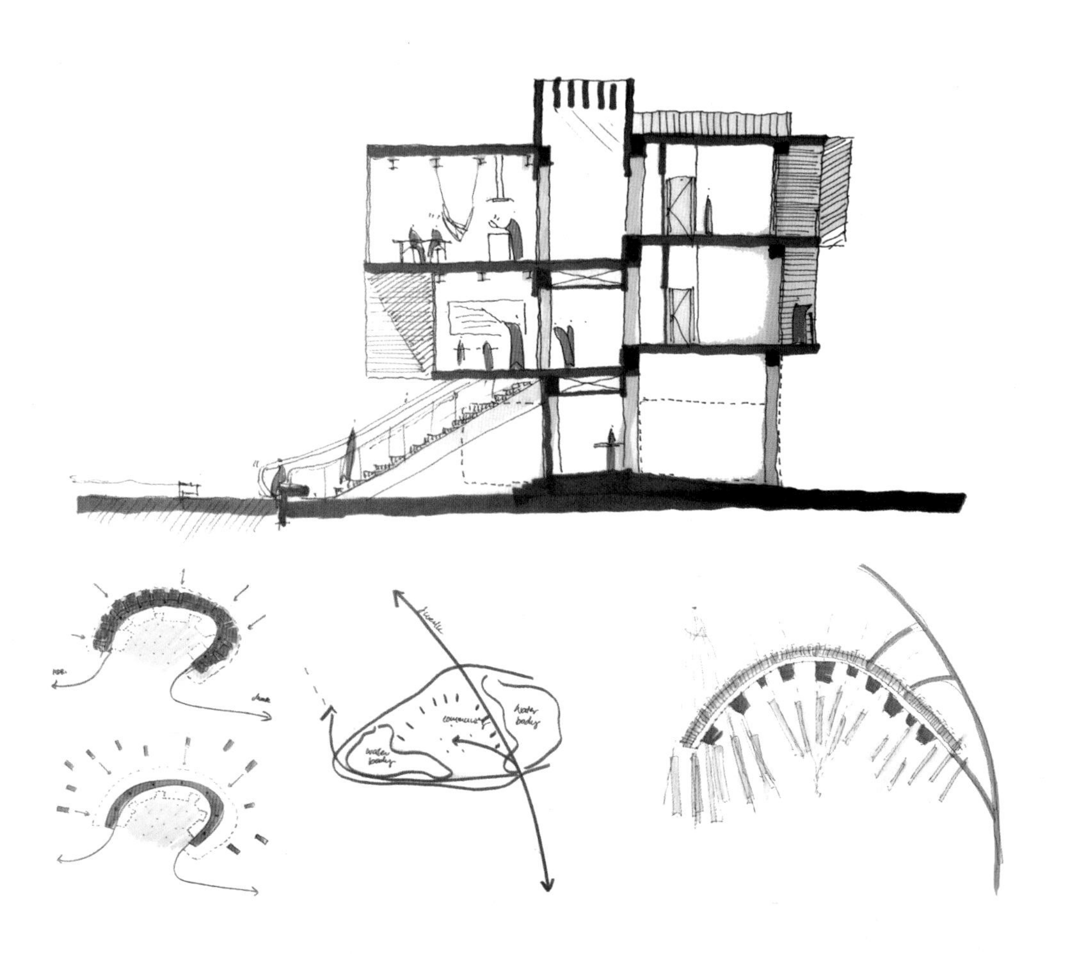

Section sketch and diagrams showing the flow of mobility pods and site response.

移动单元移动性的剖面草图与分析图。

The temporary and permanent infrastructures are the two key elements of the communication hub which caters for the diversity of programmes serving the users. Temporary spaces are self-driven mobile pods which will be docked in allocated hubs in the building.

The ground floor caters mainly for public workshop events where the mobile pods are able to access. Second floor is used for elevated public access to adjacent J-walk and is also where backpackers reside with shared amenities. The third floor houses more rooms for backpackers and various facilities.

Ground Floor Plan

底层平面图

临时的和永久的基础设施是交流中心中两项关键要素，以提供多样的项目服务不同的使用者。临时空间是自组织的移动单元，能够安置于建筑中指定的位置。

地面层便于移动单元的设置，主要服务公共活动。第二层则是背包客的住所，同时也与“裕廊步行连桥”接通。第三层拥有更多提供给背包客的房间与服务。

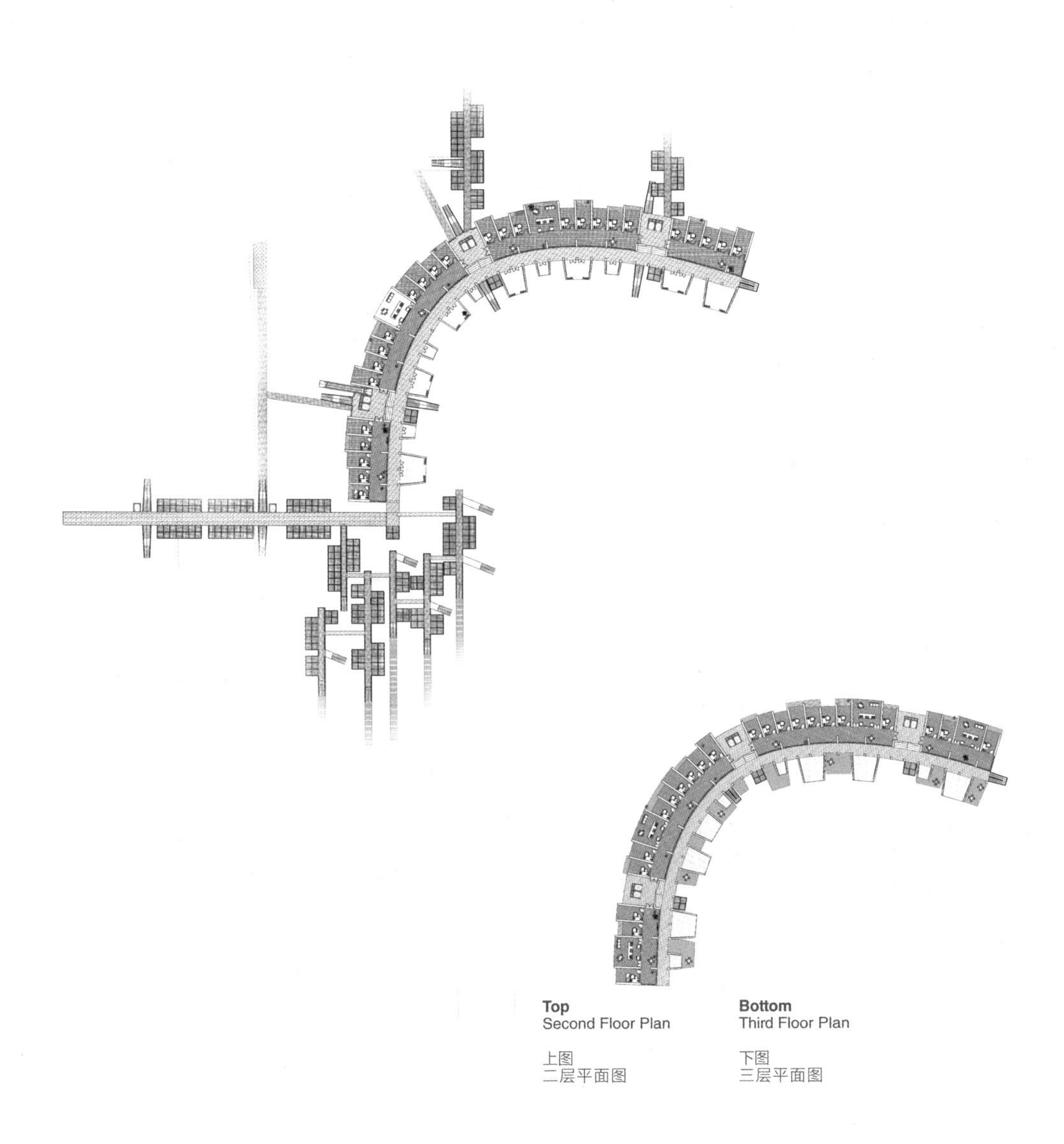

Top
Second Floor Plan

上图
二层平面图

Bottom
Third Floor Plan

下图
三层平面图

The sharing practices envisioned between the local residents and backpackers are informal and spontaneous in nature, where skills and and knowledge sharing form a part of their daily lives. They care about their standings with each other while relationships built enable them to learn from each other.

Collective learning in this instance of a shared domain thus go beyond facilitations by formal institutions, and instead occur in informal neighbourly manners. Apart from being engaged in joint activities, discussions and information sharing, more significantly they seek to look out for each other in the semblance of community formation.

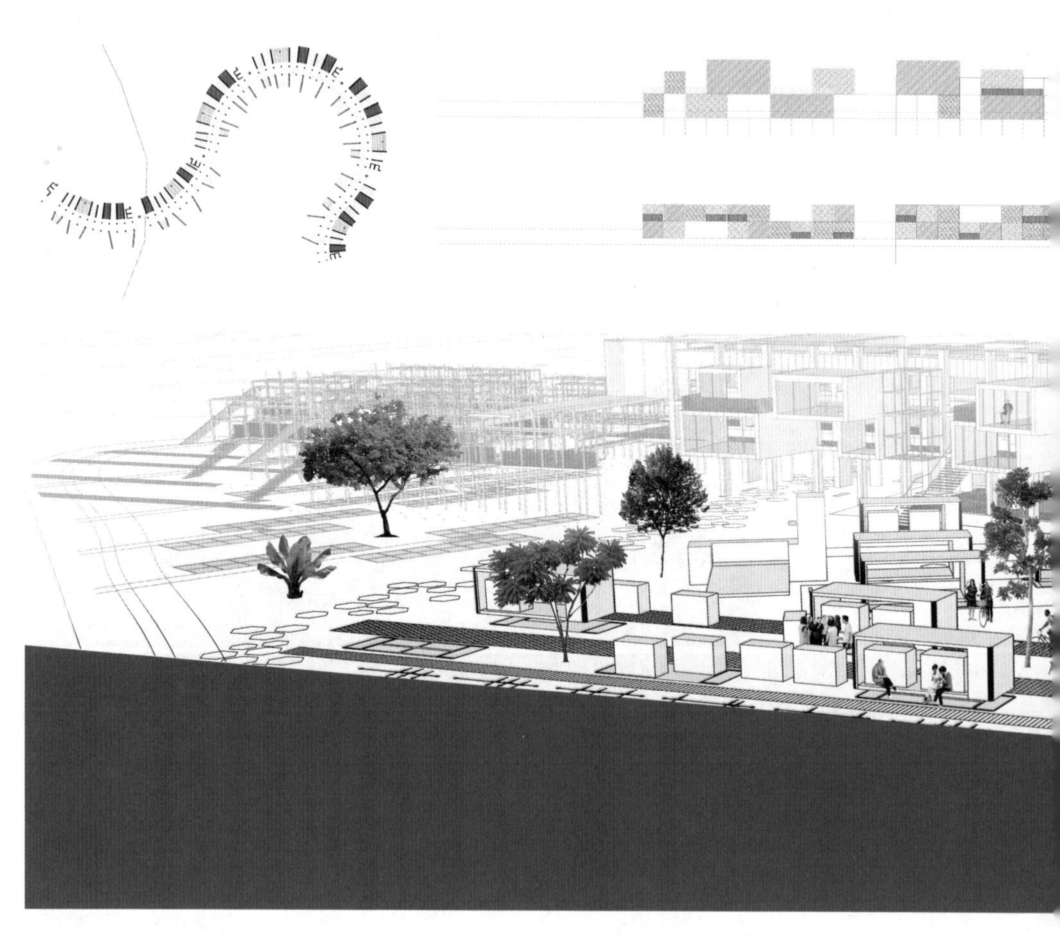

Top
Facade configuration reflecting an even mix of different types of units.

上图
不同类型单元整体混合的侧面图。

通过此设计方案，居民和背包客之间将在日常生活里参与非正式的共享活动。他们之间所建立起来的友好关系将促使他们相互学习、互相帮助。

在这种情况下，居民与背包客的学习精神并不需要正规机构的介入与刻意培养，而是通过非正式的共享活动塑造的。

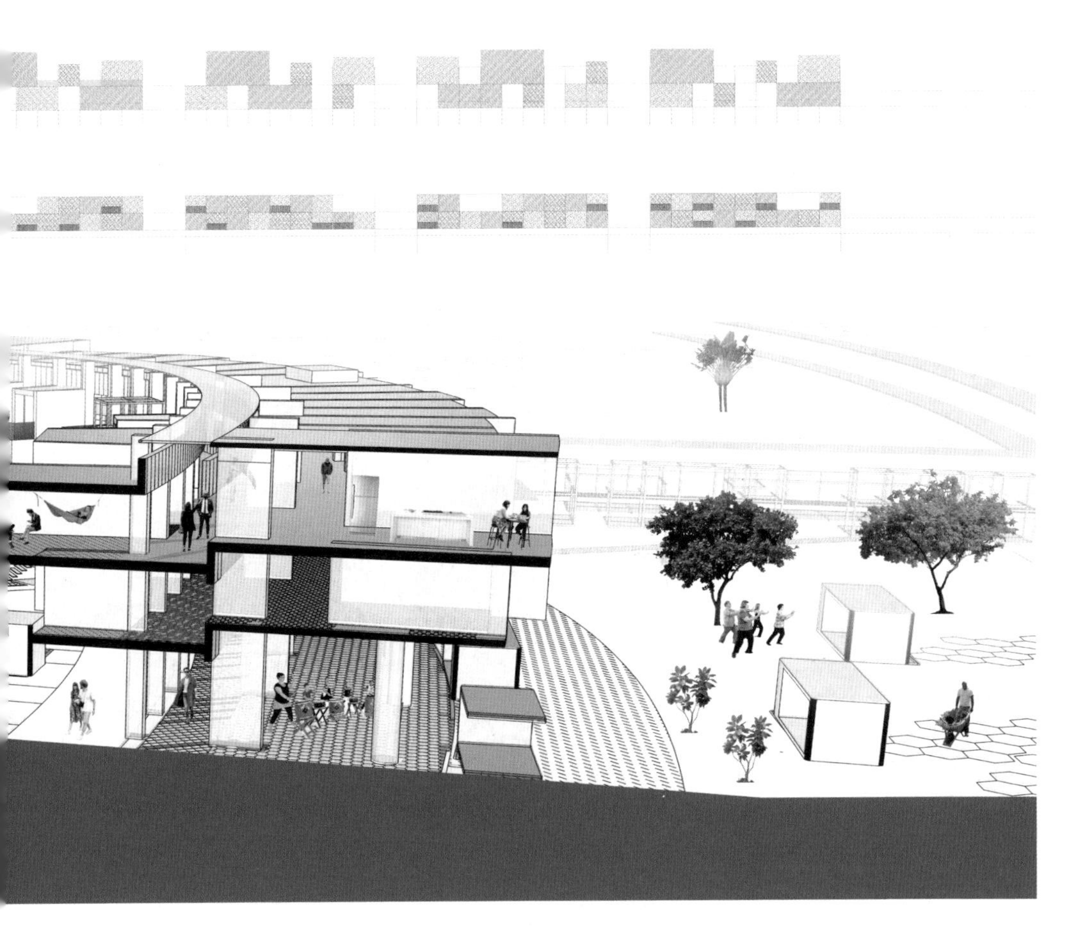

Bottom
Sectional perspective of "CC 2.0" illustrating sharing activities.

下图
“社区交流中心”里各种分享活动的剖面视角。

BIO-CYCLE

By Tan Chiew Hui

有机循环

STACKED KAMPUNGS

By Liew Yuqi

立体村落

This project imagines how urban commons may take form in view of the future challenges posed by the transformation of Jurong East into a CBD. It addresses the displacement of existing communities and inflow of new communities, specifically the low-skilled individuals and new professionals respectively, by introducing a research and living facility. Common pool resources of food, water and energy are co-produced from waste whilst living spaces are built and reconfigured by inhabitants through scaffolding structures. In short, this project hypothesises that these two implicated communities will initiate and engage in communing through their collective interests to adapt. This is fuelled by the motivation for new communities to assimilate and for existing communities to stay relevant.

该项目描绘了在裕廊东转变为CBD过程里面临的挑战之中，城市共同体如何形成的景象。它强调了新型社区潜在的对既有社区的取代可能，特别是新的专业人才对低技能个人的影响，通过研究型的活跃设施的植入来进行。食品、水和能源的共同资源池利用废弃物生成的同时，社区生活空间也在居民的搭建之中形成。简而言之，该项目设想了低技能人员和专业人士在共同兴趣驱使下，在社区中交流互动，激发了新旧社区的交融同化。

STIMULATING URBAN COMMONS

促进城市共同体的精神

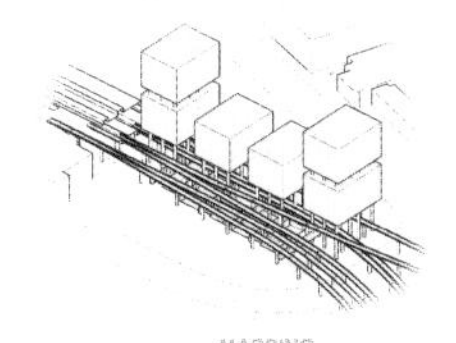
MASSING

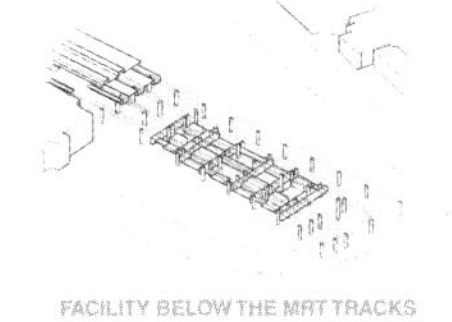
FACILITY BELOW THE MRT TRACKS

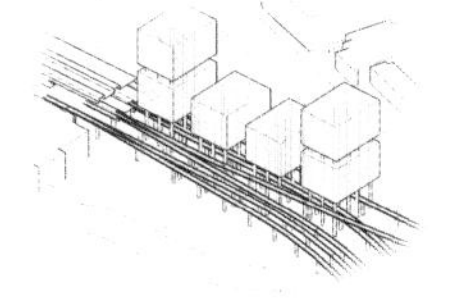
RESIDENCES ABOVE THE MRT TRACKS

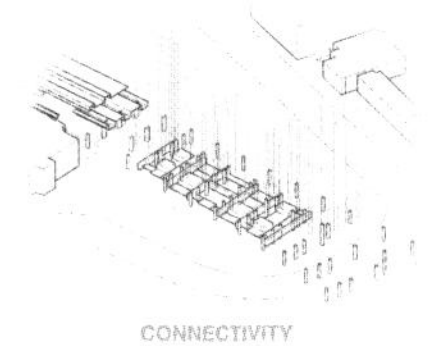
CONNECTIVITY

Top
As co-production and co-consumption occur concurrently, common pooled resources of waste are upcycled into 3D-printed bicycles, farm produce and energy, creating opportunities for collaborative living and working to take place.

上图
随着共同生产与共同消费同时进行，共享资源池的废物将被升级为 3D 打印的自行车、农业产出物与能源。这为共享居住和共享办公创造了条件。

Bottom
Underutilised spaces beneath and over the MRT tracks houses complementary co-living and co-working facilities, which are integrated through their form and circulation.

下图
捷运轨道下未充分利用的空间，设有共享居住和共享办公设施，其建筑形式与流线进行了整合。

The mechanism for the symbiotic relationship of co-existence, collaboration, and co-production between the two communities is an eco-system managed and sustained by them. Situated below and above the MRT tracks adjacent to Jurong East MRT station, the edifice is sited on "unvalued" territory. Such is a gesture in line with the principle of the sharing economy – that of identifying value in surplus resources and appropriating them for alternate uses.

The combined proposal assimilates both populations through this transition of an epoch through the design semblance of the urban commons. Embedded within the processes of co-production are spontaneous social exchanges and formation of social relations.

In addition to facilitate communing, this project perpetuates the spirit of sharing in the larger context of Jurong East.

50 PEOPLE
(~ 10 HOUSEHOLDS)

VEGETABLES
(SELF-SUSTAINING)

281 plants/person per year
3-4 months maturation
0.25 sq metre for each plant
17.5 sq metre per person (including rotation of 3 months)
17.5 X 50 persons=
875 sq metres (FARM)
(~30m X 30m)

0.0254m X 875 sq metres=
22.3 cubic metres (COMPOST)
(~5m X 5m X 1m)

ELECTRICITY
(DAILY USE)

A 3-room flat household electricity consumption averages at 250 kwh per month
10 households X 250 kWh = 2500 kWh per month
1 tonne of biomass generate 60 cubic m of biogas
60 cubic metre of biogas generate
120kWh of electricity
240kWh of heat
20 tonnes of food waste/month to be fully sufficient
food waste density = 600 kg per cubic metre
20 000 / 600 =
33 cubic metres (BIOMASS)
(~7m X 5m X 1m)

33+22.3=
55.3 cubic metres (FOOD WASTE COLLECTION)
(~8m X 7m X 1m)

WATER

151 litre/person per day
20 litres/shower X 2
2 litre drinking water
109 litre for other uses
(laundry, washing etc)
151 X 50persons = 7550 litre
7550 / 1000kg m^-3 =
7.55 cubic metres (RAINWATER)
(~4m X 2m X 1m)

Based on the average amount of food, electricity and water comsumption of an average person in Singapore, the amount of resources required to provide for 50 people is determined. This informs the size and type of space requirements in the design.

根据新加坡人均食物、水和电消耗量的数据，50 人所需的资源量被估算出，进而得到了设计中所需空间的类型与大小。

位于裕廊东捷运轨道下的空间将给予两个社群一起共同生活、合作以及达成共识的机会。此设计方案找出了未充分利用空间的价值，从而以替代用途来促使城市共同体的产生,毫无疑问，这符合共享经济的原则。

这两项设计方案通过城市共同体，提供给两个社群成员生存及适应的机会。与此同时，他们也能在协同合作的过程中，彼此了解并增强社群意识。通过这项计划，我们也希望裕廊东居民能见证共享精神所带来的好处。

100 PEOPLE/DAY
(50 LUNCH, 50 DINNER)

14.9g
per aluminum can
14.9g X 10 persons =
149 g per day

12.7g
per plastic bottle
12.7g X 10 persons =
127 g per day

aluminum frame - 1.2kg
aluminum seatpost - 220g
8 days
to accumulate

carbon handlebar - 200g
carbon bike fork - 300g
carbon seatpost - 200g
carbon fiber - 200g
7 days
to accumulate

The time needed to produce parts for a completed bicycle is determined by estimations of plastic and metal recyclables that can be collected each day.

通过塑料和金属可回收物的收集量，自行车各个零件所需要的生产时间也被计算出。

有机循环

BIO-CYCLE

Beneath the tracks are collaborative spaces and training facilities for bicycle manufacturing. A pneumatic waste system running along the length of the tracks collects waste from neighbouring residential estates.

While food waste is turned into biogas energy and compost for urban agriculture; plastics and metals waste are up-cycled into raw materials for the 3D printing of rental bicycles.

Perspective view of the ground floor, which is publicly accessible.

底层公共空间的透视图。

位于轨道下方是为自行车制造提供给两个社群的合作空间和培训设施。沿地铁轨道设置的气动垃圾回收系统，把邻近住宅区的废物输到制造厂，以进行再循环处理与利用。

食物残渣首先被转化为沼气能以及农业堆肥，可循环使用的塑料和金属将通过机器处理成为3D打印自行车零件所需的原材料。

Perspective from the second floor, displaying the bike manufacturing facility.

自行车制作空间的透视图。

In view of its adjacency to the MRT station, the ground floor is left publicly accessible, providing spaces for eating, waiting, eating and gathering which serve commuters, pedestrians, the lunch time crowd and shoppers.

Programmatically, it locates the manual collection point for food waste, composting area, farming and communal kitchen in close proximity to each other, affording a visible cycle of food waste into agricultural products and collective activities.

While electricity and biogas for the kitchen are produced in-house from food wastes collected, heat for composting and liquid slurry for farming are by-products of these production from the machineries above.

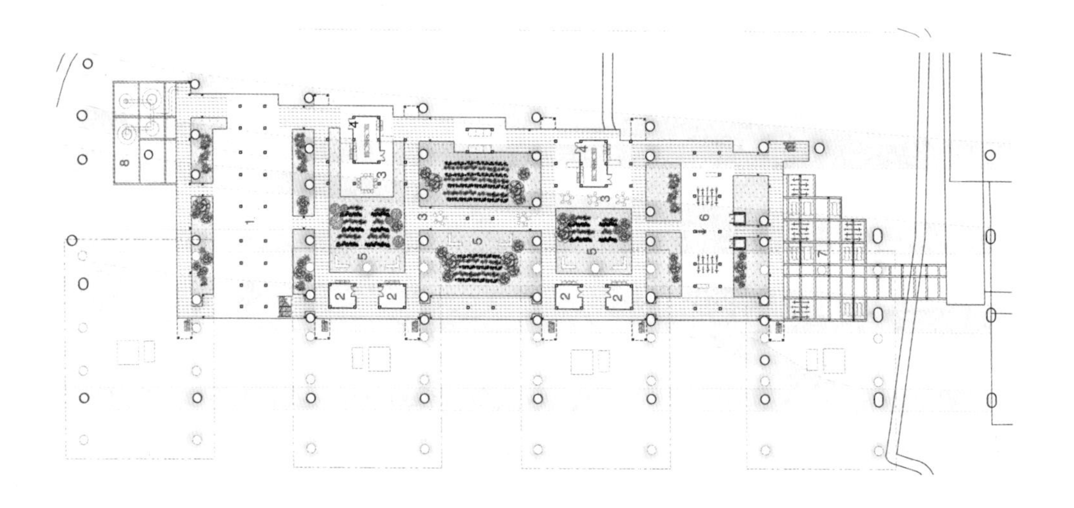

Ground Floor Plan
(1) Multi-purpose Communal Area; (2) Gardening Tool Shed; (3) Common Dining; (4) Common Kitchen; (5) Horticulture & Farming; (6) Bike Repair; (7) Integrated Bike Rental & Public Eating; (8) Water Storage

一层平面图
（1）多用途公共空间；（2）园艺工具棚；（3）公用餐饮处；（4）公用厨房；（5）园艺农业区；（6）自行车修理室；（7）自行车出租和饮食处；（8）储水处

从其临近地铁车站的角度来看，建筑底层的空间留给大众，好让乘客、行人、午餐人群和购物者能方便聚集或休息。

设计方案也把食物残渣收集处、堆肥区、小型农业区和公用厨房安排在同一个范围内，目的是给予公众了解食物来源和再循环的机会，从而让他们见证城市共同体的产生。

公用厨房所用的沼气都是通过食物残渣再循环产生的，而液体浆料和热量是发电机所产生的副产品。在这个再循环的系统里，所谓的废物都不再被浪费。

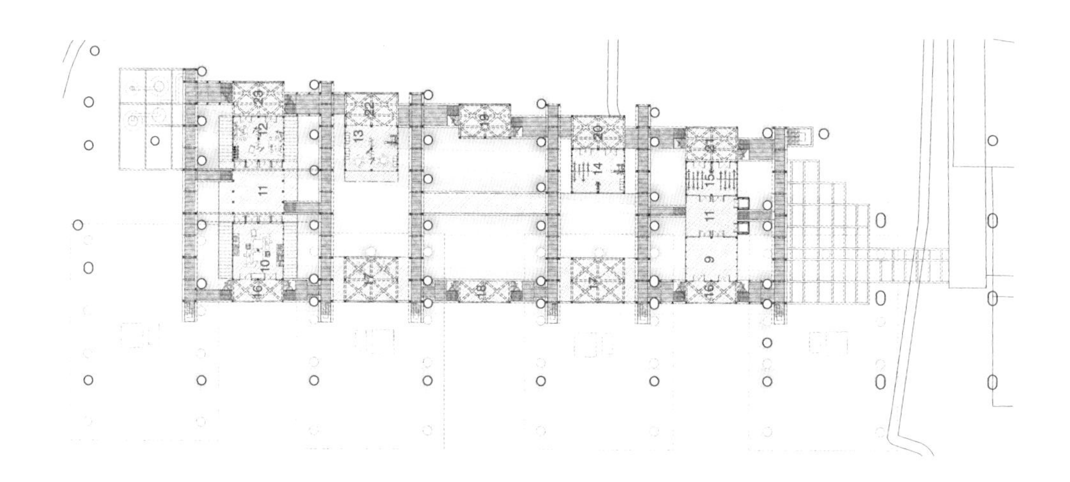

Second Floor Plan

(9) Material Processing Laboratory; (10) 3D Printing Workshop; (11) Shared Space Extension; (12) Primary Bike Assembly Area (13) Secondary Bike Assembly Area; (14) Bike Finishing Laboratory; (15) Bike Repair Workshop; (16) Primary Digestor; (17) Secondary Digestor; (18)Biogas Scrubber; (19) Biogas Storage; (20) Steam Engine; (21) Generator; (22) Boiler (Water Distillation); (23) Condensor (Water Distillation)

二层平面图

（9）材料加工实验室；（10）3D 印刷间；（11）活动空间；（12）自行车主装配区；（13）自行车次装配区；（14）自行车整修实验室；（15）自行车修理间；（16）主要消化器；（17）二次消化器；（18）沼气洗涤器；（19）沼气储存；（20）蒸汽机；（21）发电机；（22）锅炉（水蒸馏）；（23）冷凝器（水蒸馏）

The six main circulation cores connecting "Bio-Cycle" to "Stacked Kampungs" leverage on the strength of the structural columns of the MRT tracks. The openness of each space is also determined by the extent of shade provided by the MRT tracks above.

The corresponding bicycle manufacturing spaces form a looped spatial relationship sequentially, from material collection, material processing, 3D-printing, bike parts furnishing, to bike assembly.

Biogas, electricity and water distillation machineries are located on the third mezzanine floor, directly connected to the pneumatic waste pipes attached to the MRT tracks.

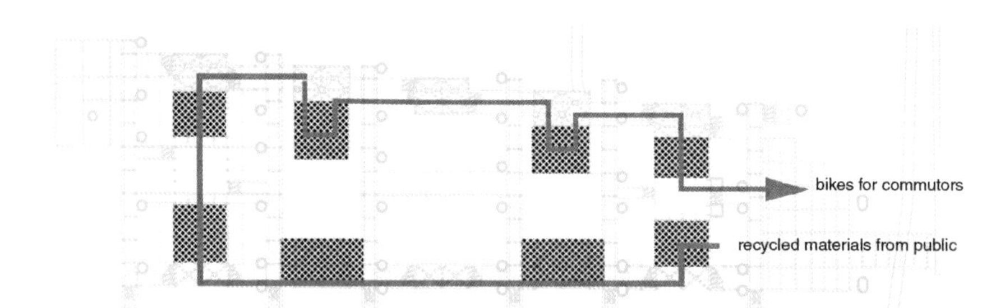

Cycle of Waste Plastics and Metals into Bicycle Parts
回收塑料和金属再循环成自行车零件的图解

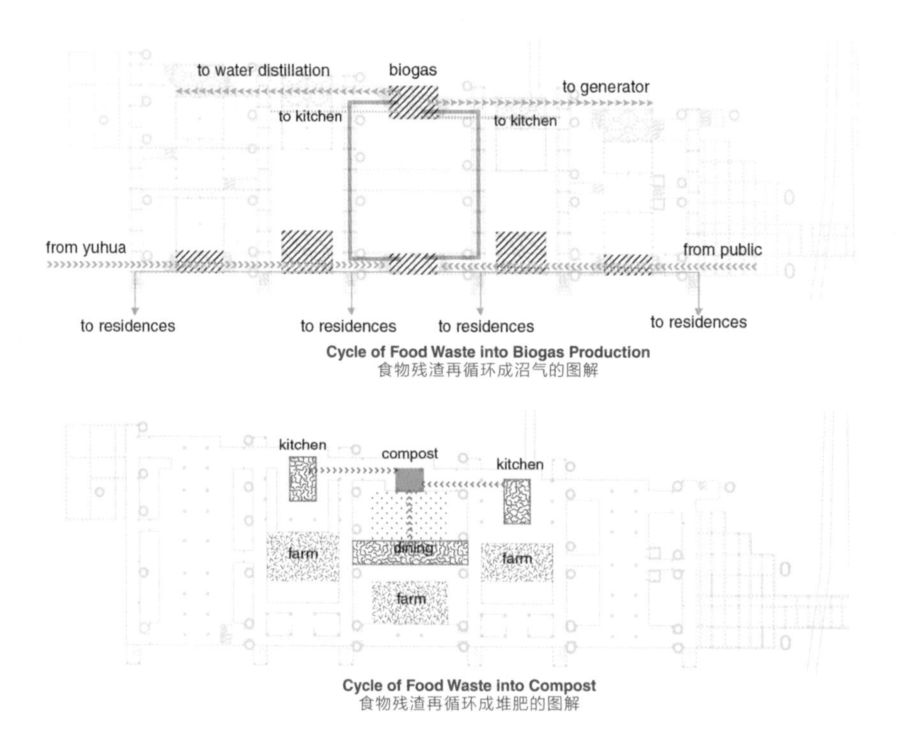

Cycle of Food Waste into Biogas Production
食物残渣再循环成沼气的图解

Cycle of Food Waste into Compost
食物残渣再循环成堆肥的图解

连接“有机循环”和“立体村落”的六个主要人行通路利用了捷运轨道结构柱作为支撑力量。每个空间的开敞性也由上侧的轨道所提供的阴凉处而决定。

自行车制造厂的材料收集处、材料处理室、3D印刷室、自行车组装室，也形成环形的空间序列。

沼气发电机和其他机械设备则位于第三层，与捷运轨道下的气动废料管道直接相连。

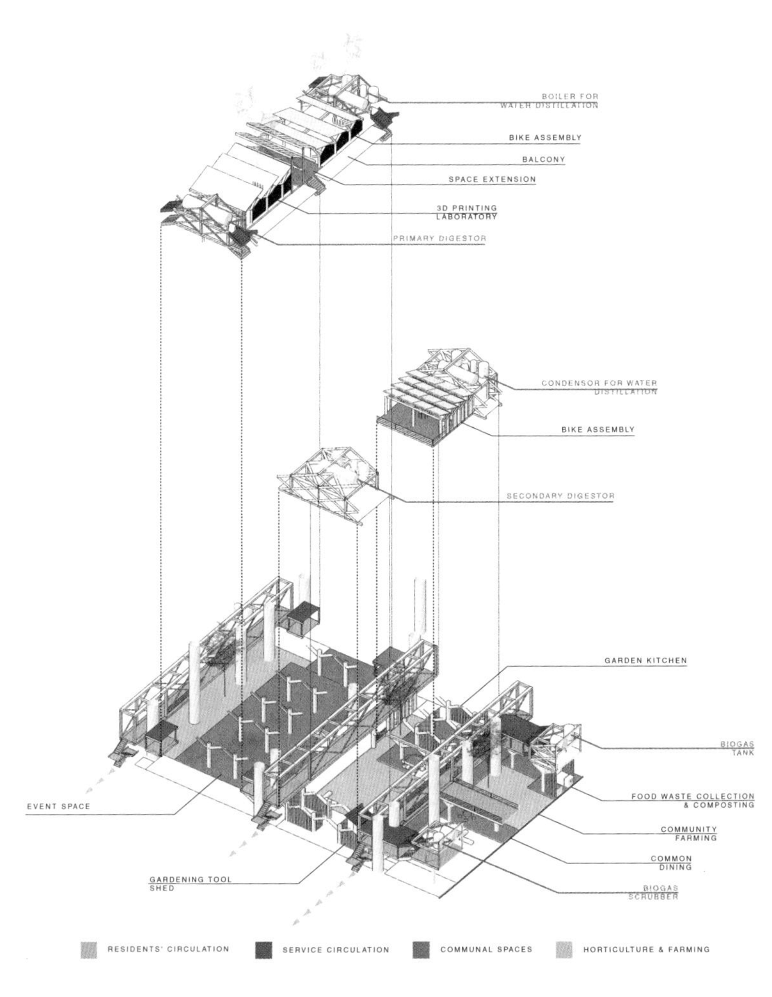

Axometric Diagram Showing Spatial Configuration & Circulation
空间布局与流线的轴测图

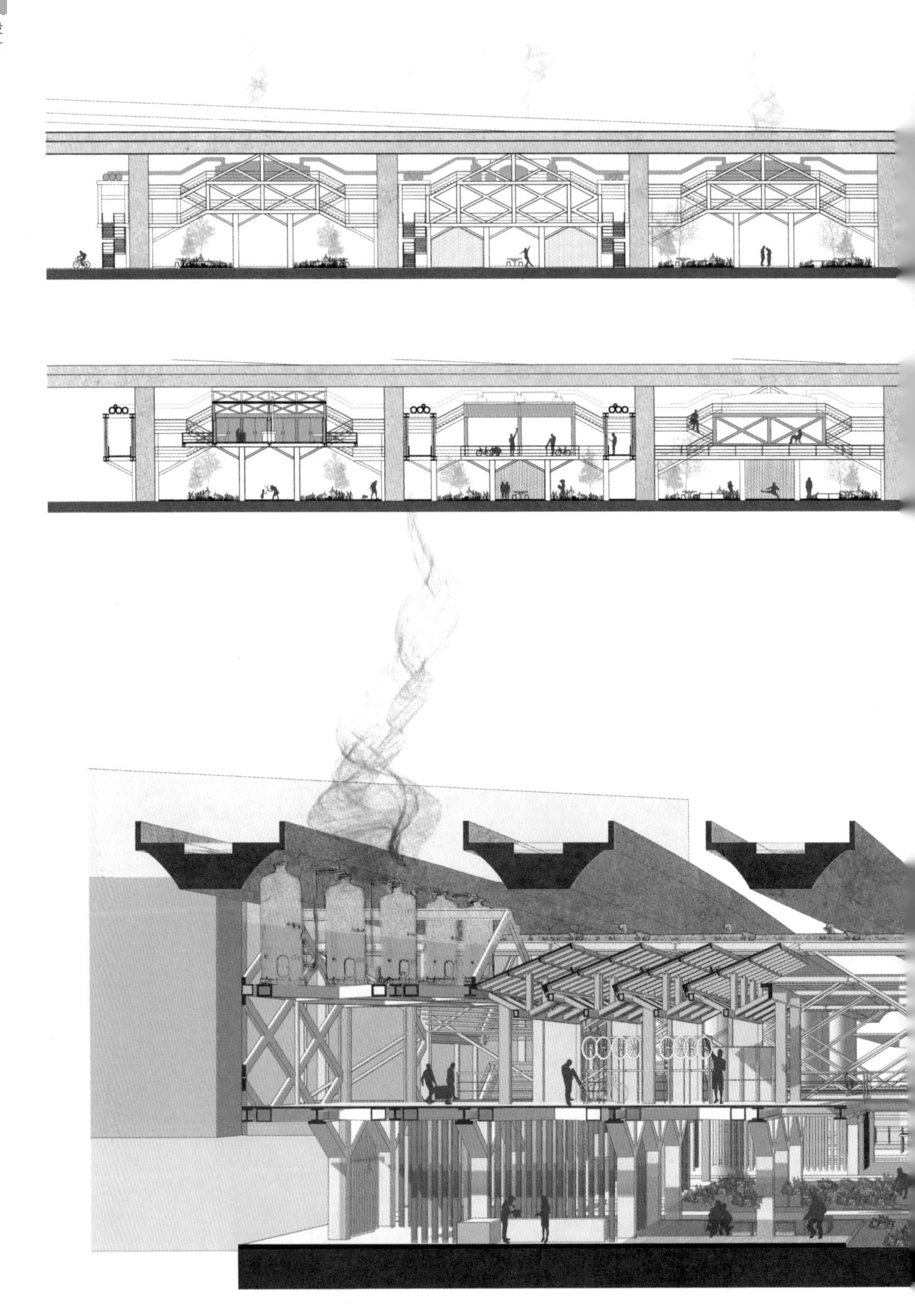

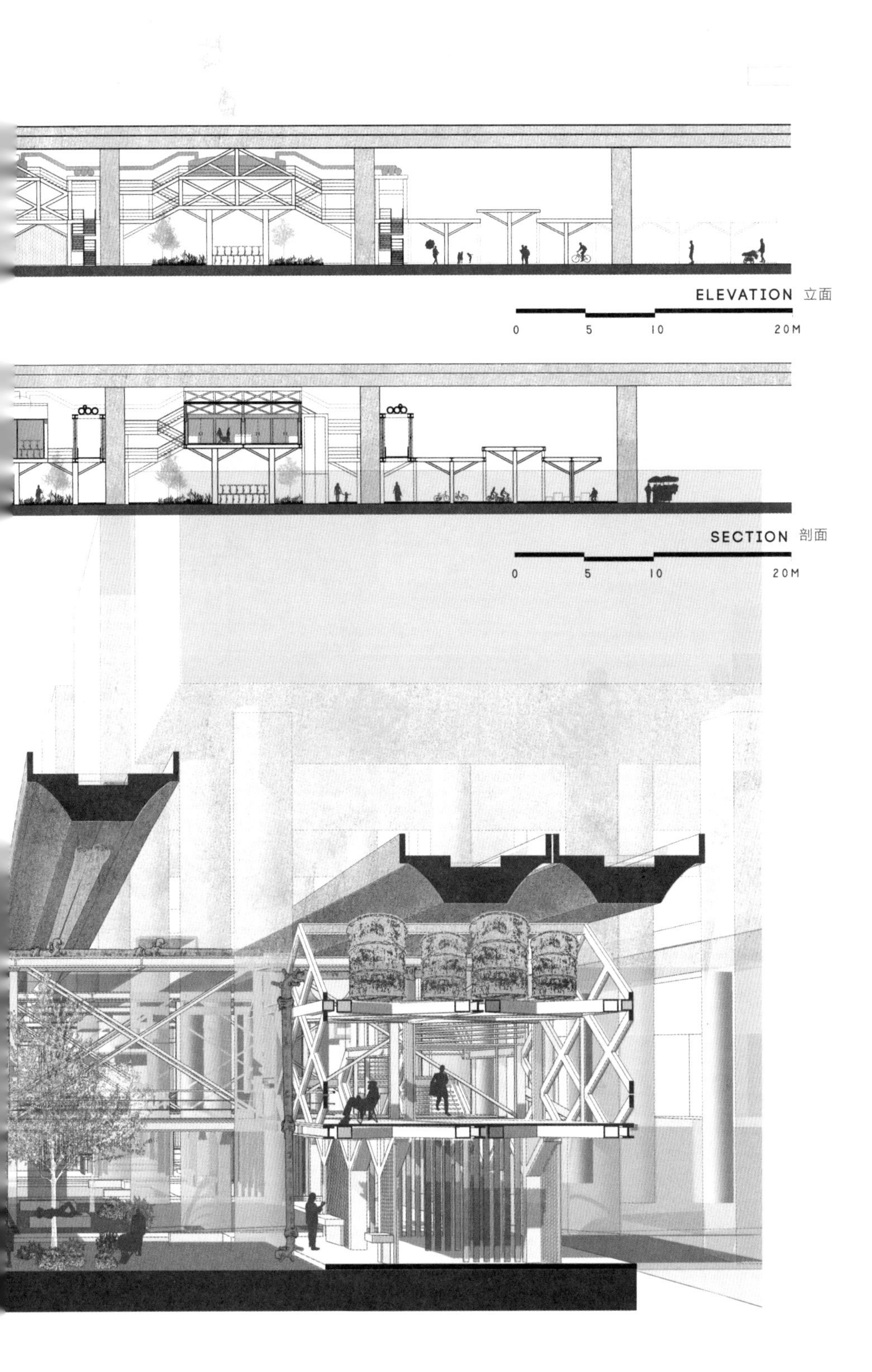
ELEVATION 立面
0 5 10 20M
SECTION 剖面
0 5 10 20M

立体村落

Above the tracks are residential and communal spaces housing both communities connected by greenery. With an aquaponics system that also serves as a façade, the architecture represents a new living typology.

STACKED KAMPUNGS

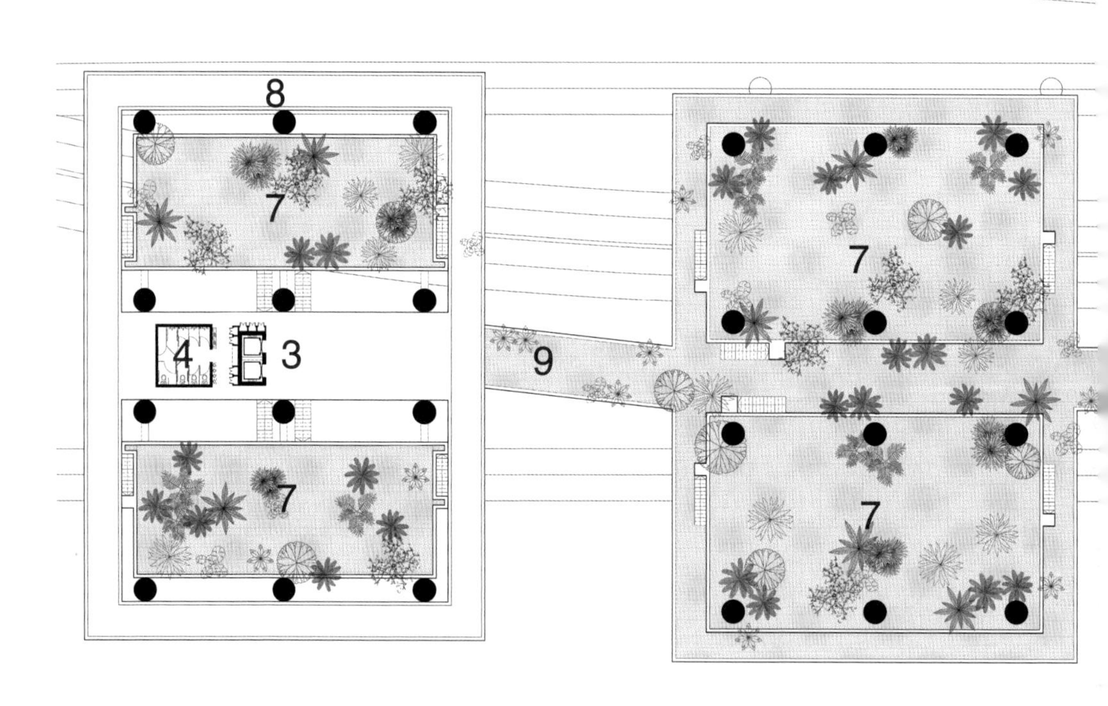

Typical Residential Floor Plan
(1) Living area; (2) Aquaponics; (3) Lift lobby; (4) Communal toilets; (5) Communal laundry area; (6) Communal kitchen; (7) Roof Garden; (8) Running track; (9) Garden bridge; (10) Running bridge

The scaffold housing system is co- produced through accessible assemblage and flexible reconfiguration. Communal and shared amenities are at the core of the building, maximising informal sharing and interaction between neighbours.

位于轨道上方承载不同社群的居住空间和公共空间以绿化空间相连接。鱼菜共生系统的架构将成为建筑的立面，此建筑的整体设计体现了新兴的生活形式。

具有灵活性的脚手架外壳系统能让居民轻易地重整与设计住宅空间。公共设施将位于建筑空间的中心，好让大家随时方便地聚集在一起，增加邻居之间的非正式的共享与互动。

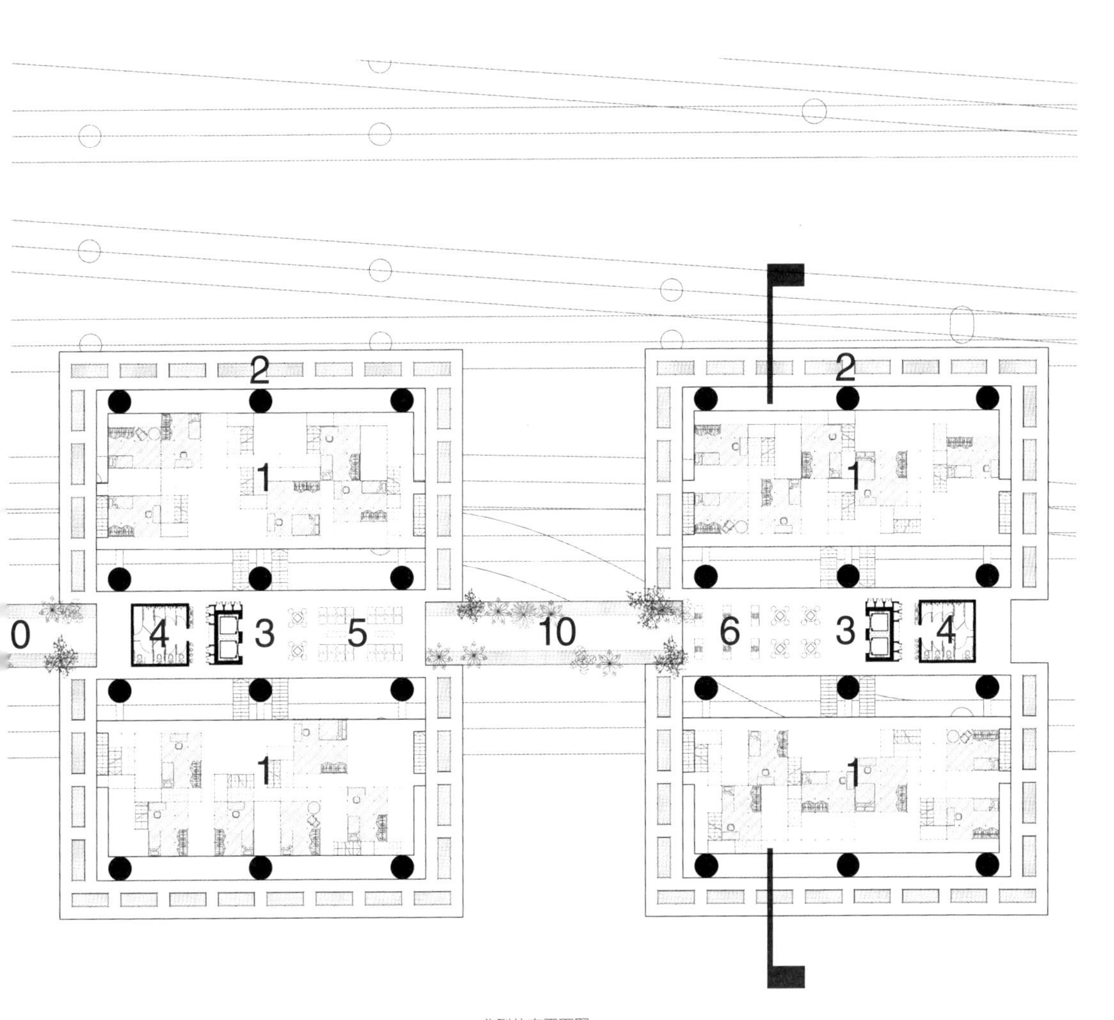

典型住宅平面图
（1）生活区；（2）鱼菜共生；（3）电梯大厅；（4）公共厕所；（5）公共洗衣区；（6）公用厨房；（7）屋顶花园；（8）跑道；（9）花园桥；（10）跑桥

Alternate families unrestricted to the boundaries of a nuclear family are expected to form, from which they are served by complementary capabilities of each community.

由陌生人聚集居住在一起而形成的“替代家庭”中的每位成员，将为居住社群提供不同的协助与服务。

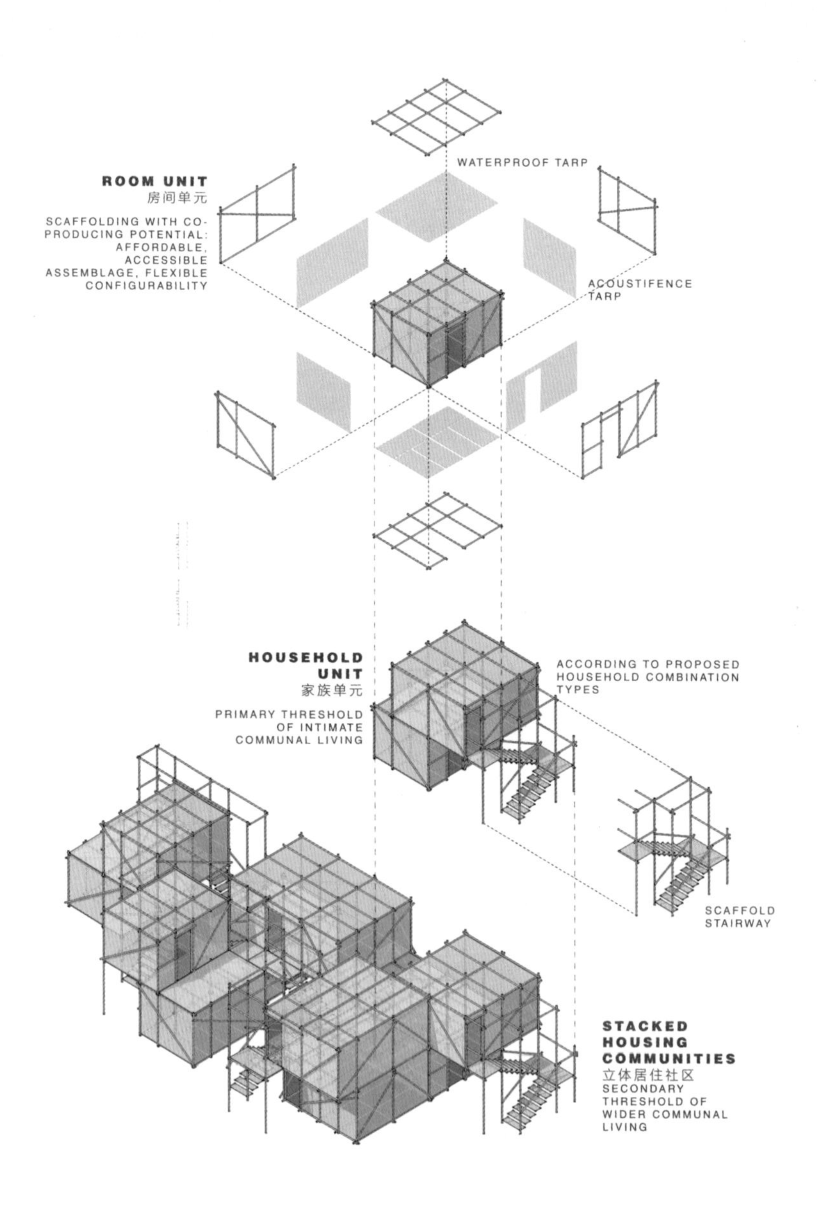

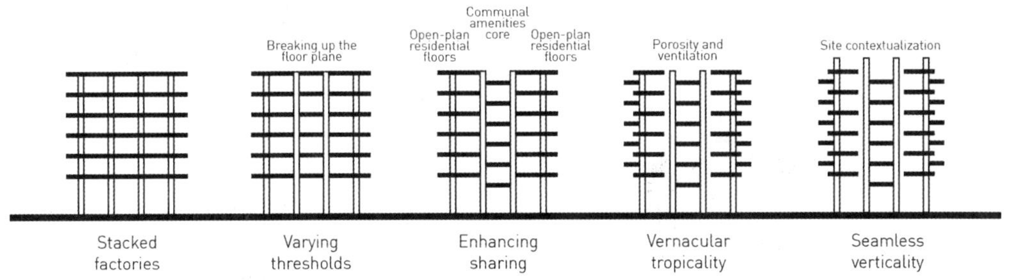
Communal amenities core
Breaking up the floor plane
Open-plan residential floors
Open-plan residential floors
Porosity and ventilation
Site contextualization
Stacked factories
Varying thresholds
Enhancing sharing
Vernacular tropicality
Seamless verticality

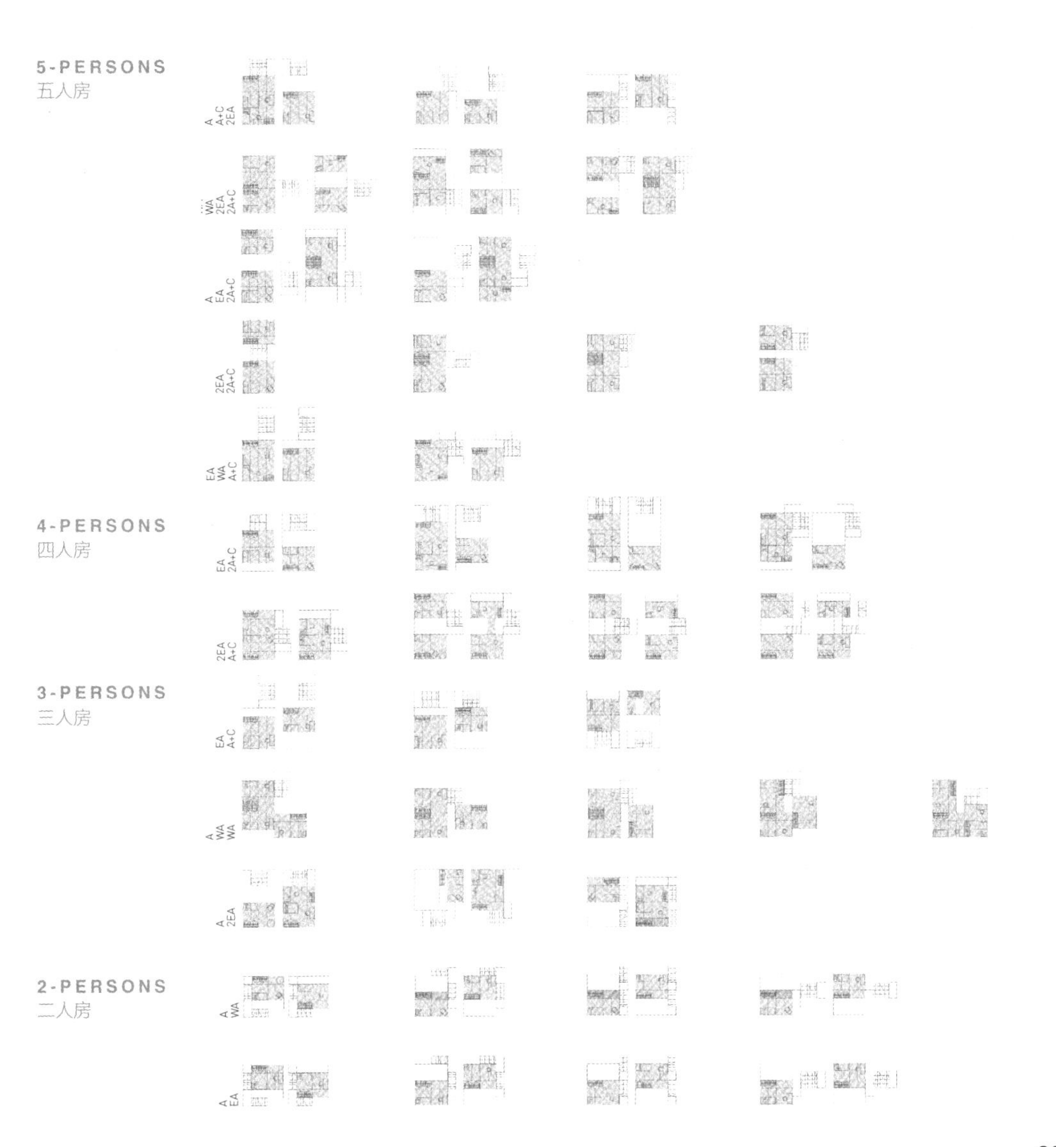
5-PERSONS
五人房
A
A+C
2EA
WA
2EA
2A+C
A
EA
2A+C
2EA
2A+C
EA
WA
A+C
4-PERSONS
四人房
EA
2A+C
2EA
A+C
3-PERSONS
三人房
EA
A+C
A
WA
WA
A
2EA
2-PERSONS
二人房
A
WA
A
EA

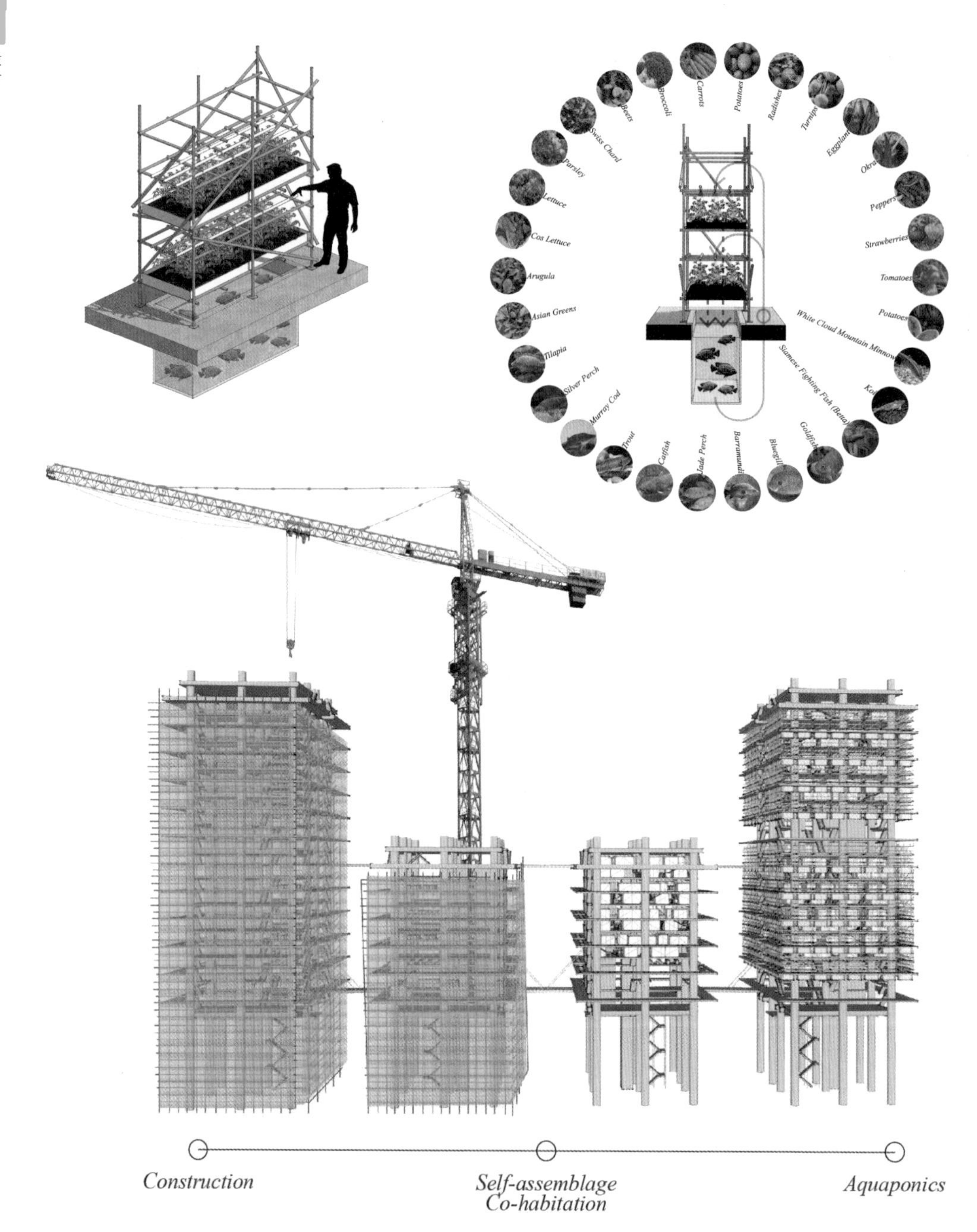

Top Left
Aquaponics is a sustainable source of agriculture and fish as food. The system can be created out of scaffold. This food production system becomes the façade of the architecture.

左上图
鱼菜共生系统可提供具有持续性的食物来源，它的结构是由脚手架制成的。由脚手架搭建而形成的接头系统也将因此成为建筑的侧面。

Right
Waste water is nutrients to the plants while purified water may be reused for the fishes. A variety of plants and fishes can be grown with varying growth conditions.

右图
鱼菜共生系统能给予各种植物和鱼类适合的生长环境。植物的营养物质来自于鱼的粪水，而经过植物过滤的净水可以重新用于养鱼。

Bottom Left
Scaffolding used in the construction of the super structure can later be used as a resource for the co-production of liveable spaces. It is also used as a frame for the aquaponics system.

左下图
施工建造过程中的脚手架可作为除了作为住宅单位外，还能被用作鱼菜共生系统的框架。

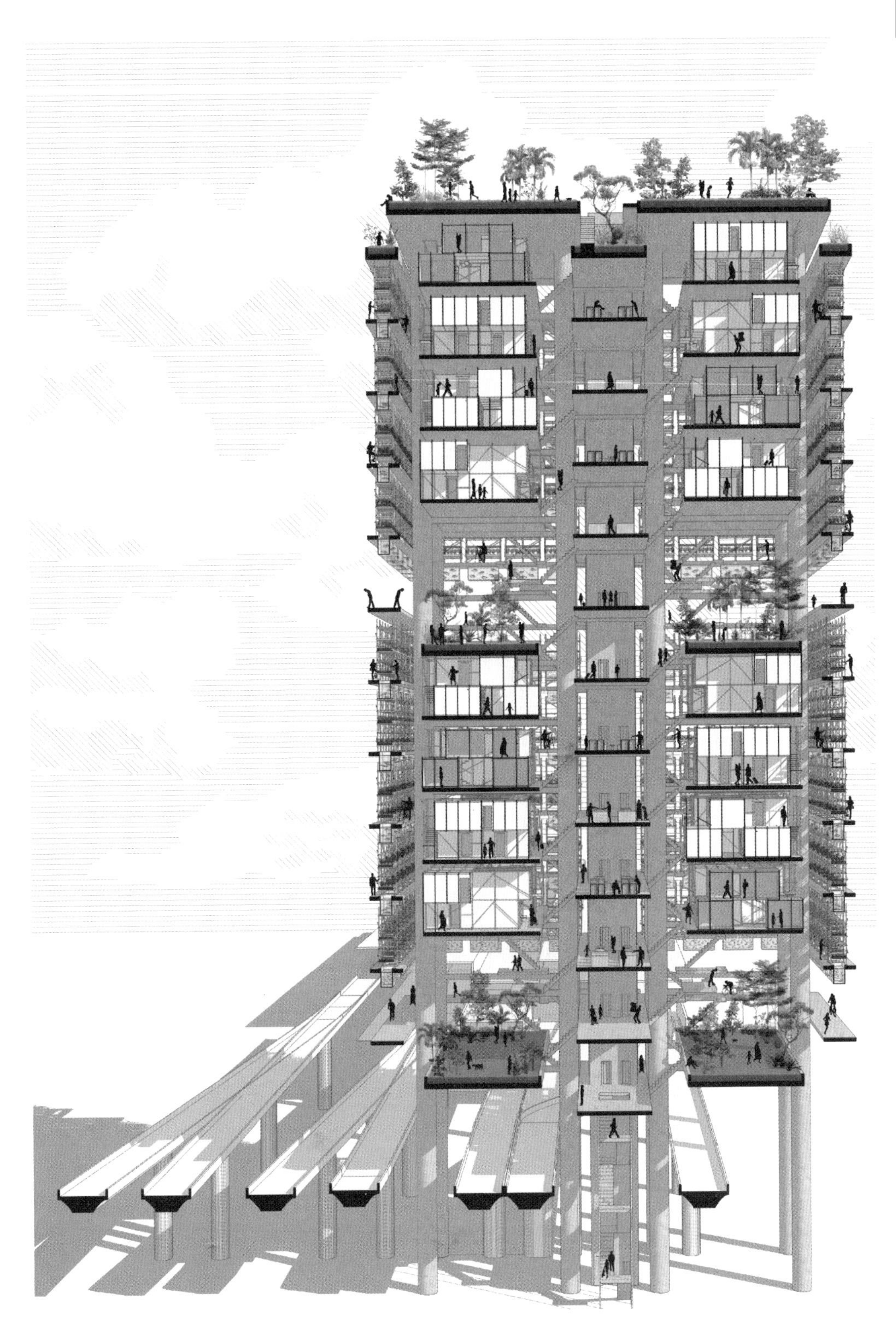

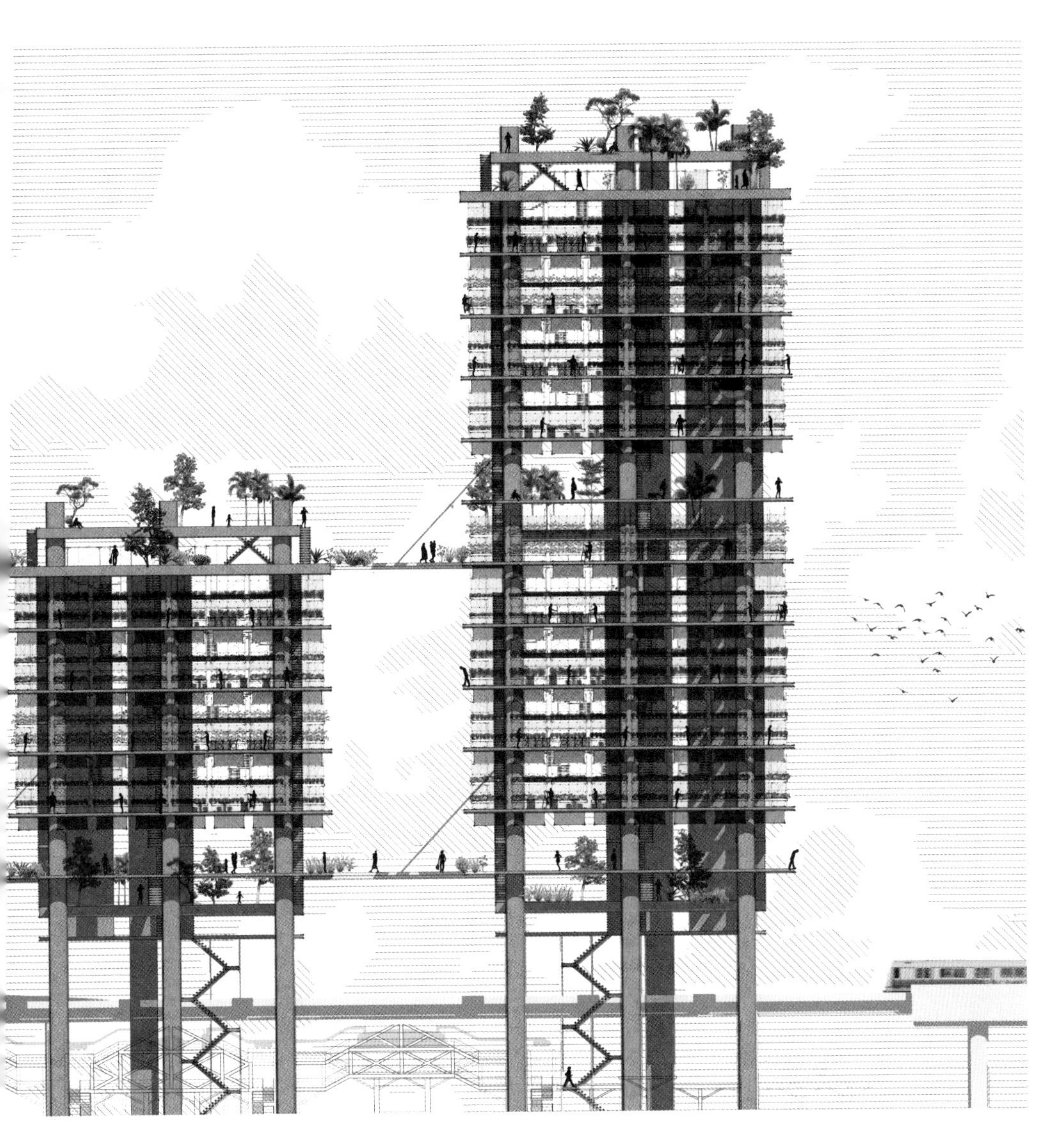

ARTSTREAM

By Gloria Neo

艺游

This project anticipates the importance of the cultural dimension in Jurong East as an emerging form of urban commons in the city. Cultural activity is one of the key producers to urban commons (McLaren and Agyeman 2015) as creative clusters emerging in the urban sphere tend to be motivated by common interests of creative individuals. The rise of the sharing economy has revolutionised the distribution methods of the arts and products of the creative industry. Distribution and advertising become just as important as the skill of making, in order for creatives to share their skills with the world.

这项设计预期文化活动将成为裕廊东城市共同体的新兴形式。当创新个体的共同兴趣驱动城市层面的创意集群出现，文化活动成为城市共同体的主要生产者之一。因此，文化产业对于城市共同体的产生将有巨大的影响。近年来，共享经济的兴起已彻底改变了艺术的分销方式和创意产业的生产方式。为了创意者在全球范围内分享他们的经验，艺术的分销与广告和艺术的生产技能变得一样重要。

ANTICIPATING EMERGING FORMS

IMPORTANCE OF MUTUAL RELATIONSHIP BETWEEN UPSTREAM & DOWNSTREAM ARTS*

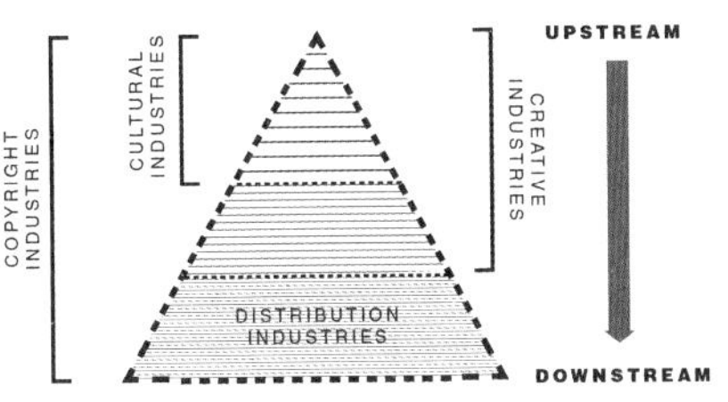

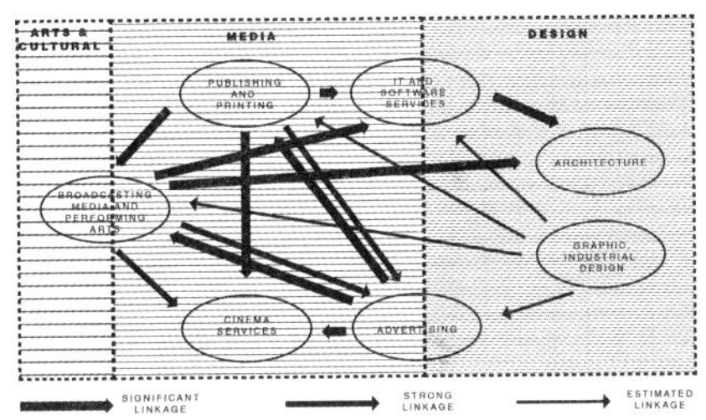

INCREASED ACCESSIBILITY AND VALUE OF ART

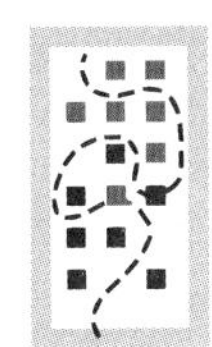

GEOGRAPHICAL SEGREGATION

PHYSICAL INTERACTION

EXISTING 10X10M GRID SYSTEM IN MASTERPLAN

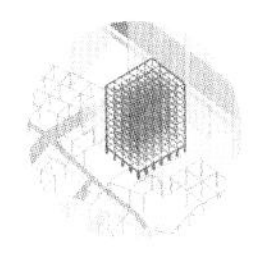

INTRODUCING 5X5M GRID

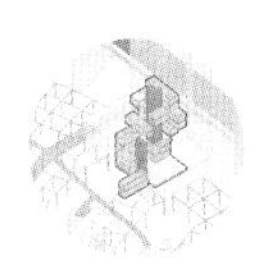

2 CORES FOR UPSTREAM AND DOWNSTREAM ARTS

INSERTING IN VOLUMES ACCORDING TO PROGRAMMATIC NEEDS

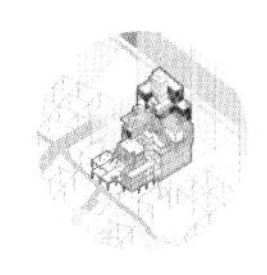

LINKING VOLUMES TO FORM CONTINUOUS CIRCULATION

Top
With a paradigm shift in the way art is distributed and made, sharing in the cultural industry evolves following the increased accessibility of the arts

上图
艺术品创作与流通方式的改变，使得共享在文化产业中也随之改变

Bottom
As site response is integrated with the spatial translation of upstream and downstream art, design decisions are made with respect to the overall massing and spatial volumes

下图
当“上游”和“下游”空间与场地结合起来，设计的整体性也随之产生

艺游：游艺

ARTSTREAM

While new media allows for increased equity of public accessibility to and appreciation of the arts, excessive distribution fuelled by the virtual commons of the Internet also creates risks of its exploitation and thus devaluation. Posed with such a dilemma, "Artstream" is a mechanism that resolves the conflicts that arise from the processes of making and distributing.

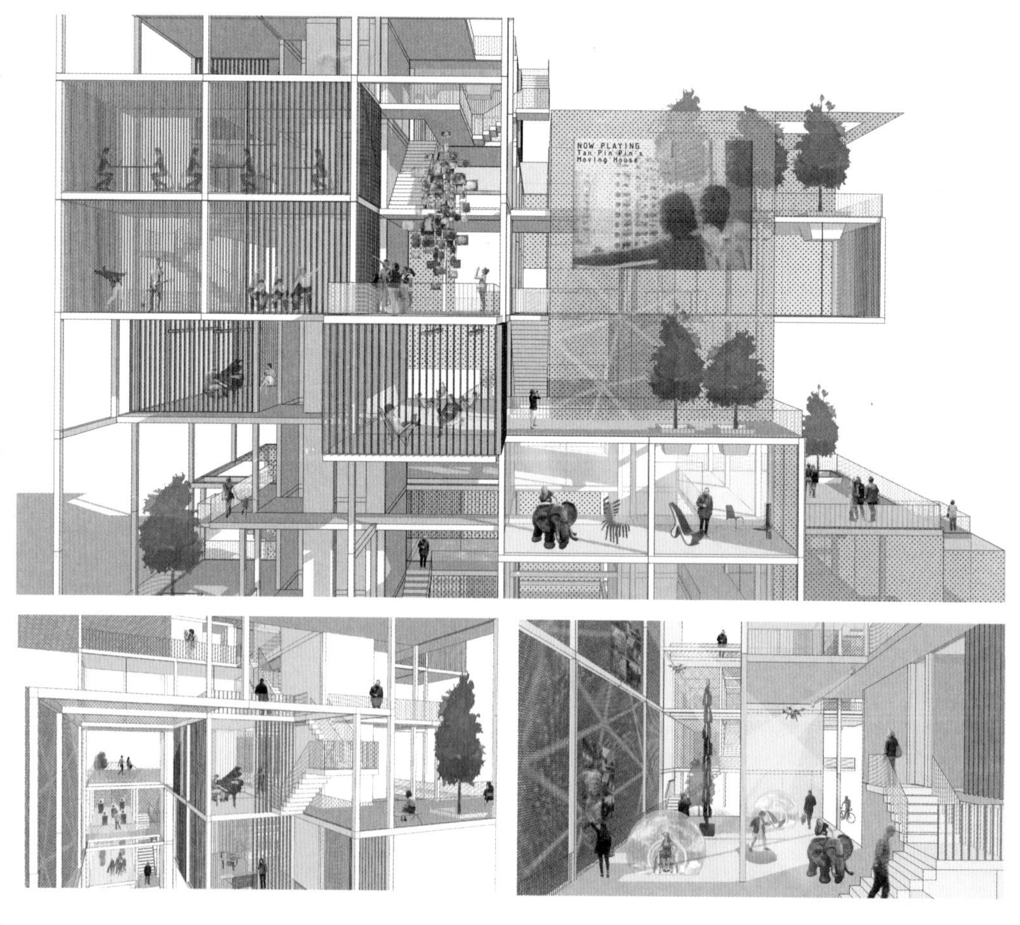

A medium for self and community expression, the cultural domain is crucial in engaging public participation and enhancing social capital. Through the intangible resources of ideas and dialogues through the artistic medium, urban commons in this project is given a different interpretation and new meaning.

虽然新媒体能让公众更方便地接触艺术，互联网也同时削弱了艺术的社会价值。“艺游：游艺”的目的则是减少艺术创造和分销过程之间所产生的冲突。

作为凝聚不同社群的媒介，艺术文化能促使社会资本的增加。通过非物质资源和艺术媒介的对话，该项目中的城市共同体被赋予了新的意义。

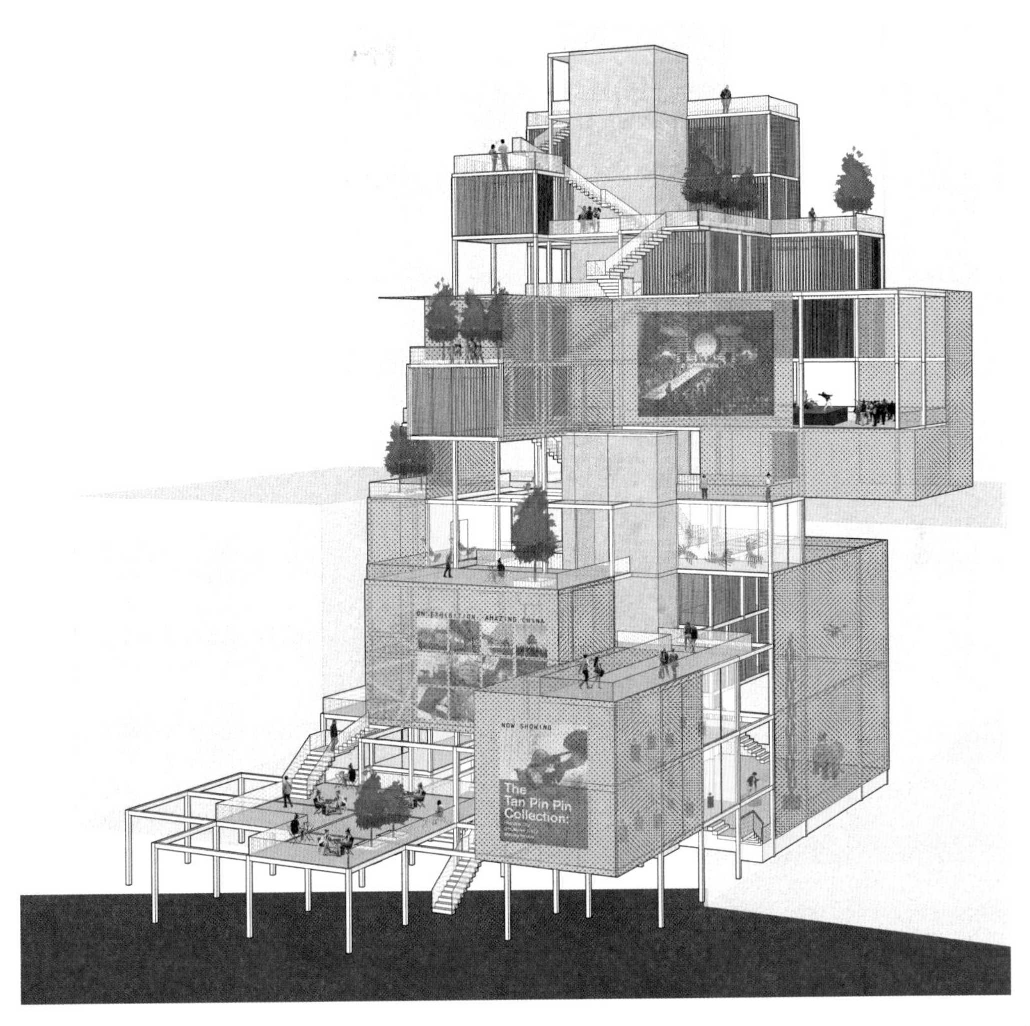

This project explores a new building typology which facilitates collaboration between the makers and the distributors (upstream and downstream arts). The building in itself also showcases the various arts to the public, playing a role in preserving the value of the arts.

Upstream making spaces are arranged around the inner core according to levels of privacy and interaction required for the specific programme. These volumes which are more intimately scaled, have a vertical corten-strip facade, as well as moveable louvred privacy screens, enabling the makers to open up or close themselves from the public when required.

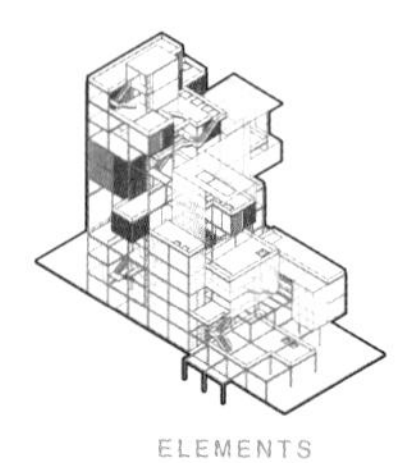

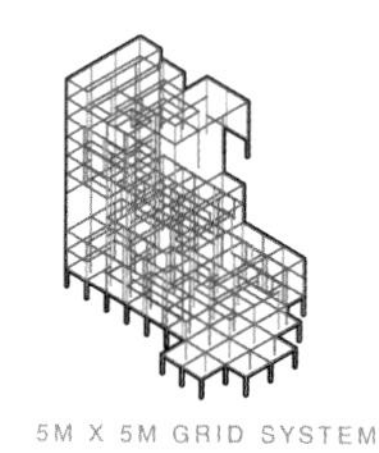

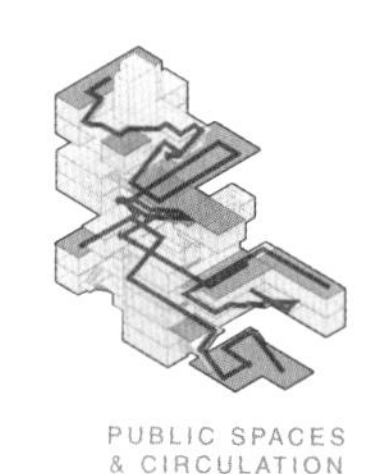

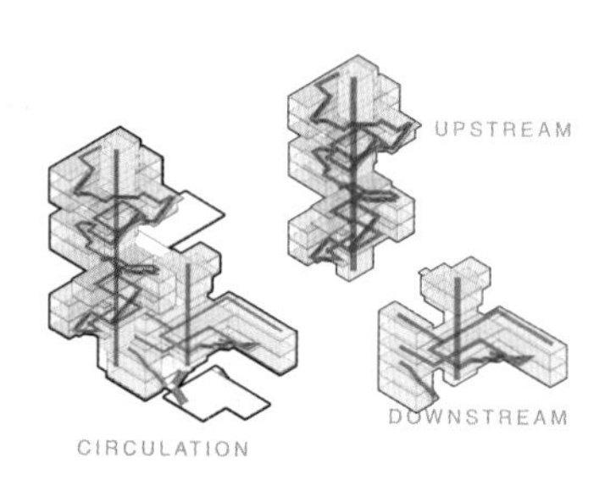

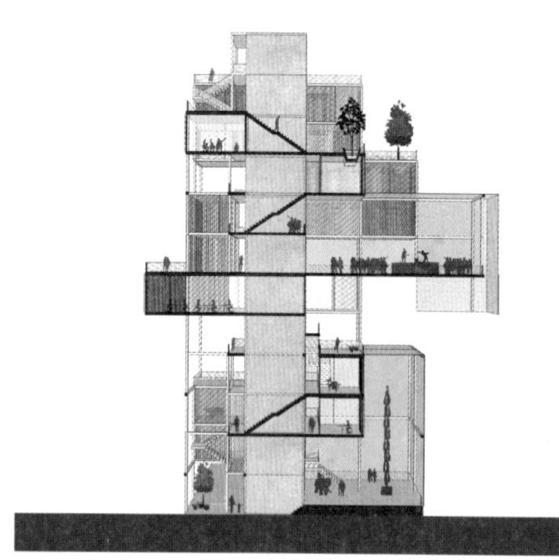

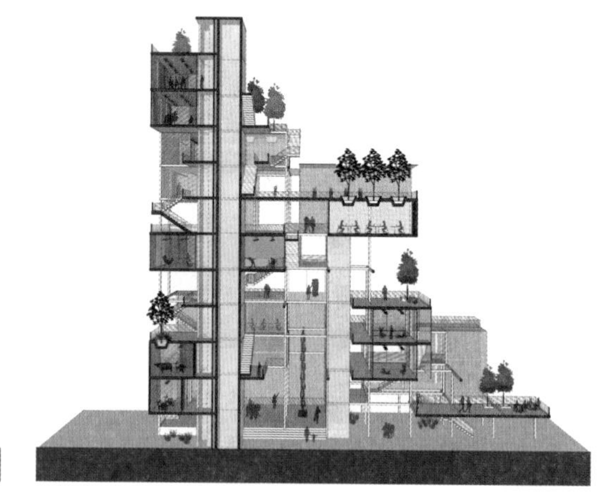

Top
Diagrams showing the design of massing and circulation.

上图
建筑体量与流线的分析图。

Bottom
Section drawings showing upstream and downstream spaces.

下图
“上游”工作室与“下游”配送空间的剖面图。

该设计方案探索一种新的建筑类型，促进艺术家（艺术创作流程的上游）和经销商（艺术创作流程的下游）之间的协作。该建筑也将通过各种艺术的展示，来保留艺术的社会价值。

考虑到不同艺术活动所需的隐私与互动交流空间不同，“上游”的工作室分布于建筑的内层核心筒周围。这些体量拥有近人尺度，可开启的百叶窗使艺术家们能够自由开关，随时展示他们的制作过程。

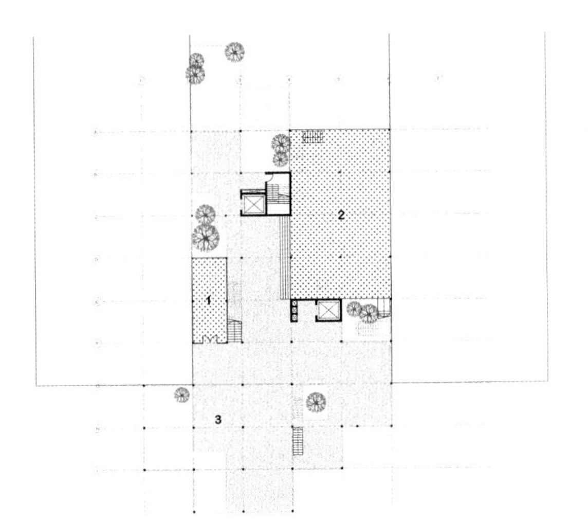

Ground Floor Plan
(1) Reception + Cafe; (2) Showcase Zone; (3) Landscaped Seating

一层平面图
（1）接待厅 + 茶馆；（2）展示区；（3）户外座位

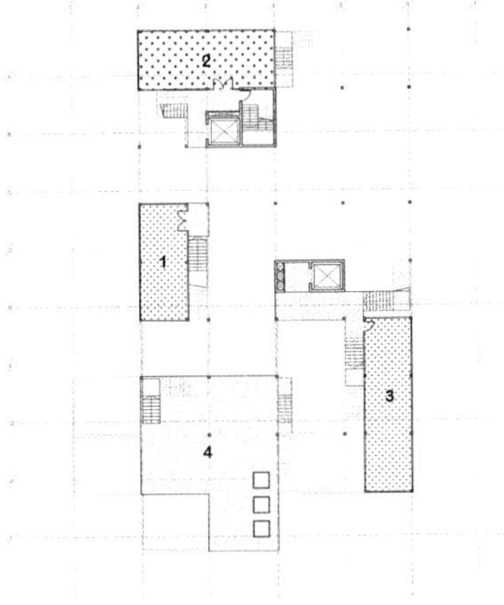

Second Floor Plan
(1) Administrative Office; (2) Makers' Workshop; (3) Gallery Space; (4) Landscaped Seating/ Showcase Space

二层平面图
（1）行政办公室；（2）艺术者工作室；（3）画廊；（4）户外座位 / 展示空间

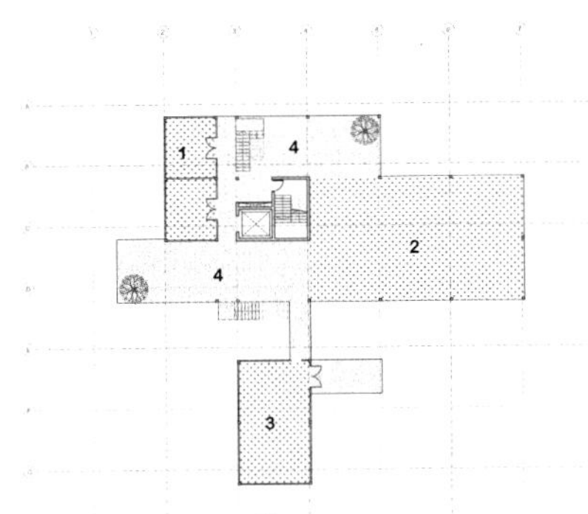

Third Floor Plan
(1) Digital Editing Labs; (2) Performance Stage; (3) Drama Room; (4) Landscaped Deck

三层平面图
（1）编辑实验室；（2）演出舞台；（3）戏剧室；（4）户外空间

Fourth Floor Plan
(1) Music Rooms; (2) Recording Room; (3) Landscape Deck / Showcase Space

四层平面图
（1）音乐室；（2）录音室；（3）户外空间 / 展示空间

Downstream distribution spaces which cater to events and the sales and showcase of arts, are similarly arranged around the outer core of the building. These volumes, located nearer to the pedestrian network of the new masterplan and with their LED glass facades, allow for projections and advertising to occur, enticing passers-by to stop by the building to take a closer look.

On plan, the upstream and downstream volumes are linked up to form various circulation pathways from ground to roof, fostering new forms of collaboration between the makers and distributors; and triggering public exploration. New pockets of spaces emerge on every floor, providing opportunities for landscaped sanctuaries or outdoor showcase spaces.

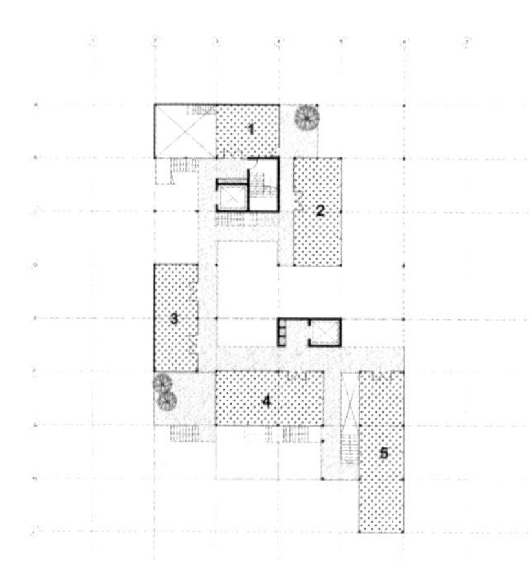

Fifth Floor Plan
(1) Pottery Studio; (2) Design Studio; (3) Gallery Space; (4) Digital Showroom; (5) Showroom

五层平面图
（1）陶艺工作室；（2）设计工作室；（3）画廊；（4）电子展厅；（5）展厅

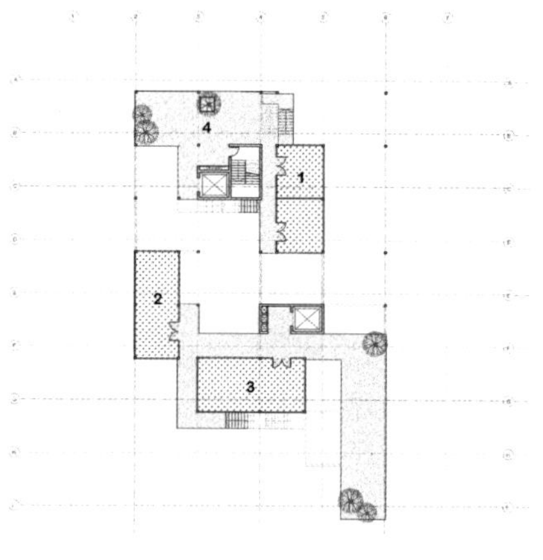

Sixth Floor Plan
(1) Administrative Office; (2) Makers' Workshop (3) Gallery Space; (4) Landscaped Seating/ Showcase Space

六层平面图
（1）行政办公室；（2）艺术者工作室；（3）画廊；（4）户外座位 / 展示空间

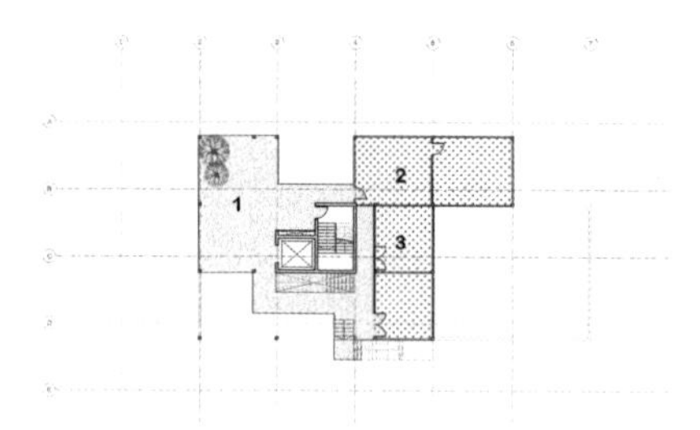

Seventh Floor Plan
(1) Landscaped Deck; (2) Music Rooms (3) Classrooms

七层平面图
（1）户外空间；（2）音乐室；（3）教室

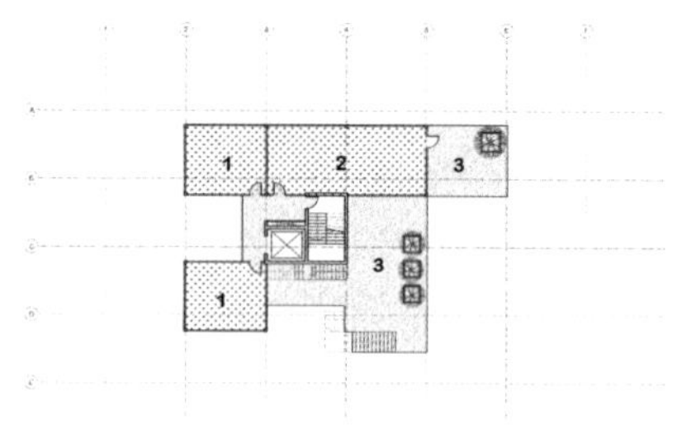

Eighth Floor Plan
(1) Art Studios; (2) Dance Studio (3) Landscaped Deck

八层平面图
（1）艺术工作室；（2）舞蹈室；（3）户外空间

主要作为展示与销售艺术品的“下游”配送空间则安排在建筑物的外层。“下游”配送空间位于公共人行道的附近，立面上的LED玻璃广告银幕吸引路人慢下脚步，来仔细观察艺术品的制造过程。

“上游”工作室和“下游”配送空间相互联系，形成接通不同楼层的循环路径。这不仅能促进艺术家和分销商之间的合作，也增加了公众参与、互动的机会。每层楼的户外空间也提供给艺术家和公众更多展示艺术及交流的机会。

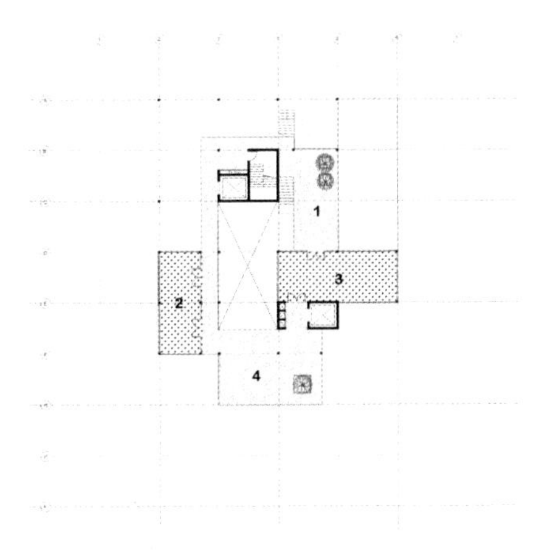

Ninth Floor Plan
(1) Landscaped Deck/Showcase Space;
(2) Office Space; (3) Collaboration Space;
(4) Landscaped Seating

九层平面图
（1）户外空间 / 展示空间；（2）办公室；
（3）合作空间；（4）户外座位

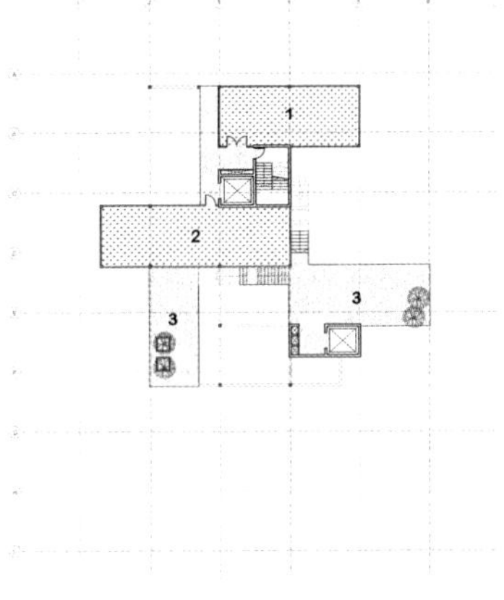

Tenth Floor Plan
(1) Animation Room; (2) Photography Studio
(3) Landscaped Deck

十层平面图
（1）动画室；（2）摄影工作室；
（3）户外空间

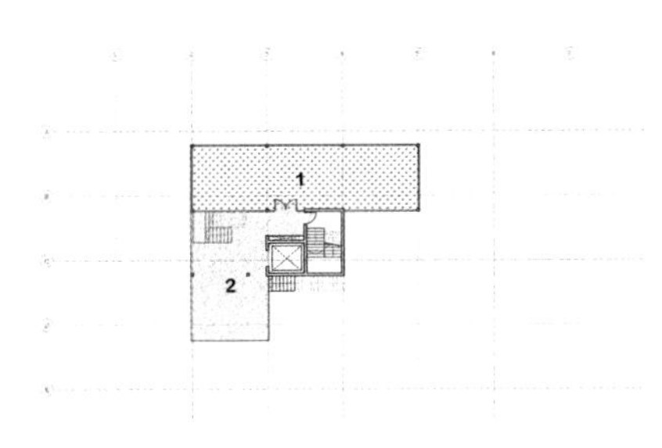

Eleventh Floor Plan
(1) Film Studio; (2) Landscaped Deck

十一层平面图
（1）电影制片厂；（2）户外空间

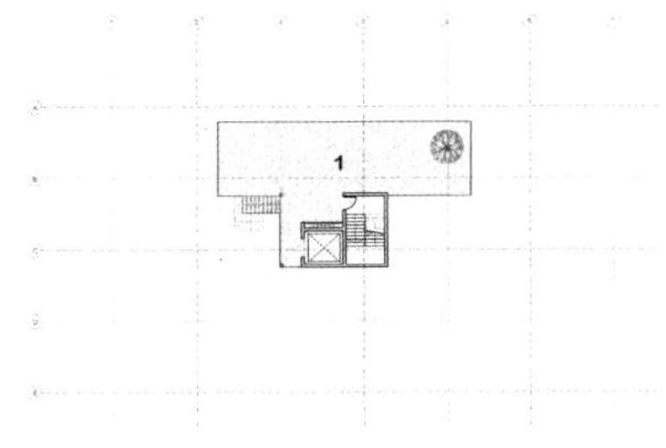

Twelfth Floor Plan
(1) Animation Room

十二层平面图
（1）动画室

5

REFLECTIONS 交流

SHARING CITY FORUM

At the end of the design studio, Tsinghua University held the final review and sharing city forums. In the “Sharing Cities: Sharing Economy and Urban Renewal” forum, Jeffery Chan Kok Hui, the assistant professor from the National University of Singapore, presented a lecture entitled “Philosophy of Sharing”, to theoretically interpret the sharing category and the role of planning designers. Former Uber China Senior Marketing Manager, W studio Project Leader Gao Wenxin shared about the “Sharing of Space under the Shared Economy”, including the introduced “Shared Confe ence Room” project implementation in Beijing, from the practical perspective. Randy Chan Keng Chong, head of Zarch's co-design company in Singapore, explains “Singapore Dream: Roaming in the Scenery of Awareness Geography”, analyzing the role of the designer under the current circumstance. From the Beijing Municipal Planning and Design Institute, Beijing Xicheng District Youth Committee Youth Work Committee leader Zhao Xing introduced the “Conservation and public participation in East Fourth South Historic District”, explaining the multi-party participation creation of “shared courtyard space”. The sharing city lectures presented in four different views stirred up a great interest of the audience towards the sharing city topic, hence followed by an active interaction session between speakers and audience.

BEIJING 2017.4.19

设计结束后，在清华大学举办了最终评图和共享城市论坛。“共享城市：共享经济与城市更新”论坛上，来自新加坡国立大学的助理教授Jeffery Chan Kok Hui发表了题为“共享的哲学”的讲演，从共享的类别和规划设计者在其中起到的作用进行理论解读。前Uber中国高级市场经理、Wstudio项目负责人高文心分享了“共享经济下的城市共享空间”，从实践层面介绍了在北京开展的“共享会议室”项目实施情况。新加坡Zarch合作设计公司的负责人Randy Chan Keng Chong以多元的视角解读了“新加坡梦：在意识地理学的景观中漫游”，分析了设计者在当前环境下的角色。来自北京市规划设计研究院、北京市西城区名城委青年工作委员会牵头人赵幸介绍了“东四南历史街区保护更新公众参与”，从规划师在旧城保护中如何协调各方参与的实际案例，分析了营造“共享院落空间”的多方参与途径。从四个不同角度解读共享城市的讲座激发了听众们极大的兴趣，此后大家进行了积极的交流。

UIA EXHIBITION

2017.5.27–2017.6.1

The achievement of the joint studio participated the exhibition and international studio in UIA 2017 held in Seoul. Team of Tsinghua University and of National University of Singapore both got involved in.

Seoul International Studio 2017 UIA 2017 的国际设计坊

SEOUL

2017.9.3-2017.9.10

联合教学的成果参加了在首尔举办的UIA 2017 展览和学生工作营，清华大学和新加坡国立大学均参加了相关活动。

Exhibition in UIA 2017 UIA 2017 的展览

DESIGN WEEK

The results of the work first was shown during Beijing Design Week 2017, on location of the studio's site, Baitasi, in a courtyard being made available by the Baitasi Remade initiative.

The exhibition consists of two parts;
- an exhibition if student work from international master students from Tsinghua University's School of Architecture
- an inhabitable installation to facilitate the activities of the local community, based on the students work from the studio

The exhibition is made up of separate furniture elements, that can be re-used in the school afterward.

此设计课题作品在2017北京国际设计周期间，在此设计课的基地白塔寺历史地区内，于一个院落中举办展览。

展览由两部分组成；
– 清华大学建筑学院的国际硕士生的作品展览
– 学生设计的装置用于促进当地社区的活动

这次展览的作品是由不同的独立家具组成，以后也可以在学校里重新使用。

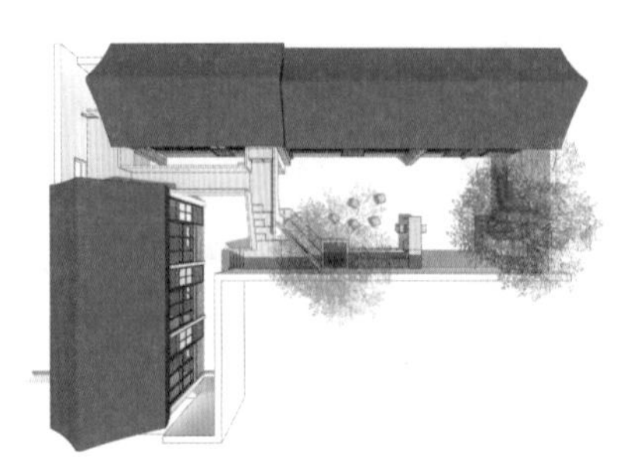

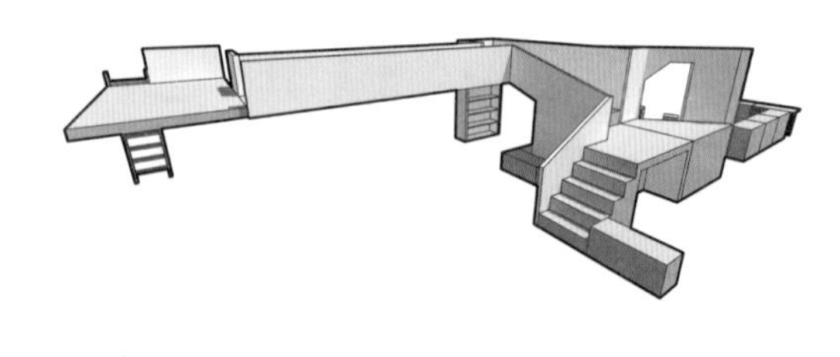

学生设计的院落改造装置 The inhabitable installation for courtyard regeneration design by Tsinghua EPMA students.

共享
城市

北京

DESIGN WEEK

Views of exhibition in Beijing design week
设计周展览实景

BEIJING

6

ACKNOWLEDGEMENTS 致谢

TEAM

The sharing cities joint studio set up by Tsinghua University and National University of Singapore, with chosen sites from Beijing and Singapore, has gained strong support from both institutions and organizations. For Singapore, million thanks to Ng Teng Fong Charitable Foundation for supporting this joint studio. For Beijing, Beijing Huarong Investment Development Co., Ltd. has provided positive help in the old city of the White Pagoda Temple site for site survey and research. Thanks to the Sustainable Settlements Joint Research Centre of Tsinghua University (School of Architecture) - Cifi Holdings (Group) for supporting the "Sharing Cites" forum. And also, special thanks to the "World Architecture" magazine, the White Pagoda Temple area management agencies and related organizations that gave support in the Beijing Design Week 2017.

Thank you to all the guest speackers and experts that participated and gave lectures during the joint studio - Professor Wang Caiqiang, Professor Rosemann Jurgen, assistant professor, Jeffrey Chan Kok Hui, Zarch cooperative design company and chief architects, Randy Chan Keng Chong, and The Press Room founder, Kelley Cheng from Singapore and META- the cross-border Research Institute founder Mr. Wang Shuo, W studio project manager Miss Gao Wenxin, and Miss Zhao Xin from the Beijing planning design and research institute from Beijing.

Finally, special thanks to students from the National University of Singapore, Kenny Chen Han Teng, Grace Koh Kah Shin, Tan Chiew Hui and Liew Yuqi in editting and design the layout for NUS part, Zhang Lu and Loo Hui Xin from Tsinghua University in monitoring overall layout and verification work. Thanks to the editors of the Tsinghua University Press for their hard work in publishing this book.

ACKNOWLEDGEMENTS

本次清华大学—新加坡国立大学的“共享城市”联合教学选取了北京和新加坡的地段，获得了双方机构和组织的大力支持。新加坡方面，感谢Ng Teng Fong Charitable Foundation (黄廷方慈善基金)对本次联合设计教学项目的大力支持。北京方面，在旧城地区的白塔寺地段的调研设计环节中，北京华融金盈投资发展有限公司提供了积极的帮助。感谢清华大学(建筑学院)—旭辉控股(集团)可持续住区联合研究中心为“共享城市”论坛举办所给予的大力支持；在北京设计周的相关活动组织中，《世界建筑》杂志社、白塔寺地区管理机构给予了大力支持。

感谢在联合教学和北京设计周展览过程中，参与授课、讨论的所有专家嘉宾——新加坡国立大学的王才强教授、Jurgen Rosemann教授、Jeffrey Chan Kok Hui 助理教授，Zarch合作设计公司的主持建筑师Randy Chan Keng Chong，The Press Room 的创始人 Kelley Cheng 和来自北京的META-跨界研究院创始人王硕先生、Wstudio项目负责人高文心女士、北京市规划设计研究院的赵幸女士。

最后要感谢新加坡国立大学的Kenny Chen Han Teng, Grace Koh Kah Shin, Tan Chiew Hui 以及Liew Yuqi同学为本书的出版进行的版面设计，清华大学的张璐和吕蕙欣同学为本书进行的统筹排版以及校核工作。感谢为本书出版做出辛勤贡献的清华大学出版社的各位编辑。

TEAM

清华大学

清华大学团队的教师为：张悦教授、黄鹤副教授、Martijn de Geus 博士。

Zhang Yue 张悦

Huang He 黄鹤

Martijn de Geus 和马町

Laurene 罗南

Zhang Xiaowen 张潇文

Clementine 安吉

Katja Toivola 云英

Cheng Fung 征峰

Gyoungmin Ko 高庚旼

Ying Yan Boey 梅颖彦

Hui Xin Loo 吕蕙欣

Deandrea 週殿

Nein 哈斯乃

Vijay Rathod 卫杰

Alessandra 丽桑

Laura 劳拉

David Vargas 大卫

Jamar Rock 洛克

Ahmed 谢强

Ricardo Simmons 希瑞

ACKNOWLEDGEMENTS

新加坡国立大学团队的教师为：
张烨助理教授、Tan Teck Kiam 副教授。

Zhang Ye 张烨

Tan Teck Kiam 陈德钦

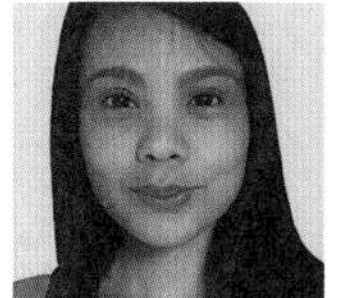
Cherie Chan Wan Qing 陈琬晴

Grace Koh Kah Shin 许家欣

Hung Yu Shan 洪馨珊

Jeshuren Moses

Kenny Chen Han Teng 管瀚霆

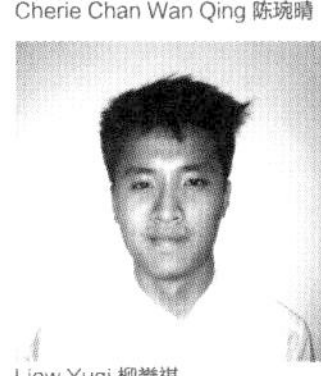
Liew Yuqi 柳誉祺

Lin Lei 林蕾

Neo Xin En Gloria 梁馨恩

Neoh Tse Wei 梁世玮

Ong Cheng Siang 王证翔

Tan Chiew Hui 陈秋慧

Tan Jia Yu 陈家裕

Ye Baogen Josiah 叶保根

Yeong Shilin Krista 杨诗琳

Yin Aiwei 尹爱惟

Yuan Yijia 袁懿嘉

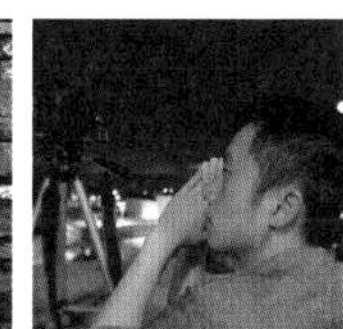
Zhang Hanfei 张翰飞